Exploration seismology
Volume 2
*Data-processing
and interpretation*

Exploration SEISMOLOGY Volume 2

Data-processing and interpretation

R. E. SHERIFF
Professor of Geophysics,
University of Houston

L. P. GELDART
Coordinator, Canadian International
Development Agency Program for Brazil

The right of the
University of Cambridge
to print and sell
all manner of books
was granted by
Henry VIII in 1534.
The University has printed
and published continuously
since 1584.

Cambridge University Press
Cambridge
New York   New Rochelle
Melbourne   Sydney

Published by the Press Syndicate of the University of Cambridge
The Pitt Building, Trumpington Street, Cambridge CB2 1RP
32 East 57th Street, New York, NY 10022, USA
10 Stamford Road, Oakleigh, Melbourne 3166, Australia

First published 1983
Reprinted 1985 (twice), 1987

Printed in Great Britain at the University Press, Cambridge

Library of Congress catalogue card number: 81 – 18176

British Library Cataloguing in Publication Data

Exploration seismology
Vol. 2: Data-processing and interpretation

I. Seismology
I. Sheriff, R.E.   II. Geldard, L.P.
551.2′2   QE539

ISBN 0 521 25064 1

v

Contents

Contents

Preface to volume 2

This volume deals mainly with seismic data-processing and interpretation, the two areas in which the most startling advances in applied seismology have been made in recent years. We have attempted to give a systematic approach, being consistent with the conventions established in volume 1. We have written for both students and for those engaged in seismic exploration and we have endeavored to make our work understandable to those who are not mathematically inclined as well as to those who are. Much of the tremendous improvement in seismic data quality today results from processing. Information theory has been used to extract geologically significant information buried in noise. The digital computer has become an essential tool in our analysis. Chapter 8 discusses data-enhancement procedures and how data-processing is carried out while chapter 10 gives the mathematical background more completely. In general chapter 8 deals with digitized data whereas chapter 10 deals with both digital and continuous functions.

The end product of exploration seismology is geologic understanding. Seismic velocity, one of the main factors relating seismic waves to geology, is discussed in chapter 7. Chapter 9 takes up the geologic meaning of seismic features. It includes some review of structural principles, structure mapping being the principal objective of most seismic work, but it also deals with interpretation procedures, velocity anomalies, stratigraphic interpretation and hydrocarbon indicators.

As with volume 1, terms are italicized where they are defined, Sheriff (1973) being the standard for definitions. Each chapter begins with an overview to aid the reader to see the aim of the various sections, and each chapter ends with problems which illuminate aspects not treated in the text as well as develop proofs. Occasional references to chapters 1 to 6 refer to volume 1.

We acknowledge the assistance of many people in the preparation of this book and we specifically thank Margaret Sheriff, Leslie Denham, Howard L. Taylor, Thomas L. Thompson and Willis H. Reed.

Mathematical conventions and symbols in volume 2

(a) General rules and definitions

(i) General functions

$g(t)$ function of a continuous variable t

g_t function of the discrete variable $t = n\Delta, n = 0, \pm 1, \pm 2, \ldots$

$g(t)*h(t)$ convolution of $g(t)$ with $h(t)$

$G(v)$ transform of $g(t)$ to a function of frequency v; the arguments v or ω indicate Fourier transform, s Laplace transform, z z-transform

$\phi_{gh}(\tau)$ correlation of $g(t)$ with $h(t)$ as function of the time shift τ; a cross-correlation if $g \neq h$, $\phi_{gg}(\tau) =$ autocorrelation

$\hat{g}(\zeta), \hat{G}(\omega)$ functions involving the cepstrum transform: log $G(\omega) = \hat{G}(\omega) \leftrightarrow \hat{g}(\zeta)$

(ii) Special functions

$\text{box}_a[t]$ boxcar of unit height and width a, centered at $t = 0$

$\text{comb}[t]$ series of equally-spaced unit impulses

$\text{sgn}[t]$ sign of $t = -1$ for $t < 0$, $+1$ for $t > 0$

$\text{sinc}[t]$ $(1/t)\sin t$

$\text{step}[t]$ unit step function, $\text{step}[t] = 0$ for $t < 0$, $+1$ for $t > 0$

$\delta[t], \delta_t$ unit impulse at $t = 0$

(iii) Mathematical conventions

$\approx$ approximately equal to

$\leftrightarrow$ denotes corresponding functions in different domains, the arguments v, ω, s, z indicating the type of transform; thus $g(t) \leftrightarrow G(v)$ and $g(t) \leftrightarrow G(\omega)$ indicate Fourier transforms, $g(t) \leftrightarrow G(s)$ a Laplace transform, $g_t \leftrightarrow G(z)$ a z-transform; lower-case letters indicate time-domain, capitals the frequency domain.

$[a, \overset{\downarrow}{b}, c, d]$ denotes a time sequence consisting of the elements a, b, c, d, with the superscribed arrow indicating the value associated with $t = 0$ (b in this instance); values not otherwise specified are zero.

$\mathbf{A}$ vector quantity, magnitude is $|\mathbf{A}|$

$\mathbf{A} \cdot \mathbf{B}, \mathbf{A} \times \mathbf{B}$ scalar and vector products of $\mathbf{A}$ and $\mathbf{B}$

∇ del, the vector operator $\mathbf{i}(\partial/\partial x) + \mathbf{j}(\partial/\partial y) + \mathbf{k}(\partial/\partial z)$.

∇^2 Laplacian operator, $\partial^2/\partial x^2 + \partial^2/\partial y^2 + \partial^2/\partial z^2$.

$\nabla\phi$ gradient of ϕ (grad ϕ)

$\nabla\cdot\mathbf{A}$ divergence of $\mathbf{A}$ (div $\mathbf{A}$)

$\nabla \times \mathbf{A}$ curl of $\mathbf{A}$ (curl $\mathbf{A}$)

$\det(a)$ determinant with elements a_{ij}

$\mathscr{A}$ matrix with elements a_{ij}

$\mathscr{A}^{\mathrm{T}}$ transpose of matrix $\mathscr{A}$

$|w|, |\mathbf{W}|$ absolute value or modulus of $w, \mathbf{W}$

$\mathrm{Re}\{g(t)\}, \mathrm{Im}\{g(t)\}$ real, imaginary parts of $g(t)$

$\arg(\omega)$ argument of ω

$\overline{G(z)}, G(z^{-1})$ complex conjugate of $G(z)$

$\displaystyle\prod_{i=0}^{n} a_i$ product $a_0 a_1 a_2 a_3 \ldots a_n$

$\displaystyle\sum_{i=0}^{n} g_i$ sum $g_0 + g_1 + g_2 + \ldots + g_n$

$\displaystyle\sum_{k} g_k, \sum g_k$ sum of g_k over appropriate values of k

$\exp(x)$ e^x

$g(0+), g(0-)$ value of g when approaching 0 from right, left

$\mathfrak{P}$ principal value of

$P(e_j)$ Probability of e_j

$\mathfrak{D}$ difference operator

$\mathfrak{R}$ derivative operator

$\mathfrak{I}$ integration operator

$\mathfrak{Z}$ delay operator

(b) Latin symbols

a, b constants

a_n, b_n, c_n Fourier series coefficients

$A(t)$ amplitude of envelope, amplitude attributed to reflectivity

$A(v), B(v)$ amplitude spectrum

A, B constants

D_{s} depth of shot

e_i error in ith output

e_t impulse response of sequence of reflectors

E sum of errors squared, Young's modulus

$f(t), f_t$ filter in time domain

$f_{\mathrm{L}}(t), f_t^{\mathrm{L}}$ low-pass filter

$f_{\mathrm{H}}(t), f_t^{\mathrm{H}}$ high-pass filter

$f_{\mathrm{u}}(t)$ response of filter to unit step in time domain

F constant, magnitude of force

$\mathbf{F}$ force

$F(v), F(\omega), F(z)$ filter in frequency domain

$F_{\mathrm{u}}(s)$ transform of response to unit step

g acceleration of gravity

$g(t), g_t$ seismic trace in time domain, input trace

$g_\perp(t)$ quadrature trace

$\hat{g}(\zeta)$ cepstrum of $\hat{G}(\omega)$ or $\hat{G}(z)$

$G(v), G(\omega), G(z)$ seismic trace in frequency domain

$h(t), h_t$ output in time domain

$\hbar(t), \hbar_t$ desired output

$H(v), H(\omega), H(z)$ output in frequency domain

$\mathscr{H}(v), \mathscr{H}(\omega), \mathscr{H}(z)$ desired output in frequency domain

$\mathbf{i}, \mathbf{j}, \mathbf{k}$ unit vectors in $x-, y-, z$-directions

i_t inverse filter in time domain

$I(v)$ transform of inverse filter

j $(-1)^{\frac{1}{2}}$

k constant

K effective elastic modulus

l, m, n direction cosines relative to $x-, y-, z$-axes

ℓ_1, ℓ_2, ℓ_4 criteria in optimum filtering

L prediction lag

L_k, M_k delay at location k due to structure, normal-moveout error

m, n constants

n integer, number of layers

n_t impulse response of near-surface zone

p raypath parameter

$\mathscr{P}$ pressure

r, s receiver, source coordinates

r_t additive noise

r, R radii

R, R_i reflection coefficient of ith interface

R_k, S_k delay due to geophone, source at location k

$R(\omega)$ real part of Fourier transform = cosine transform

R' resistivity

s Laplace transform parameter = $\sigma + \mathrm{j}\omega$, distance

s_t impulse response of source

S semblance, entropy density

S_{w} water saturation

$\mathscr{S}$ area, surface

t time, traveltime

t_0 traveltime for geophone at the shotpoint

t_{ij} time shift between traces i and j

Δt time interval

Δt_n normal moveout

T period

v_t Vibroseis ™ input to ground

V velocity

$\bar{V}$ equivalent average velocity

$\bar{\bar{V}}$ rms velocity

V_{s} stacking velocity

V_{a} apparent velocity

$\mathscr{V}$ volume

w weighting factors

w_t equivalent wavelet, impulse response of water layer, downgoing waveform

x source-to-geophone (offset) distance

$X(\omega)$ imaginary part of Fourier transform = $-$(sine transform)

z depth, z-transform parameter

$Z(t)$ P-wave acoustic impedance

(c) Greek symbols

α	P-wave velocity
α_n	Fourier-series coefficients
β	S-wave velocity
γ	phase
$\gamma(v), \gamma(\omega)$	phase spectrum
$\gamma(t)$	instantaneous phase
Γ	measure of simplicity
Δ	sampling interval, interval of independent variable
ε	strain
$\boldsymbol{\varepsilon}$	error matrix
ζ	quefrency
$\boldsymbol{\eta}$	outward-drawn unit normal
θ	angle, argument of complex quantity
κ	$2\pi(\text{wavenumber}) = 2\pi/\lambda$
κ_a	$2\pi(\text{apparent wavenumber})$
κ_N	$2\pi(\text{Nyquist wavenumber})$
λ	wavelength, Lamé constant, length ratio, constant, weighting factor

λ_a	apparent wavelength
λ_N	Nyquist wavelength
μ	rigidity modulus, mass ratio
v	frequency $= \omega/2\pi = 1/T$
$v_i(t)$	instantaneous frequency
v_N	Nyquist frequency
ξ	dip
ρ	density
σ	Poisson's ratio, convergence factor, standard deviation
τ	delay, time ratio
ϕ	porosity, P-wave displacement function
$\phi_{gh}(\tau)$	correlation of $g(t)$ with $h(t)$ as function of time shift τ
$\Phi_{gh}(\omega)$	cross-energy spectrum (transform of $\phi_{gh}(\tau)$)
$\psi(x,z,t)$	wavefunction in two dimensions
$\psi^*(x,z,t^*)$	wavefunction in moving coordinate system
$\Psi(\kappa_x, \kappa_z, \omega)$	transform of wavefunction
$\Psi^*_{xz}(\kappa_x, \kappa_z, t)$	transform of wavefunction with respect to x, z
ω	angular frequency $= 2\pi v$
ω_0	natural frequency
ω_N	Nyquist frequency

7
Seismic velocity

Overview

Knowledge of velocity values is essential in determining the depth, dip and horizontal location of reflectors and refractors, in determining whether certain things like head waves and velocity distortions occur, and in ascertaining the nature of rocks and their interstitial fluids from velocity measurements.

While lithology is the most obvious factor affecting velocity, the ranges of velocity of different rock types overlap so much that it does not provide a good basis for distinction by itself. Porosity appears to be the most important single factor, and the dependence of porosity on depth of burial and pressure relationships makes velocity sensitive to these factors also. Velocity is generally lowered when gases or oil replace water as the interstitial fluid, sometimes by so much that amplitude anomalies result from hydrocarbon accumulations.

The near-surface layer of the Earth usually differs markedly from the remainder of the Earth in velocity and some other properties. This makes the near-surface low-velocity layer (LVL) especially important; our determinations of depths, attitudes and continuity of deeper events are affected as reflections pass through this layer. In arctic areas a zone of permanently-frozen earth, permafrost, distorts deeper events because of an exceptionally high velocity. Gas hydrates which form in the sediments just below the seafloor in deep water also produce velocity change.

In-situ measurement of velocity is accomplished directly by conventional velocity surveys where the traveltime is measured between a source on the surface and a geophone in a deep borehole. The velocity over short intervals is measured by sonic-logging sondes, but time–depth relationships determined from them are subject to errors resulting from integration and other factors unless calibrated by conventional velocity measurements. Most velocity information is determined by the variation of arrival time with offset distance, the normal-moveout (NMO) dependency, because deep boreholes are usually not available.

7.1 Factors affecting velocity

7.1.1 *Introduction*

Velocity can be determined from measurements (*a*) *in situ* (see §7.3) or (*b*) on samples in a laboratory. Press (1966) lists measurements of both types. Care has to be taken that measurements on samples are not distorted by changes in the sample conditions; many early measurements gave misleading values because they were made on desiccated or otherwise altered samples. Gregory (1977) discusses laboratory measurements and gives a number of references from outside the usual geophysical literature. Reports of velocity measurements in the literature are numerous and in the following sections we cite only those believed to be representative and which give insight into the interrelationship of factors.

Equations (2.52) and (2.53) can be written

$$\alpha^2 = (\lambda + 2\mu)/\rho, \quad \beta^2 = \mu/\rho \text{ (solid media)}$$
$$\alpha^2 = \lambda/\rho, \quad \quad \beta = 0 \text{ (fluid media)};$$

hence, in general,

$$V = (K/\rho)^{\frac{1}{2}} \qquad (7.1)$$

where K = effective elastic parameter. Thus, the dependence of V upon the elastic constants and density appears to be straightforward. In fact, the situation is much more complicated because K and ρ are interrelated, both depending to a greater or lesser degree upon lithology, porosity, the properties of interstitial fluids, pressure, depth, cementation, degree of compaction, etc.

7.1.2 *Effect of lithology*

Lithology is probably the most obvious factor affecting velocity (fig. 7.1). Some rocks extend outside the ranges shown in the figure. The most impressive aspect of this chart is the tremendous overlap of velocity values for different lithologies, which suggests that velocity is not a good criterion for determining lithology. High velocity for sedimentary rocks generally indicates carbonates and low velocity generally indicates sands or shales, but intermediate velocity can indicate either.

Distinguishing between sandstone and shale is often of considerable importance and velocity measure-

Fig.7.1. P-wave velocity for various lithologies. The graphs for shale, sandstone, limestone and dolomite show also the dependence on porosity (indicated in percentages). Based on tables and graphs by Press (1966), Gardner *et al.* (1974), and Lindseth (1979).

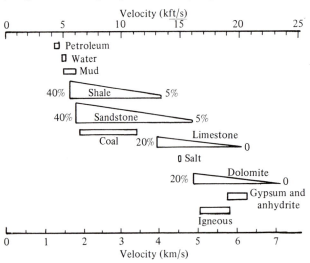

ments are sometimes the basis for prediction. Figure 7.2 shows a portion of a well log where sands are distinguished from shales on the basis of spontaneous potential (SP). While there is a difference between the best-fit line through the sand velocity values and one through the shale velocity values, the scatter of individual values exceeds this. Statistical predictions sometimes are satisfactory when based on data in an area but predictions for specific samples are often extremely far off.

While data for the velocity of S-waves are more sparse than for P-waves, the ratio of velocity for the two types of waves appears to be indicative of lithology; this is illustrated by fig. 7.3.

7.1.3 *Effect of density*

The density of a rock depends directly upon the densities of the minerals making up the rock (ignoring for the moment the effect of porosity). Table 7.1a shows that the densities of those minerals which constitute most sedimentary rocks vary over a range of about 20%. In table 7.1b the range of density variations within a rock type is low for igneous rocks (about 10%), intermediate for metamorphics and limestone (12–18%) and relatively high for clastic sediments (25–30%). Density variations play a significant role in velocity variations and high densities usually correspond to high velocities (fig. 7.4). Equation (7.1) which implies an inverse relationship, is oversimplified since density also affects K in the numerator.

The data of Gardner *et al.* (1974) suggest the relationship

$$\rho = aV^{\frac{1}{4}} \qquad (7.2)$$

Table 7.1a. *Density of representative sedimentary rock minerals (after Robie et al., 1966).*

calcite	$CaCO_3$	2.71 g/cm^3
dolomite	$CaMg(CO_3)_2$	2.87
anhydrite	$CaSO_4$	2.96
halite	$NaCl$	2.16
quartz (α)	SiO_2	2.68
albite	$NaAlSi_3O_8$	2.62
orthoclase	$KAlSi_3O_8$	2.55
kaolinite	$Al_2Si_2O_5(OH)_4$	2.60
muscovite	$KAl_2(AlSi_3O_{10})(OH)_2$	2.83

Many natural minerals vary in composition and hence in density. Kaolinite and muscovite are included as representative of clay minerals.

Table 7.1b. *Density of representative rocks (after Daly et al., 1966)*

granite	2.51–2.81 g/cm^3	mean: 2.67 g/cm^3
diorite	2.68–2.96	2.84
gabbro	2.85–3.12	2.98
diabase	2.80–3.11	2.96
gneiss	2.59–3.06	2.71
schist	2.70–3.03	2.80
sandstone	2.17–2.70	2.42
limestone	2.37–2.77	2.60
shale	2.06–2.66	2.38

Fig.7.2. Portion of SP- and velocity logs for a well in the US Gulf Coast. The SP-values distinguish sands from shales. (After Sheriff, 1978.)

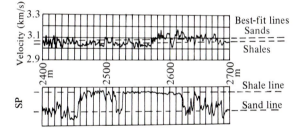

Fig.7.3. Relation between S- and P-wave velocities (β and α) for various lithologies. Data from Pickett (1963). Shales are expected to lie below the limestone region but data are not available to verify this. Porosity generally decreases to the right.

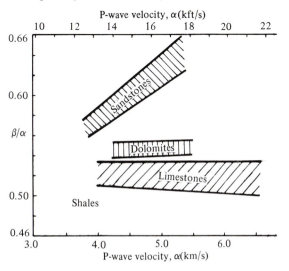

where ρ is in g/cm^3, V in m/s when $a = 0.31$ and in ft/s when $a = 0.23$. The graph of this equation is shown in fig. 7.4. Salt, anhydrite and coal do not fit (7.2).

7.1.4 *Effect of porosity*

Table 7.1a shows that the minerals which make up most sedimentary rocks have densities in the range of $2.7 \pm 4\%$ whereas table 7.1b shows that sandstones, for example, have densities of $2.4 \pm 10\%$. The discrepancy is due largely to the effect of porosity. Sedimentary rocks are of two broad classes: clastic and chemically-deposited. *Clastic rocks* are composed of fragments of minerals, other rocks, shells, etc., mainly made up of the minerals shown in table 7.1a, and hence have appreciable void space. Chemically-deposited rocks may have been subjected to recrystallization and/or the effects of percolating solutions, which also often result in appreciable void space. In both cases the voids are usually filled with fluids and the bulk density ρ is given exactly by

$$\rho = \phi\rho_f + (1 - \phi)\rho_m \qquad (7.3)$$

where ϕ = porosity, ρ_f = fluid density and ρ_m = matrix density.

In addition to affecting the velocity through the bulk density, porosity also has a direct effect on the velocity since a part of the wave path is in low-velocity fluids. The empirical *time-average equation* developed by Wyllie *et al.* (1958) is often used to relate velocity V and porosity ϕ (fig. 7.5); it assumes that the traveltime per unit path length in a fluid-filled porous rock is the average of the traveltimes per unit path length in the matrix material, $1/V_m$, and in the fluid, $1/V_f$, the traveltimes being weighted in proportion to the respective volumes:

$$\frac{1}{V} = \frac{\phi}{V_f} + \frac{(1 - \phi)}{V_m}. \qquad (7.4)$$

This relationship is used extensively in well-log interpretation; its form is similar to that of (7.3) except that (7.4) is statistical and empirical. The values of V_f and V_m used in (7.4) are often those which give the best fit over a range of interest and the fit may be poor outside the range, for example for poorly-consolidated, high-porosity sediments.

Random packs of well-sorted particles have porosities in the range of 45–50% but under pressure the particles deform at the contacts, and as a result the density increases and the porosity decreases (Sheriff, 1977a) (the elastic constants also change – see §7.1.5). Very few of the various things which can happen to rocks increase porosity (see fig. 7.6), hence porosity generally decreases with increases in depth of burial, cementation and age, as sorting becomes poorer, etc. Porosity is usually the most important factor in determining the velocity in a sedimentary rock.

7.1.5 *Effects of depth of burial and pressure*

Porosity generally decreases with increasing depth of burial (or overburden pressure) and hence velocity

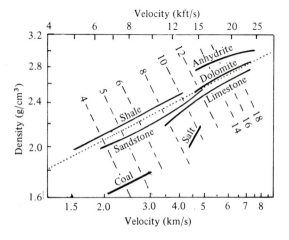

Fig.7.4. P-wave velocity–density relationship for different lithologies (the scale is log–log). The dotted line shows eq. (7.2) and the dashed lines show constant acoustic impedance (kg/s m^2 × 10^6). (After Gardner *et al.*, 1974, and Meckel and Nath, 1977.)

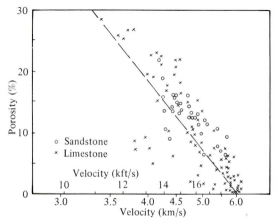

Fig.7.5. Velocity–porosity relationship. The horizontal scale is linear in transit time (proportional to $1/V$). The dashed line is the time-average equation (7.4) for $V_m = 5.94$ km/s (19.5 kft/s) and $V_f = 1.62$ km/s (5.3 kft/s). (After Wyllie *et al.*, 1956.)

increases with depth. The elastic constants also depend upon the pressure. These effects are attributable to the structure of sedimentary rocks, which are not homogeneous as simple elasticity theory assumes.

Fig.7.6. The effect of various processes on porosity of a clastic rock. Porosity decreases with depth of burial (compaction), cementation or poorer sorting. Porosity is essentially unchanged by uplift. (After Zieglar and Spotts, 1978.)

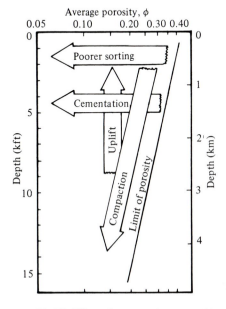

Fig.7.7. Effects of compression on a cubic packing of spheres. (After White, 1965.) (a) Cubic packing; (b) force causes centers to move closer together; (c) force causes point contact to become circle of contact; (d) effect of change in force.

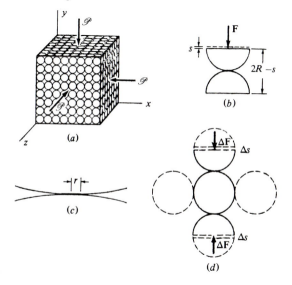

The simplest rock model consists of identical spheres arranged in a cubic pattern (fig. 7.7a) with the matrix subjected to a compressive pressure $\mathscr{P}$. If the radius of the spheres is R, the force F pressing two adjacent spheres together is the total force acting on a layer of $n \times n$ spheres (that is, $(2Rn)^2 \mathscr{P}$) divided by the number of spheres (n^2), or $F = 4R^2 \mathscr{P}$. This force causes a point of contact to become a circle of contact of radius r and the centers to move closer together a distance s (see figs. 7.7b, c), r and s being related to R, F and the elastic constants E, σ of the spheres by Hertz' equations (see Timoshenko and Goodier, 1951, p. 372–7):

$$\left. \begin{array}{l} r = \{3(1 - \sigma^2)RF/4E\}^{\frac{1}{3}}, \\ s = \{9(1 - \sigma^2)^2 F^2/2RE^2\}^{\frac{1}{3}}. \end{array} \right\} \quad (7.5)$$

When a P-wave passes, $\mathscr{P}$ changes by $\Delta\mathscr{P}$, resulting in changes $\Delta F = 4R^2 \Delta\mathscr{P}$ and $\Delta s = -2R\varepsilon$, where ε is the strain in the direction of F (see fig. 7.7d). Thus the effective elastic modulus K is given by

$$K = -\Delta\mathscr{P}/\varepsilon = \frac{1}{2R}\frac{\Delta F}{\Delta s} = \{3E^2\mathscr{P}/8(1 - \sigma^2)^2\}^{\frac{1}{3}}.$$

on differentiating (7.5). The average density is the weight of a sphere divided by the volume of the circumscribed cube, that is, $\bar{\rho} = (\frac{4}{3}\pi R^3 \rho)/(2R)^3 = \frac{1}{6}\pi\rho$, ρ being the density of the material of the spheres. Thus we get for the P-wave velocity, V_{cubic},

$$V_{\text{cubic}} = (K/\bar{\rho})^{\frac{1}{2}} = \{81E^2\mathscr{P}/(1 - \sigma^2)^2\pi^3\rho^3\}^{\frac{1}{6}}. \quad (7.6)$$

Gassmann (1951) calculated the velocity for a hexagonal packing of identical spheres (fig. 7.8b) under a pressure produced by the weight of a thickness z of overlying spheres; he obtained for a vertical ray

$$V_{\text{hex}} = \{128E^2gz/(1 - \sigma^2)^2\pi^2\rho^2\}^{\frac{1}{6}}, \quad (7.7)$$

where g is the acceleration of gravity. Since $\mathscr{P}$ is nearly proportional to z, (7.6) and (7.7) give the same variation of velocity with depth. Faust (1953) found an empirical formula for velocity in terms of depth of burial z and formation resistivity R' which is consistent with (7.6) and (7.7):

$$V_{\text{p}} = 900(zR')^{\frac{1}{6}}, \quad (7.8)$$

V_{p} being in m/s, z in m and R' in Ωm. However, the deviations of individual measurements were very large, indicating the presence of other factors which have not been taken into account.

The section in the Louisiana Gulf Coast consists of relatively undistured clastic rocks whose condition is similar to the foregoing rock models. Gregory (1977) gives velocity versus depth data for Gulf Coast sands and shales under normal pressure conditions (fig. 7.9). The data fit a $\frac{1}{4}$ exponent more nearly than the $\frac{1}{6}$ exponent shown by (7.8).

In actual rocks the pore spaces are filled with a fluid under a pressure which is usually different from that resulting from the weight of the overlying rocks. In this situation the effective pressure on the granular matrix is the difference between the overburden and fluid pressures.

Where formation fluids are under abnormal pressure, the differential pressure becomes that appropriate to a shallower depth and the velocity also tends to be that of the shallower depth. Such a lowering of velocity can be seen in the well log of fig. 7.10, although the change is often gradational rather than abrupt as shown here. Laboratory measurements (Gardner *et al.*, 1974) also show that velocity is essentially constant when the overburden and fluid pressures are changed, provided the differential pressure remains constant. Abnormal fluid pressure constitutes a severe hazard in drilling wells and one use of

Fig.7.8. Close packing of spheres. (*a*) Cubic packing (as in fig. 7.7*a*), an arrangement which is not gravitationally stable; (*b*) hexagonal packing, which is gravitationally stable; (*c*) first layer of a hexagonal stack, showing the two classes of sites (A and B), adjacent sites of which cannot both be occupied at the same time (e.g., the two dashed locations); (*d*) second layer of spheres in place showing how occupying A sites excludes some B sites; (*e*) hexagonal pack with left portion using A sites and right portion B sites; the consequence is more porosity than otherwise expected from all A sites alone; a random initial choice of A or B sites leads to a completely random pack after a few layers.

(*a*)

(*b*)

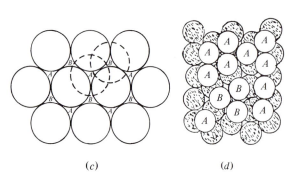

(*c*)

(*d*)

(*e*)

seismic velocity measurements is in predicting such zones (see §7.2.4).

The variation of velocity with depth, usually referred to as the velocity function (§3.2.4), is frequently a reasonably systematic increase as we go to greater depths. Velocity versus depth relationships for several areas are shown in fig. 7.11.

A contradiction to the concept that compaction and consequent loss of porosity are primarily responsible for the increase of velocity with depth is provided by non-porous igneous rocks, which also show an increase in velocity with depth (Press, 1966), although at a slower rate. Gardner *et al.* (1974) postulated that 'microcracks' exist in such rocks which delay seismic waves, and that these microcracks close with pressure so that they exert less influence. They subjected rocks to shock treatment to induce more microcracks, after which the velocity decreased and the dependence of velocity on pressure increased.

7.1.6   *Effects of age, frequency and temperature*

An early form of Faust's law (Faust, 1951) included the age of the rock as a factor in determining velocity. Figure 7.12 is taken from Faust's paper; the data points are each averages of many values. Older rocks generally have higher velocities than younger rocks, but most geophysicists agree that age is probably merely a measure of the net effect of many geologic processes, that is, older rocks have merely had longer time to be subjected to various factors (cementation, tectonic stresses, etc.) which decrease porosity. Since the history of rocks varies so much in time and space, the time factor must be only approximate. Time-dependent strain may play some part but its nature is not known.

Experimental data generally support the contention that velocity does not depend on frequency, over the range from hertz to megahertz, and hence there is no clear evidence that dispersion of P- or S-waves exists.

Velocity appears to vary slightly with temperature (fig. 7.13), decreasing by 5–6%/100°C.

Fig. 7.9. Velocity–depth relationship for Gulf Coast sands (x) and shales (o). Best-fit quadratic curves are also shown. Also shown by the step graph are data for Offshore Venezuela where the sediments are under similar conditions. The dotted graph shows average velocity to various depths for the Venezuelan data. (Data from Gregory, 1977.)

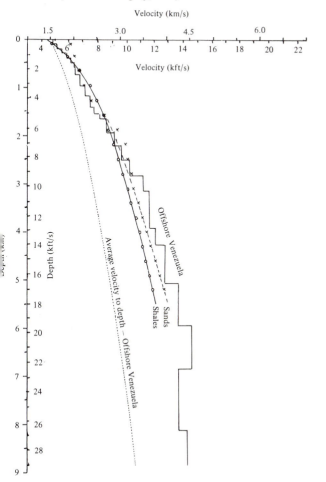

Fig. 7.10. SP- and velocity log for a US Gulf Coast well which encountered abnormal pressure. The lithology is sand-shale, the two being distinguished by the SP-log. Depths are in kilometers. (After Sheriff, 1978.)

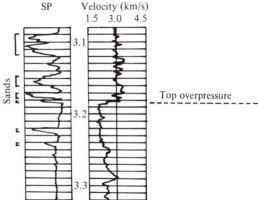

7.1.7 *Effect of interstitial fluid*

Porous rocks are almost always saturated with fluids, generally salt water, the pores in oil and gas reservoirs being filled with varying amounts of water, oil and gas. The replacement of water by oil or gas changes the bulk density and the elastic constants, and hence also the P-wave velocity and the reflection coefficient. These changes are sometimes sufficient to indicate the presence of gas or oil. Horizontal variations in reflection amplitude, velocity, frequency and other factors are sometimes important indicators of oil and gas accumulations (see §9.8). The low velocities when gas fills the pore space at least partially explain the low velocities observed in the

weathered (LVL) layer (§7.2.2) and why its lower boundary is so often the water table.

The nature of the interstitial fluid does not change the shear modulus appreciably and hence S-wave velocity changes only slightly (mainly because the density changes). The ratio of P- to S-wave velocity (α/β) has been proposed as a method of distinguishing the fluid filling the pore space (fig. 7.14).

Some early successes in locating hydrocarbons by increased reflection amplitude led to expectations that every amplitude anomaly was associated with a commercial gas or oil field. Domenico (1974), applying the formula of Geertsma (1961), showed that only a small amount of

Fig.7.11. Velocity–depth relationships from selected wells. (*a*) Data from Gulf of Alaska; Cost-B2 well, Offshore US East Coast; wells in Tyler (#1) and Dewitt (#2) Counties in Texas Gulf Coast; and in Illinois Basin; (*b*) data from Sacramento Valley (Yolo Co.,

Calif.); Wind River Basin (Fremont Co., Wyo.); Williston Basin (Divide Co., N.D.); and the Java Sea; (*c*) average velocity to various depths for the data in (*a*) and (*b*).

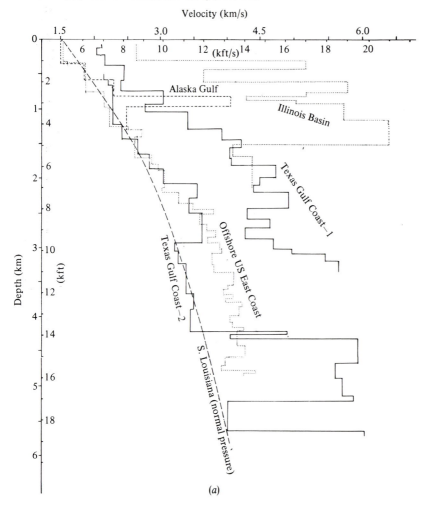

(*a*)

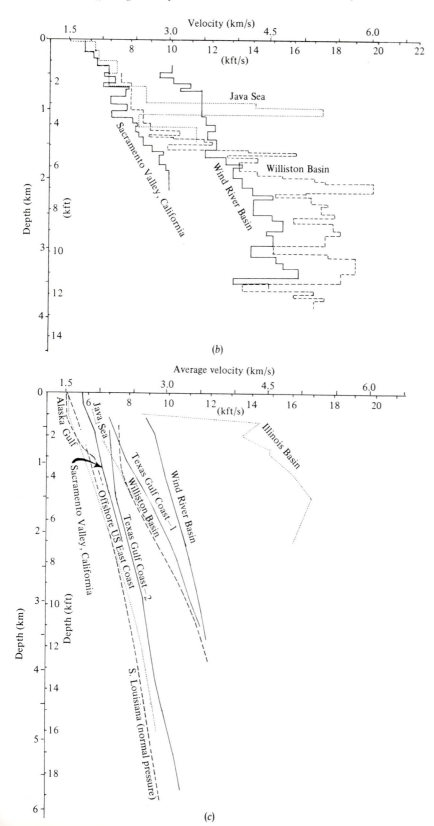

(b)

(c)

gas in the pore space produced a large decrease of velocity (fig. 7.15*a*) and a large change in reflectivity (fig. 7.15*b*). The Geertsma formula allows for the fluid compressibility as well as the density and the elastic moduli of the matrix material. Domenico (1976, 1977) partially verified the theoretical results with laboratory experiments.

7.2 Application of velocity concepts
7.2.1 *Introduction*

An understanding of the factors affecting velocity helps us foresee the kind of velocity variations to expect in an area and hence the velocity distortions to expect in seismic data (see §9.5). Areas of moderately uniform

Fig.7.12. Velocity versus age and depth of burial. (From Faust, 1951.)

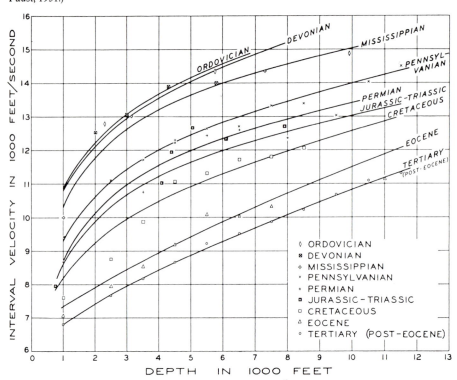

Fig.7.13. Velocity in brine-saturated Berea sandstone as a function of temperature and pressure. (After Timur, 1977.)

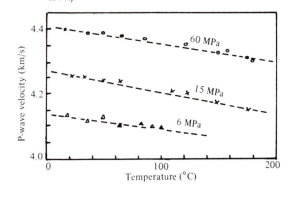
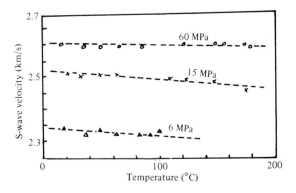

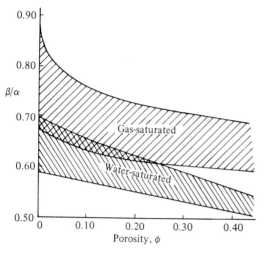

Fig.7.14. Relation of S- and P-wave velocity and porosity for gas- and water-saturated rocks. Data from Gregory (1976).

geology, such as the US Gulf Coast, exhibit little variation in the velocity function from area to area. Because of the seaward regional dip in the Gulf Coast area, as one goes seaward younger section is encountered at a given depth but the velocity function does not vary greatly; the maximum pressures to which the rocks have been subjected are the existing pressures which depend mainly on depth, not age. On the other hand, areas subject to recent structural deformation and uplift, such as California, exhibit rapid variation of velocity function from area to area. Many of the California rocks have been buried to greater depths and subjected to greater stresses than exist at present. The result is rapid lateral changes in velocity which profoundly affect seismic interpretation.

Empirical data suggest that the maximum depth to which a rock has been buried is a measure of the irreversible effect on porosity and is therefore an important parameter in determining porosity. In summary, porosity is determined principally by the existing differential pressure and the maximum depth of burial.

Fig.7.15. Effect of gas or oil saturation on velocity. Solid curves are for gas, dashed for oil. (After Domenico, 1974.) (a) P-wave velocity versus percentage of pore space filled with water (*water saturation*) with gas or oil

filling the remaining pore space, for various depths; (b) reflection coefficient for oil or gas sands overlain by shale.

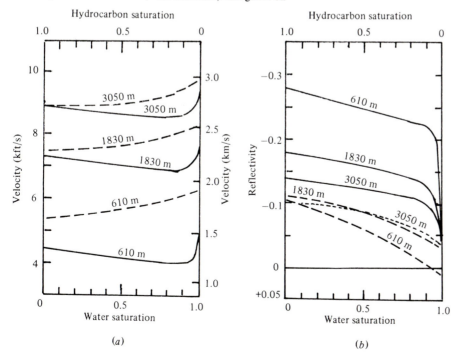

(a)   (b)

The irreversible change in porosity (and consequently in velocity) with depth of burial has been used to determine the maximum depth at which a section formerly lay. If the velocity–depth relationship for a given lithology can be established in an area not subjected to uplift, the maximum depth of burial can be ascertained from the observed velocity–depth relationship and hence the amount of the uplift can be inferred. In fig. 7.16 the shale and limestone regression lines (curves *A* and *C*) represent measurements on 'pure' shales and limestones which are believed to be at their maximum depth of burial. Curve *B*, which is obtained from these curves by interpolation, is the predicted curve based on the relative amounts of shale and limestone actually present and assuming the rocks to be at their maximum depth of burial. The displacement in depth required to fit this curve to the actual measurements is presumed to indicate the amount of uplift which has occurred. This technique can sometimes be used to determine if rocks have ever been buried deeply enough to acquire the high temperatures required for hydrocarbon generation (see §9.1.1).

7.2.2 *The weathered or low-velocity layer*

Seismic velocities which are lower than the velocity in water usually imply that gas (air or methane resulting from the decomposition of vegetation) fills at least some of the pore space (Watkins *et al.*, 1972). Such low velocities

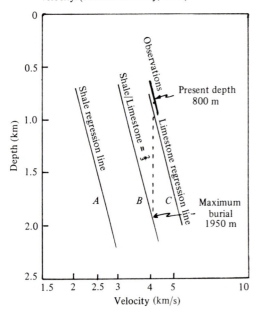

Fig.7.16. Finding maximum depth of burial from velocity. (From Jankowsky, 1970.)

are usually seen only near the surface in a zone called the *weathered layer* or the *low-velocity layer*, often abbreviated *LVL*. This layer, which is usually 4–50 m thick, is characterized by seismic velocities which are not only low (usually between 250 and 1000 m/s) but at times highly variable. Frequently the base of the LVL coincides roughly with the water table, indicating that the low-velocity layer corresponds to the aerated zone above the water-saturated zone, but this is not always the case. In areas of seasonal fluctuation of the water table, leaching and redeposition of minerals may produce the effect of double weathering layers. Double weathering effects sometimes result from a perched water table or changes at the base of glacial drift which is at a different depth than the water table. In desert areas where there may be no definite water table the LVL may grade continuously into sediments with normal velocity. In subarctic areas muskeg swamp is mushy with low velocity in summer and frozen with high velocity in winter (see also §7.2.3). In other areas the nature of the low-velocity layer and the problems associated with it change considerably with season. Obviously the term 'weathering' as used by geophysicists differs from the geologist's 'weathering' which denotes the disintegration of rocks under the influence of the elements.

The importance of the low-velocity layer is five-fold: (1) the absorption of seismic energy is high in this zone, (2) the low velocity and the rapid changes in velocity have a disproportionately large effect on traveltimes, (3) because of the low velocity, wavelengths are short and hence much smaller features produce significant scattering and other noise, (4) the marked velocity change at the base of the LVL sharply bends seismic rays so that their travel through the LVL is nearly vertical regardless of their direction of travel beneath the LVL, and (5) the very high impedance contrast at the base of the LVL makes it an excellent reflector, important in multiple reflections. Because of the first factor, records from shots in this layer often are of poor quality and efforts are made to locate the shot below the LVL. Methods of investigating the low-velocity layer are discussed in §5.3.6 and methods of correcting for it in §5.6.2.

In some areas where there is significant compaction with depth within the low-velocity layer, the velocity increase with depth z approximates

$$V = az^{1/n} \tag{7.9}$$

where a and n are empirically-derived constants. Blondeau and Swartz developed the *Blondeau method* for determining the vertical traveltime to a datum when the velocity obeys (7.9)(Duska, 1963; Musgrave and Bratton,

1967). If the first-break time–distance curve is a straight line when plotted on log–log paper, the method is applicable. The line's slope gives *n*. The calculation procedure is discussed in problem 7.11. This method has been applied mainly in glacial-drift areas.

7.2.3 *Permafrost*

The temperature of near-surface rocks is usually about the mean annual temperature for the location, and in arctic and some subarctic areas this temperature is below the freezing point. Seismic velocity generally increases markedly when the pore fluid in a rock freezes. In muskeg areas where the near-surface material is essentially swamp when not frozen and rich in undecayed vegetation, the velocity may increase from 1.8 km/s or lower to 3.0–3.8 km/s upon freezing. Timur (1968) reports Berea sandstone velocity changing from 3.9 to 5.2 km/s, Spergen limestone from 4.4 to 5.7 km/s, and black shale from 3.6 to 3.9 km/s upon freezing. The amount by which the velocity changed was roughly proportional to the porosity.

The portion of the section which is frozen year-round is called *permafrost*. There is usually a layer above it which thaws in the summer and the general increase of temperature with depth imposes a lower limit. Permafrost thickness varies from tens of centimeters to a kilometer. Where it is very thick, the velocity near its base may decrease with depth gradually until velocities are normal for the rock type. Where the permafrost is relatively thin, the decrease in velocity at its base may be fairly abrupt.

A body of water on the surface usually does not freeze deeper than a few meters and the water insulates the sediments lying below it from the cold so that permafrost is often absent under water bodies. The lateral change from normal velocities under lakes and rivers to high permafrost velocities on adjacent land areas can be very abrupt and can produce the appearance of major fictitious structures deeper in the section.

Whereas refraction at the base of a low-velocity layer tends to make raypaths traverse the layer more nearly vertically, refraction at a permafrost boundary makes raypaths in the permafrost more oblique and increases the traveltime spent in the permafrost. Furthermore, this effect is greater with longer offset traces which generally have more horizontal component of travel, and consequently some of the assumptions of the models on which static corrections, velocity analysis, CDP stacking, etc., are based may not apply. The result is that our ability to correct for permafrost effects is often poor. To complicate the problem, we usually cannot

determine accurately permafrost thickness and velocity variations.

Another phenomenon associated with permafrost is *frost breaks* (ice breaks) which result from cracking of the ice outward from the shotpoint (fig. 7.17). These sudden energy releases occur abruptly at various times after the shot and involve appreciable energy release, so that their effect is that of repeated erratic shots which may obscure reflections from the primary shot. Frost breaks are less likely to occur as the source energy decreases, so one may have to use smaller charges than otherwise desirable and increase the amount of stacking to compensate (Rackets, 1971).

7.2.4 *Abnormal-pressure detection*

'Normal' pressure for rocks is the situation where the pressure of the fluid in the rock's pore space is that of a hydrostatic head equal to the depth of burial. If the density of the fluid is ρ_f, the fluid pressure $\mathscr{P}_f$ is $\mathscr{P}_f = \rho_f z$ where *z* is the depth. Drillers often speak of the pressure gradient, $d\mathscr{P}_f/dz = \rho_f$, which is about 10 kPa/m or 0.45 psi/ft for $\rho_f = 1.04$ g/cm^3 (gradients between 0.48 and 0.43 psi/ft are usually regarded as 'normal'). The pressure exerted by the rock overburden is about $\mathscr{P}_m = 22.5$ kPa/m or 1.0 psi/ft (for $\rho_m = 2.3$ g/cm^3). The effective stress on a rock (as discussed in §7.1.5) is the differential pressure, $\Delta\mathscr{P} = \mathscr{P}_m - \mathscr{P}_f = 12.5$ kPa/m or 0.55 psi/ft.

Abnormal or overpressure situations (subnormal pressures are also occasionally encountered) result from a sealing of formations as they are buried so that the formation liquid cannot escape and allow the formation to compact under the increased overburden pressure (Plumley, 1980). In effect, part of the weight of the overburden is transferred from the rock matrix to the fluid in the pore spaces. Consequently the rock 'feels' that it is under a differential pressure appropriate to some shallower depth and the velocity of the rock is that of the shallower depth.

The deeper portions of many depositional sequences involve fine-grained sediments where the permeability was not sufficient to allow the interstitial water to escape during compaction and abnormal pressures are common. This is especially true in young Tertiary basins where deposition has been fairly rapid, such as the US Gulf Coast, the Niger and Mackenzie deltas and along the continental slopes in many areas. Abnormal-pressure formations may behave as viscous fluids lacking shear strength and become involved in diapiric flow (see fig. 9.19), or become weak detachment zones for faulting (Gretener, 1979).

Fig.7.17. Seismic records from the Arctic showing frost
breaks. (Courtesy Petrocanada.)

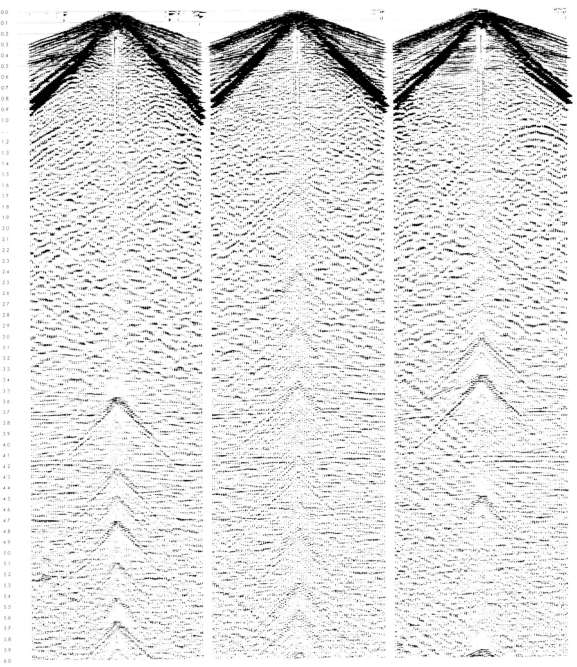

Where reflectors lie within or below abnormally-pressured section, velocity analyses (§8.2.3) may permit calculating interval velocities and not only detecting overpressure but also determining the amount (Reynolds, 1970). The tendency in picking velocity analyses is to honor only velocity data which show a monotonic increase of stacking velocity with depth, which excludes the velocity inversions which evidence abnormal-pressure zones. Because data in abnormal-pressure zones are usually stacked with a velocity which is too high, reflections within such zones usually appear very poor and sometimes the deterioration of reflections can be used to indicate such zones. Multiples, of course, are also usually evidenced by low stacking velocity and hence may make the interpretation of abnormal-pressure zones difficult. The prediction of abnormal pressure is of considerable importance in drilling plans to minimize the possibilities of blowouts and other drilling problems. It is also of importance in predicting gas reserves since gas reservoirs which lie within abnormally-pressured zones can contain exceptionally large amounts of gas for the reservoir volume.

When searching for abnormal pressures, velocity surveys (§8.2.3) are usually run with smaller increments than when the objective is primarily to determine the stacking velocity. Aud (1976) contends that velocity-scan increments should be 50 ft/s and time increments 10 ms; he also argues that data should be at least 12-fold with offsets sufficient to give at least 100 ms of normal moveout. He often picks events spaced only 100 ms apart. The technique of averaging data for a number of adjacent midpoints generally leads to less noise in measurements but also to poorer identification of velocity values with specific events. Because of the uncertainties in the values determined from any single velocity analysis, weighted averaging of the results of several adjacent analyses improves reliability.

7.2.5 *Gas-hydrate effects*

Reflections which cut across the bedding are sometimes seen on deep-water seismic data a short distance below the seafloor, as in fig. 7.18. These are often attributed to gas hydrates, ice-like crystalline lattices of water molecules in which gas molecules are trapped physically. These are stable under the temperature and pressure conditions found just below the seafloor in deep water. They apparently can form where the gas concentration exceeds that necessary to saturate the interstitial water. The velocity in methane-hydrate sediments is about 2.0–2.2 km/s (Tucholke *et al.*, 1977). A 'base of gas-hydrate reflection' usually is roughly conformable to the seafloor in a region of landward-dipping bedding, and the depth of the reflection beneath the seafloor corresponds roughly to the limit of stability of methane hydrate (fig. 7.19). The reflection is thus interpreted as marking

Fig.7.18. Seismic line on the Blake Outer Ridge, Offshore South-eastern US, showing gas-hydrate reflection. (From Shipley *et al.*, 1979.)

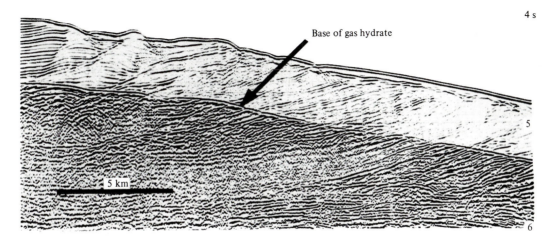

the interface between hydrate and gas trapped by the overlying hydrate. The gas trapped in this way may someday be an energy resource.

7.3   Measurement of velocity

7.3.1   *Conventional well surveys*

The most direct methods of determining velocity require the use of a deep borehole. Two types of *well surveys* are used: the 'conventional' method of shooting a well and sonic logging (or continuous velocity survey) which is discussed in the next section.

Shooting a well consists of suspending a geophone or hydrophone in the well by means of a cable and recording the time required for energy to travel from a shot fired near the well down to the geophone (see fig. 7.20). Airguns in the mud pit or in the water for marine wells are also used as energy sources. The geophone is specially constructed to withstand immersion under the high temperatures and pressures encountered in deep oil wells. A mechanical arm presses the geophone against the borehole wall to assure coupling. The cable has a threefold role: it supports the geophone, it serves to measure the depth of the geophone and it carries electrical conductors which bring the geophone output to the surface where it is recorded. Shots are fired at one or more points near the wellhead. The geophone is moved between shots so that the results are a set of traveltimes from the surface down to various depths. The geophone depths are chosen to include the most important geological markers, such as tops of formations and unconformities, and also intermediate locations so that the interval between successive measurements is small enough to give reasonable accuracy (often 200 m apart).

Results of a typical well survey are shown in fig. 7.21. The vertical traveltime, t, to the depth, z, is obtained by multiplying the observed time by the factor $\{z/(z^2 + x^2)^{\frac{1}{2}}\}$ to correct for the actual slant distance. The average velocity between the surface and the depth z is then given by the ratio z/t. Figure 7.21 shows the average velocity V and the vertical traveltime t plotted as functions of z. If we subtract the depths and times for two shots, we find the *interval velocity* V_i, the average velocity in the interval $(z_m - z_n)$, by means of the formula

$$V_i = \frac{z_m - z_n}{t_m - t_n} \qquad (7.10)$$

Shooting a well gives the average velocity with good accuracy of measurement. It is, however, expensive

Fig. 7.20. Shooting a well for velocity.

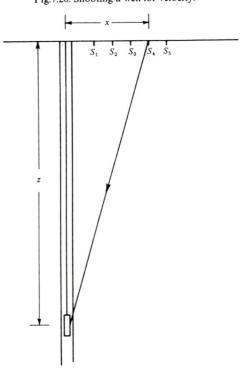

Fig. 7.19. Limit of stability of methane hydrate in water containing $3\frac{1}{2}\%$ NaCl. The horizontal distance between the dotted vertical line (indicating seafloor temperature of 3 °C) and the stability-limit line is roughly proportional to the thickness of the gas-hydrate zone. The 0.6 s thickness and 2.0 km/s velocity of fig. 7.18 are consistent with measured temperature gradients of about 0.03 °C/m. (After Tucholke *et al.*, 1977.)

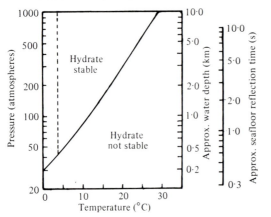

since the cost includes not only the one-half to one day's time of the seismic crew but also the cost of standby time for the well (which often far exceeds the seismic cost). Potential damage to the well is another factor which discourages shooting wells; while the survey is being run, the well must stand without drill stem in the hole and

hence is vulnerable to cave-in, blowout, or other serious damage. A further disadvantage in new exploration areas is that seismic surveys are often completed before the first well is drilled.

7.3.2   *Velocity logging*

The continuous-velocity survey makes use of one or two pulse generators and two or four detectors, all located in a single unit called a *sonde* which is lowered into the well. Figure 7.22a shows the borehole-compensated sonic-logging sonde developed by Schlumberger. It consists of two sources of seismic pulses, S_1 and S_2, and four detectors, R_1 to R_4, the 'span' distances from R_1 to R_3 and from R_2 to R_4 being 61 cm (2 ft). The spacing of the 'borehole-compensated' sonde is 1.22 m (4 ft), on the 'long-spaced' sonde 2.44 m (8 ft). With the longer spacing there is greater likelihood that the velocity measured will be that of unaltered formation. The velocity is found by measuring the traveltime difference for a pulse traveling from S_1 to R_2 and R_4, similarly for a pulse going from S_2 to R_3 and R_1, then taking the average of the differences. The sonde is run in boreholes filled with drilling mud which has a seismic velocity of roughly 1500 m/s; however, the first energy arrivals are the P-waves which have travelled in the rock surrounding the borehole. Errors arising from variations in borehole size or mud cake thickness near the transmitters are effectively eliminated by measuring the

Fig.7.21. Plot of well-velocity survey.

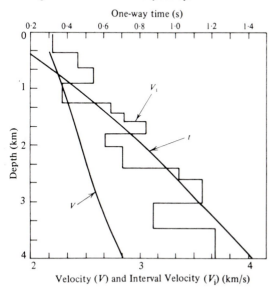

Fig.7.22. Sonic logging. (Courtesy Schlumberger.)
(*a*) Borehole-compensated logging sonde; (*b*) sonic log.

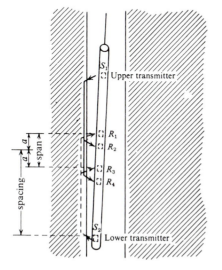

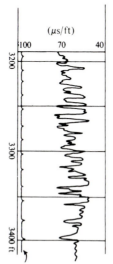

Integrated traveltime showing ms ticks

(*a*)   (*b*)

difference in arrival time between two receivers; errors resulting from such variations near the receivers are reduced by averaging the results from the two pairs of receivers. The sonic log (fig. 7.22*b*) shows as a function of depth the transit time divided by the span (expressed in μs/ft), the result being the reciprocal of the P-wave velocity in the formation.

The traveltime interval between sonic log receivers is measured by a device which automatically registers the arrival of the signal at each of the two receivers and measures the time interval between the two. Since the signal at the receiver is not a sharp pulse but instead is a wavetrain, the detector is actuated by the first peak (or trough) which exceeds a certain threshold value. At times the detector is not actuated by the same peak (or trough) at the two receivers and hence the increment of traveltime will be in error. This effect, called *cycle skip*, usually can be detected and allowed for since the error is exactly equal to the known interval between successive cycles in the pulse (Kokesh and Blizard, 1959).

The accuracy of sonic log values is often rather poor, as evidenced by frequent disagreements between regular and long-spaced logs. This fact should be realized by geophysicists, who are inclined to believe sonic and distrust seismic data when faced with disagreements. Sonic logs often suffer from inadequate penetration, hole caving, alterations with time after drilling the borehole, and other factors. Depending on the use to which sonic log data are to be put, editing and correcting may be desirable. Editing involves detailed comparisons of different logs to locate improbable values and replace these with values believed to be more reasonable. Editing is often essential if good synthetic seismograms (see §9.4.4) are to be made from log data.

The sonic log is automatically integrated to give total traveltime which is then shown as a function of depth by means of ticks at intervals of 1 ms (see fig. 7.22*b*). There is a tendency for small systematic errors to accumulate in the integrated result. Check shots can be made at the base and top of the sonic log so that the effect of the cumulative error can be reduced by distributing the difference in a linear manner. The sonde often includes a well seismometer of the type used in shooting a well to facilitate taking check shots.

The instantaneous velocity fluctuates rapidly in many formations, as seen in fig. 7.22*b*. While the velocity distribution, if considered in detail, is an extremely irregular function, the wavelengths used in seismic exploration are so long (generally greater than 30 m) that the rapid fluctuations are not significant in determining the path of waves.

Sonic logs are used for porosity determination because porosity appears to be the dominant factor in seismic velocity. Although sonic logs are of great value to the geophysicist, they are usually not run with the geophysical uses in mind and hence often do not produce all the information which the geophysicist wants. For example, check shots are not necessary for porosity determination and, therefore, are often omitted; the log usually does not cover the entire hole depth and sonic log data are rarely available for the shallow part of the borehole. Thus the sonic log data are usually incomplete so that using such logs for velocity control involves assumptions where the data are missing. This is especially true in the applications of sonic logs in preparing synthetic seismographs with which we shall deal in §9.4.4.

Several variations of acoustic borehole logs are used sometimes though they have little application in seismic work. If the entire waveform is recorded, we get in addition to the first-arrival P-wave in the rock, an S-wave in the rock, a P-wave in the mud, and a tube wave (see §2.2.10*d*). The different waves usually differ sufficiently in arrival times that their amplitudes can be measured by presetting time windows. The amplitude of the S-waves is sometimes displayed on a 'fracture-finder log,' high amplitude indicating the absence of fractures. The P-wave amplitude when the sonde is inside casing is displayed in the 'cement-bond log,' the amplitude being high when the casing is hanging freely (because the energy is trapped in the steel casing), low when the casing is bonded to the formation by cement. The full waveform is sometimes displayed in a '3-D log,' which is also used to measure the quality of bonding casing to formation.

7.3.3 *Measurements based on variations of traveltime with offset*

(*a*) *X^2–T^2 method.* The arrival time of reflected energy depends not only on the reflection depth and the velocity above the reflector but also on offset distance. Several methods (including the velocity-analysis methods which will be described in §8.2.3) utilize this dependence on offset as a means of measuring the velocity. Two classical methods, $X^2 - T^2$ and $T - \Delta T$, are central to surface velocity-measurement methods, even though both methods have fallen into disuse.

The X^2–T^2 *method* is based upon (3.21). We write

$$t^2 = x^2/V_s^2 + t_0^2. \tag{7.11}$$

When we plot t^2 as a function of x^2, we get a straight line whose slope is $1/V_s^2$ and whose intercept is t_0^2, from which we can determine the corresponding depth. The quantity V_s is the velocity assumed in CDP stacking (§8.2.5) and hence is called *stacking velocity*. When we have horizontal velocity layering and horizontal reflectors, V_s is the same as the rms velocity $\bar{V}$ in (3.22):

$$V_s^2 \approx \bar{V}^2 = \sum_{i=1}^{n} V_i^2 \Delta t_i / \sum_{i=1}^{n} \Delta t_i, \qquad (7.12)$$

where V_i is the velocity in a layer through which the two-way traveltime is Δt_i. Under different circumstances,

$V_s = V$ for constant velocity and a horizontal reflector,
$V_s = V/\cos \xi$ for constant velocity and reflector dip ξ,
$V_s = \bar{V}$ for horizontal velocity layering and reflectors,
$V_s = \bar{V}/\cos \xi$ for dipping but parallel velocity layering and reflectors.

In the general case, there is no simple relationship between V_s and $\bar{V}$. The stacking velocity V_s is used in correcting CDP data before stacking (hence its name) even where its relation to the velocity distribution is very complicated or not known. We sometimes use the equivalent average velocity $\bar{V}$(§3.2.2) for depth determination:

$$\bar{V} = \sum_{i=1}^{n} V_i \Delta t_i / \sum_{i=1}^{n} \Delta t_i; \qquad (7.13)$$

this equation also assumes horizontal velocity layering (and vertical raypath).

Waters (1978) shows seismic data plotted where the scales are linear with x^2 and t^2 rather than linear with x and t, as an aid in picking V_s.

When the regular seismic profile does not have a sufficiently large range of x-values to enable us to find V with the accuracy required, special long-offset profiles are shot. Dix (1955) described an arrangement of profiles which involves the same portion of the reflecting bed and which eliminates the effect of dip in the calculations and provides time-ties that can be used as checks that the picking of the reflection events has not involved a jump of one leg, that is, one cycle. With the longer offsets used in CDP work, special shooting for velocity is less common although still sometimes employed, especially for deep reflectors. An X^2–T^2 survey can give velocities accurate within a few per cent if: (1) the records are of good quality and have at least a moderate number of reflections, (2) accurate near-surface corrections are applied, (3) the field work and the interpretation are carefully done, and (4) the velocity distribution is simple (that is, no lateral variation of velocity or complexity of structure).

Once the velocities have been determined to two successive parallel reflectors using (7.12), the interval velocity, V_n, can be found from the Dix formula. Writing $\bar{V}_L$ for the velocity to the nth reflector and $\bar{V}_U$ for the velocity to the reflector above it, (7.12) gives

$$\sum_{1}^{n} V_i^2 \Delta t_i = \sum_{1}^{n-1} V_i^2 \Delta t_i + V_n^2 \Delta t_n = \bar{V}_L^2 \sum_{1}^{n} \Delta t_i$$

$$\sum_{1}^{n-1} V_i^2 \Delta t_i = \bar{V}_U^2 \sum_{1}^{n-1} \Delta t_i.$$

Subtracting and dividing both sides by Δt_n then gives the *Dix formula*,

$$V_n^2 = \left(\bar{V}_L^2 \sum_{1}^{n} \Delta t_i - \bar{V}_U^2 \sum_{1}^{n-1} \Delta t_i \right) / \Delta t_n. \qquad (7.14)$$

Note that the Dix formula implies that the travel paths to the $(n-1)$th and nth reflectors are essentially identical except for the additional travel between the two reflectors. When the two reflectors are not parallel or when the offset is large, this condition is not satisfied and the Dix formula may give meaningless results.

(b) *T–ΔT method*. The *T–ΔT method* is based upon (3.7) which can be written in the form

$$V = \frac{x}{(2t_0 \, \Delta t_n)^{\frac{1}{2}}} \qquad (7.15)$$

With symmetrical spreads Δt_n can be calculated from the arrival times of a reflection event at the shotpoint (t_0) and at the outside geophone groups, t_1 and t_k. Dip moveout is eliminated by averaging the moveouts on the opposite sides of the shotpoint:

$$\Delta t_n = \tfrac{1}{2}\{(t_1 - t_0) + (t_k - t_0)\} = \tfrac{1}{2}(t_1 + t_k) - t_0. \qquad (7.16)$$

The values of Δt_n given by this equation are subject to large errors, mainly because of uncertainties in the near-surface corrections. To get useful results, large numbers of measurements must be averaged in the hope that weathering variations and other uncertainties will be sufficiently reduced (Swan and Becker, 1952).

(c) *Best-fit approaches*. Most velocity determination is done in data-processing, which is discussed in §8.2.3. These methods are based on either (1) finding the hyperbola which best fits coherent events assumed to be primary reflections within some given space and time window, or (2) finding which stacking velocity produces the 'best' stacked section. Such measurements are generally sufficiently accurate for stacking but not always for the lithologic conclusions sometimes drawn from them.

7.3.4 *Other sources of velocity information*

At least in principle, several other sources of velocity information might be used. The curvature of diffractions (see fig. 4.3) depends on velocity and occasionally is used for velocity determination where the orientation of the diffraction-generating feature is known (or assumed). Automatic migration methods (of which the collapsing of diffractions is part) also involve velocity, and maximizing the coherency of a migrated section conceptually provides velocity information. The amplitude of reflections provides information about changes in the velocity-density product which is used in the generation of seismic logs (§9.4.5), and the variation of reflection amplitude with angle of incidence also depends on velocity. The accuracy of velocity values derived by these various methods is however usually poor.

Velocity information can also be obtained from types of measurements which do not depend on reflection travel paths, such as of head-wave velocity.

7.4 Interpretation of velocity data

With common-depth-point data which have a high degree of redundancy (that is, which sample the same depth point many times) and velocity-analysis programs (§8.2.3*a*), interval velocities can be calculated from (7.14) for the intervals between parallel reflectors at many points on the section – in fact, almost on a continuous basis. After due allowance for the uncertainties involved in the measurements, systematic variations might be interpreted in stratigraphic terms (Hofer and Varga, 1972). Carbonate and evaporite velocities are sufficiently higher than clastic velocities (especially in Tertiary basins) that they can often be distinguished.

The analysis of velocity data constitutes an important interpretation problem. As with other interpretation problems, some interpretations can be ruled out because they imply impossible or highly improbable situations. From experience we know that velocity does not vary in a 'capricious' manner. Thus it would be unreasonable to expect the velocity to vary from place to place in other than a slow systematic way unless the seismic section shows significant structural or other changes which suggest a reason why the velocity should change rapidly. Thus one might expect two velocity analyses to show

some differences between portions of the section which are separated by a fault whereas one expects little variation in portions which appear to be continuous. Likewise, interpreting an increase in velocity as a reef buildup in an otherwise clastic section might be warranted if the increase is accompanied by changes in reflection character and by structural evidence such as diffractions, or if the increase occurs over deeper structure which could have caused this area to be relatively high during deposition so that a reef might have grown here rather than elsewhere.

The desire to extract stratigraphic information from velocity data sometimes results in interpretations which exceed the limitations of the data. Small errors in normal-moveout measurements can produce sizeable errors in stacking velocities, especially for deep reflections, and these in turn cause large errors in the calculation of interval velocities when the intervals are small. Where the reflectors are not parallel, interval-velocity calculations are meaningless (Taner *et al.*, 1970). Velocity measurements sometimes are severely distorted by various factors, such as interference effects, noise of various kinds, distortions produced by shallow velocity anomalies or weathering variations, and great care must be exercised that these effects are not taken as indications of actual changes in the velocity of the rocks.

Problems

7.1. Sandstone is generally expected to have a higher velocity than shale. From fig. 7.1, what percent difference should be expected at depths of 1, 2 and 4 km, assuming that both sandstone and shale have porosities as given by the limit curve in fig. 7.6. How does the difference shown in fig. 7.2 compare?

7.2. (*a*) Assume that sandstone is composed only of grains of quartz, limestone only of grains of calcite, and shale of equal quantities of kaolinite and muscovite. For salt-water ($\rho = 1.03$ g/cm^3) saturated sandstone, limestone and shale, what porosities are implied by the lower and upper limits of the density ranges and mean values shown in table 7.1*b*? (*b*) What velocities would be expected for these nine values according to Gardner's rule? Where do these values plot on fig. 7.1?

7.3. Based on the reasoning in §7.1.5, how would you expect velocity to depend on grain size?

7.4. Assume that the velocity of calcite is 6.86 km/s and of quartz is 5.85 km/s; what velocities should be expected of 10, 20 and 30% porosity (*a*) limestone composed only of calcite; (*b*) sandstone composed only of quartz? Where do these values plot on figs. 7.1 and 7.5?

7.5.   What physical fact determines the 'limit of poros-ity' line in fig. 7.6? What is implied for measurements which fall to the right of this line?

7.6.   Figure 7.9 shows velocity versus depth for normally pressured shales. (*a*) How do the velocities shown in fig. 7.10 above and below the 'top overpressure' compare with the curve? What depth would correspond to normal pressure for the overpressured shale? What porosity would you expect for the overpressured shale? (*b*) Plot the velocities for 100% water-saturation from fig. 7.15 on the curve; how do they compare? (*c*) Why do the velocity–depth curves for the various areas depart from this curve? Incorporate your knowledge of the geology of the various areas in your answer. (*d*) Plot the shale and limestone values from fig. 7.16 for depths of 1000 and 2000 m on this curve; how do they compare?

7.7.   What shale velocities are consistent with the oil–sand data shown in fig. 7.15*b*? (Determine velocities for two values of water-saturation for each of the three depths, neglecting differences between sand and shale densities.)

7.8.   Assume that raypaths have an angle of approach of $10°$, $20°$ and $30°$ in the sub-weathering with a velocity of 2400 m/s. (*a*) For a weathered layer 10 m thick with a velocity of 500 m/s, how do traveltimes through the weathering compare with that for a vertically-traveling ray? What is the horizontal component of the raypath in the weathering. (*b*) For permafrost 100 m thick with a velocity of 3600 m/s, answer the questions in (*a*).

7.9.   In the early days of refraction exploration for saltdomes sketches were drawn indicating that the angle of approach to the surface should have a large horizontal component, but measurements with 3-component seismographs showed very little horizontal component and controversy arose therefore over whether the travelpaths could be as drawn. Explain the apparent discrepancy based on your concept of the actual Earth.

7.10.   (*a*) Assume a subsiding area without uplift activity. A shale is normally pressured until it reaches a depth of burial of 1400 m, at which point it becomes cut off from fluid communication, that is, interstitial fluid can no longer escape. If it is found at depth of 2000 m, what

velocity and what fluid pressure would you expect? If at a depth of 3000 m? (*b*) Assume a shale buried to 3000 m and then uplifted to 2000 m, being normally pressured all the time; what velocity and fluid pressure would you expect? (*c*) Assume the shale in part (*a*) is buried to 3000 m and then uplifted to 2000 m, without fluid communication being established; what velocity and fluid pressure would you expect? What if uplifted to 1000 m?

7.11.   The Blondeau method of making weathering corrections starts with a curve of emergent arrival times versus offsets, t versus x in fig. 7.23 plotted on log – log graph paper. The slope gives $B = (1 - 1/n)$ where $1/n$ is the exponent in (7.9). To find t_v, the vertical traveltime to a given depth z_m, we find F, a tabulated function of B (see Musgrave and Bratton, 1967, p. 233). Then, $x = Fz_m$ so that we can get t from the $x - t$ curve. Finally, $t_v = x/F$.

  Verify the above procedure by deriving the follow-ing relations. (*a*) $z = z_m \sin^n i$ where i is the angle of in-cidence measured with respect to the vertical at depth z; (*b*) $x = Fz_m$ where $F = 2n\int_0^{\frac{1}{2}\pi} \sin^n i \, di = $ function of n, hence of B also; (*c*) $t = (G/a)(x/F)^B$ where $G = 2n\int_0^{\frac{1}{2}\pi} \sin^{n-2} i \, di$; (*d*) $dx/dt = x/Bt = $ horizontal com-ponent of apparent velocity at point of emergence; (*e*) $dx/dt = $ horizontal component of apparent velocity of wavefront at any point of the trajectory $= V_m = x/Bt$; (*f*) $t_v = \int_0^{z_m} dz/V = t/F$.

7.12.   Figure 7.24 shows data from a well velocity survey tabulated on the standard calculation form. (*a*) Plot time–depth, average velocity versus depth, and interval velocity versus depth graphs using a sea-level datum. (*b*) How much error in average and interval velocity values would result from: (i) time-measurement errors of 1 ms; (ii) depth-measurement errors of 1 m? (*c*) Determine V_0, a for a velocity function fit to these data assuming the functional form $V = V_0 + az$ where V is interval velocity and z depth.

7.13.   Velocity analysis usually results in a plot of stack-ing velocity against arrival time. Bauer (private communi-cation) devised a 'quick-look' method of determining the interval velocity, assuming that the stacking velocity is average velocity and horizontal layering. The method is shown in fig. 7.25. A box is formed by the two picks between which the interval velocity is to be picked and the diagonal which does not contain the two pick points when extended to the velocity axis gives the interval velocity. (*a*) Prove the validity of this method, or show its limitations. (*b*) This method is useful in seeing the influence of measurement errors; discuss the sensitivity of interval-velocity calculation to: (i) error in picking velocity values from the graph; (ii) error in picking time; (iii) picking events very close together; (iv) picking each event late.

Fig.7.23. Illustrating Blondean weathering corrections.

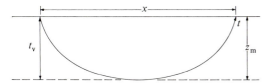

Fig.7.24. Data from a well-velocity survey.

Location / header fields: Elevation 107 m · Total Depth 3395 m

Shothole information:

	Elevation	Distance	Direction from Well
1	78 m	125 m	W
2	—	125	33
3	78	125	33
4	78	125	33

Data table:

Record No.	Shothole No.	Dgₚ	tₘ	t_c	T Reading	Pola (up/grade)
1	1	3395	026	088	0.990	+
2	2	3395	027	088	0.990	+
3	—	3305	024	089	0.976	+
4	1	3200	024	088	0.95	+
5	—	3100	023	086	0.934	+
6	—	2970	023	088	0.907	+
7	1	2810	024	089	0.868	+
8	1	2690	023	088	0.840	+
9	—	2535	023	086	0.805	+
10	2	2400	027	086	0.769	+
11	2	2270	024	089	0.752	+
12	2	2180	024	089	0.728	+
13	2	2110	024	085	0.709	+
14	2	2045	023	089	0.697	+
15	3	1925	024	088	0.671	+
16	3	1750	024	086	0.649	+
17	3	1670	027	088	0.617	+
18	3	1560	023	087	0.586	+
19	3	1355	023	086	0.546	+
20	3	1265	023	088	0.509	+
21	3	1150	023	085	0.477	+
22	3	1040	023	085	0.426	+
23	3	845	023	089	0.373	+
24	4	690	027	083	0.310	+
25	4	625	023	085	0.288	+
26	4	470	024	085	0.226	+
27	4	320	024	089	0.165	+
28	4	21.5	023	024	0.129	+

The remaining body of the form is a blank columnar worksheet with the following column headers (reading across): Dgₛ · H · tan i · cos i · Tgₛ · Δsd · Δsd/V · Tgd · Tgd Average · Dgd · ΔDgd · ΔTgd · Vᵢ Interval Velocity · Vₐ Average Velocity.

Definitions panel (right side):

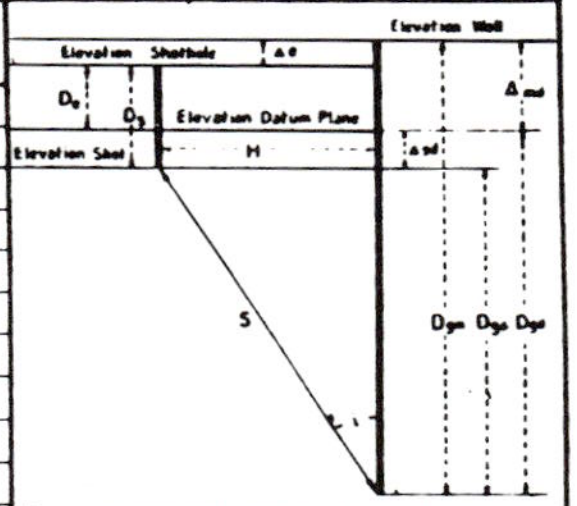

D_{gm} = Geophone depth measured from well elevation
D_{gs} = shot
D_{ge} = datum
D_s = Depth of shot
D_e = Shothole elevation to datum plane
H = Horizontal distance from well to shotpoint
S = Straight line travel path from shot to geophone
t_u = uphole time at shotpoint
T = Observed time from shotpoint to well geophone
t_r = in reference geophone
Δe = Difference in elevation between well (datum) and shot (datum) plane
Δsd = $D_s - D_e$
Δsd = $D_s - D_e$
D_{gs} = D_{gm} = tan i · $\dfrac{H}{D_{gs}}$
T_{gs} = cos i · T = true travel time from shot to geophone
T_{gd} = $T_{gs} \pm \dfrac{\Delta sd}{V}$
D_{gd} = $D_{gs} \pm \Delta e$
V_i = $\dfrac{\Delta D_{gd}}{\Delta T_{gd}}$
V_a = $\dfrac{D_{gd}}{T_{gd}}$ = Average

Surveyed by _____
Date _____
Master velocity data _____

x	t_A	t_B	x	t_A	t_B	x	t_A	t_B
0.0 km	0.855 s	0.906 s	1.4 km	1.005 s	0.977 s	2.8 km	1.330 s	1.202 s
0.1	0.856	0.902	1.5	1.017	0.991	2.9	1.360	1.234
0.2	0.858	0.898	1.6	1.037	1.004	3.0	1.404	1.253
0.3	0.864	0.898	1.7	1.068	1.019	3.1	1.432	1.272
0.4	0.868	0.899	1.8	1.081	1.037	3.2	1.457	1.296
0.5	0.874	0.902	1.9	1.105	1.058	3.3	1.487	1.304
0.6	0.882	0.903	2.0	1.118	1.066	3.4	1.513	1.334
0.7	0.892	0.909	2.1	1.151	1.083	3.5	1.548	1.356
0.8	0.904	0.916	2.2	1.166	1.102	3.6	1.580	1.377
0.9	0.906	0.922	2.3	1.203	1.121	3.7	1.610	1.407
1.0	0.930	0.932	2.4	1.237	1.127	3.8	1.649	1.415
1.1	0.945	0.943	2.5	1.255	1.158	3.9	1.674	1.438
1.2	0.950	0.950	2.6	1.283	1.177	4.0	1.708	1.459
1.3	0.979	0.965	2.7	1.304	1.195			

7.14. Determine the velocity by the X^2-T^2 method using the data given above, t_A being for a horizontal reflector while t_B is for a reflector dipping $10°$ down toward the shotpoint.

7.15. Analysis of an X^2-T^2 survey gives the following results:

i	h_i	t_i	V_s
1	1.20 km	1.100 s	2.18 km/s
2	2.50	1.786	2.80
3	3.10	1.935	3.20
4	4.10	2.250	3.64

Fig.7.25. 'Quick-look' interval-velocity determination.

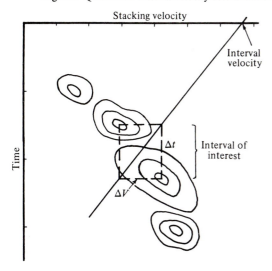

Calculate the interval velocities.

7.16. (a) Assume six horizontal layers each 300 m thick and constant velocity within each layer (fig. 7.26a), the successive layers having velocities of 1.5, 1.8, 2.1, 2.4, 2.7 and 3.0 km/s. Ray-trace through the model to determine offset distances and arrival times for rays which make an angle of incidence with the base of the 3.0 km/s layer (at A) of 0, 10, 20 and 30°. Calculate stacking velocity from each pair of values (six calculations) and compare with the average and rms velocities. (b) Repeat assuming the layers dip 20°, as shown in fig. 7.26b, for reflecting point B. (c) By trial and error, shift the reflecting point updip to achieve common midpoints. (d) Repeat by modifying the model to that shown in fig. 7.26c, for reflecting point C. (e) Shift the reflecting point C as required to achieve common midpoints.

7.17. Show that the values given in §7.2.4 for the normal pressure gradient in the Earth, namely 10 kPa/m and 0.45 psi/ft, are consistent.

7.18. By comparing figs. 7.11 and 7.21, what can you deduce about the nature of the rocks in the well for fig. 7.21?

7.19. (a) Given that the trace spacing in fig. 8.21 is 50 m, determine the stacking velocity, depth and the dip at approximately 0.5, 1.0, 1.5, 2.0, 2.3 and 2.4 s. (b) What problems or ambiguities do you have in picking the events? (c) How much uncertainty is there in your ability to pick times and how much uncertainty does this introduce into the velocity, depth and dip calculations?

7.20. (a) In fig. 8.12d, pick stacking velocity versus time pairs and calculate interval velocities; the analysis is for S.P. 100. (b) What can you tell about the lithology from this? (*c*) If the section which is present in the syncline but missing over the anticline consists of young, poorly-consolidated clastic rocks, what values would you expect a velocity function at about S.P. 45 to show? (*d*) Note the downdip thinning of the section from about 0.75 to 1.25 s at the left end of the section; suggest the explanation.

Fig.7.26. Models of 300 m thick layers (measured perpendicular to bedding), each of constant velocity.

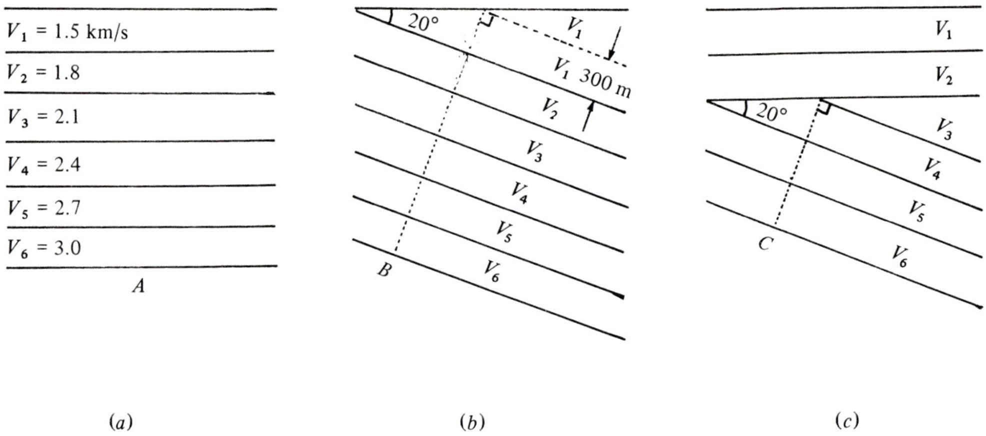
V_1 = 1.5 km/s
V_2 = 1.8
V_3 = 2.1
V_4 = 2.4
V_5 = 2.7
V_6 = 3.0
A
20°
V_1
300 m
V_2
V_3
V_4
V_5
V_6
B
C
(a)
(b)
(c)

8
Data-processing

Overview

Radar was one of the outstanding technological advances of World War II and was widely used in the detection of aircraft and ships at sea. However, noise frequently interfered with the application of radar and considerable theoretical effort was devoted to the detection of signals in the presence of noise. The result was the birth of a new field of mathematics – Information Theory. Early in the 1950s a research group at the Massachusetts Institute of Technology studied the application of the new field to seismic exploration problems (Flinn et al., 1967). Simultaneously with this development, rapid advances in digital computer technology made extensive calculations economically feasible for the first time. These two developments began to have an impact on seismic exploration in the early 1960s and before the end of the decade, data-processing (as their application is called) had changed seismic exploration dramatically, so much so that the changes are sometimes referred to as the 'digital revolution'. Most seismic recording is now done in digital form and most data are now subjected to data-processing before being interpreted.

At first, information theory was very difficult to understand because it was formulated in complex mathematical expressions and employed an unfamiliar vocabulary; however, as the number of applications has expanded, the basic concepts have been expressed more clearly and simply (Silverman, 1967; Lee, 1960; Robinson and Treitel, 1964; Anstey, 1970; Finetti *et al.*, 1971). A number of books on data-processing are now available (see §1.4); we mention specifically Kanasewich (1973), Claerbout (1976) and Robinson and Treitel (1980).

In chapter 10 we give the mathematical concepts most important in data-processing and in this chapter we show how these concepts are applied without being concerned with mathematical proofs. This chapter discusses mainly functions in digital form whereas chapter 10 treats both continuous and digital functions. Equation numbers give nearly identical equations in chapter 10.

Usually we think of seismic data as the variation with time (measured from the shot instant) of the amplitudes of various geophone outputs. When we take this viewpoint we are thinking in the *time domain*, that is, time is the independent variable. We also sometimes find it convenient to regard a seismic wave as the result of the superposition of many sinusoidal waves differing in frequency, amplitude and phase; the relative amplitudes and phases are regarded as functions of frequency and we are thinking in the *frequency domain*. The frequency-domain approach is illustrated by electrical systems which are specified by their effects on the amplitudes and phases of sinusoidal signals of different frequencies. For example, graphs of filter characteristics usually show amplitude ratios or phase shifts as ordinates with frequency as the abscissa.

Three types of mathematical operations constitute the heart of most data-processing: Fourier transforms, convolution and correlation. Fourier transforms (§8.1.1) convert from the time domain to the frequency domain and *vice versa*, and they and other types of transforms can be used to convert into and out of other domains also. The essential aspect of transforms is that in principle no information is lost in the procedure, although in actual application some degradation occurs because of the use of approximations, truncation, etc. Transforms provide alternate ways of doing things which are sometimes advantageous.

Convolution (§8.1.2) is the operation of replacing each element of an input with a scaled output function; it is the mathematical equivalent to filtering, such as occurs naturally in the passage of seismic waves through the Earth, in passing electrical signals through circuits, etc. The limitations on sampling and signal reconstitution, that there be no frequency components above half the sampling frequency, is explained using convolution concepts. Sometimes undesirable filtering can be undone by deconvolution.

Correlation (§8.1.3) is a method of measuring the similarity between two data sets. A common application is determining the time shift which will maximize the similarity. Correlation is also the means for extracting short signals of known waveshape from long wavetrains, as is used in Vibroseis processing. If a data set is correlated with itself (autocorrelation), a measure of the repetition in the data is obtained.

The objective of most data-processing is enhancing the signal with respect to the noise (§8.2). Improvements resulting from discrimination on the basis of frequency are employed in various deconvolution techniques: deterministic inverse filtering, recursive filtering, least-squares (Wiener) filtering, wavelet processing, etc. Compensation for near-surface time delays is the objective of static corrections. Use of normal moveout as a discriminator is employed in velocity analysis. Stacking, apparent-velocity filtering, and other methods also help improve the signal-to-noise ratio.

Another objective of data-processing is the re-positioning of data elements to compensate for the spatial distortions which result from the superposition of reflections coming from various directions. This is the objective of migration (§8.3). Usually the processes are applied in only two dimensions although the problem is actually three-dimensional.

A discussion of the procedures for carrying out data-processing in actual practice (§8.5) forms the concluding portion of this chapter.

8.1   The principal operations
8.1.1   *Fourier transforms*

Fourier analysis in our context involves transforming functions from the time-domain to the frequency domain and *Fourier synthesis* the inverse process of transforming from the frequency domain to the time domain. This distinction is somewhat artificial, however, and analysis and synthesis could be interchanged without making any difference in the final result. The important point with transforms is that no information is lost in transforming. We can, thus, start with a waveform in the time domain, transform it into characteristics in the frequency domain, and then transform the frequency-domain characteristics into a waveform which is identical with the original waveform. This makes it possible to do part of our processing in the time domain and part in the frequency domain, taking advantage of the fact that some processes can be executed more economically in one domain than in the other. In actual transformations we lose a small amount of information because of the truncation of series expansions, round-off errors and such, but we can make these losses as small as we wish by carrying enough terms or enough decimal places.

If we have a reasonably 'well-behaved' periodic function $g(t)$ of period T (that is, if $g(t)$ repeats itself every time t increases by T), then the function can be represented by a *Fourier series*,

$$g(t) = \tfrac{1}{2}a_0 + \sum_{n=1}^{\infty} (a_n \cos 2\pi v_n t + b_n \sin 2\pi v_n t), \quad (8.1, 10.71)$$

$$= \tfrac{1}{2}c_0 + \sum_{n=1}^{\infty} c_n \cos (2\pi v_n t - \gamma_n), \quad (8.2, 10.80)$$

$$= \sum_{n=-\infty}^{\infty} \alpha_n e^{j 2\pi v_n t}, \quad (8.3, 10.83)$$

where

$$v_n = n/T = nv_0 = n\omega_0/2\pi = \omega_n/2\pi,$$

$$\left. \begin{aligned} a_n &= (2/T) \int_{-\frac{1}{2}T}^{\frac{1}{2}T} g(t) \cos 2\pi v_n t \, dt, \\ b_n &= (2/T) \int_{-\frac{1}{2}T}^{\frac{1}{2}T} g(t) \sin 2\pi v_n t \, dt, \end{aligned} \right\} \quad (8.4, 10.77, 10.78)$$

$$\left. \begin{aligned} c_n &= (2/T) \int_{-\frac{1}{2}T}^{\frac{1}{2}T} g(t) \cos (2\pi v_n t - \gamma_n) \, dt, \\ \gamma_0 &= 0, \gamma_n = \tan^{-1}(b_n/a_n) \quad (n \neq 0), \end{aligned} \right\} \quad (8.5, 10.81)$$

$$\alpha_{\pm n} = (1/T) \int_{-\frac{1}{2}T}^{\frac{1}{2}T} g(t) e^{\mp j 2\pi v_n t} \, dt \quad (8.6, 10.84)$$

(Subscripts indicate discrete sets, as with v_n, a_n, etc., whereas functional notation, such as $g(t)$, indicates a continuous variable. Equation numbers also show equivalents in chapter 10.)

Equation (8.1) expresses $g(t)$ in terms of cosine and sine curves of amplitudes a_n and b_n, (8.2) in terms of cosine curves of amplitude c_n which have been phase-shifted by γ_n, and (8.3) in terms of complex exponentials. All three forms are equivalent, that is

$$c_n = a_n \cos \gamma_n + b_n \sin \gamma_n, = (a_n^2 + b_n^2)^{\frac{1}{2}}, \quad (8.7, 10.81)$$

$$a_n = \alpha_n + \alpha_{-n}; b_n = j(\alpha_n - \alpha_{-n}); \quad (8.8)$$

$$\alpha_{\pm n} = \tfrac{1}{2}(a_n \mp jb_n). \quad (8.9, 10.84)$$

Note that

$$\tfrac{1}{2}c_0 = \tfrac{1}{2}a_0 = \alpha_0 = \text{average value of } g(t). \quad (8.10)$$

Equations (8.1) and (8.2) show that $g(t)$ can be regarded as the sum of an infinite number of harmonic (cosine and sine) waves of frequency v_n having amplitudes a_n, b_n, or c_n (depending on which equation) and phase γ_n. These

equations thus represent the analysis of $g(t)$ into component harmonic waves.

As the period T becomes larger, it takes longer for $g(t)$ to repeat; in the limit when T becomes infinite, $g(t)$ no longer repeats. In this case we get in place of (8.3) and (8.6)

$$g(t) = \int_{-\infty}^{\infty} G(v) e^{j 2\pi v t} \, dv, \quad (8.11, 10.87)$$

$$G(v) = \int_{-\infty}^{\infty} g(t) e^{-j 2\pi v t} \, dt. \quad (8.12, 10.86)$$

The function $G(v)$ is the *Fourier transform* of $g(t)$ while $g(t)$ is the *inverse Fourier transform* of $G(v)$. Using the symbol $\leftrightarrow$ to denote equivalent expressions in different domains, we write

$$g(t) \leftrightarrow G(v).$$

We also refer to $g(t)$ and $G(v)$ as a *transform pair*.

Equations (8.11) and (8.12) can be written in several ways. In general $G(v)$ is complex,

$$G(v) = A(v) e^{j\gamma(v)}, \quad (8.13, 10.91)$$

where $A(v)$ and $\gamma(v)$ are real and $A(v)$ is also positive. We call $A(v)$ the *amplitude spectrum* and $\gamma(v)$ the *phase spectrum* of $g(t)$. Substitution in (8.11) gives

$$g(t) = \int_{-\infty}^{\infty} A(v) e^{j\{2\pi v t + \gamma(v)\}} \, dv. \quad (8.14)$$

For actual waveforms $g(t)$ is real and hence from (8.14) and problem 10.11a we get

$$g(t) = \int_{-\infty}^{\infty} A(v) \cos \{2\pi v t + \gamma(v)\} \, dv. \quad (8.15)$$

Since $G(v)$ is complex we may separate it into real and imaginary parts,

$$G(v) = R(v) + jX(v); \quad (8.16, 10.91)$$

when $g(t)$ is real,

$$\left. \begin{aligned} R(v) &= \int_{-\infty}^{\infty} g(t) \cos 2\pi v t \, dt, \\ -X(v) &= \int_{-\infty}^{\infty} g(t) \sin 2\pi v t \, dt. \end{aligned} \right\} \quad (8.17, 10.90)$$

The integrals $R(v)$ and $-X(v)$ are called the *cosine* and *sine transforms*, respectively. When $g(t)$ is real, $R(v)$ and $X(v)$ are respectively even (symmetrical, i.e., $R(v) = R(-v)$) and odd (antisymmetrical, i.e., $X(v) = -X(-v)$) functions. When $g(t)$ is real and even, $X(v) = 0$; when $g(t)$ is real and odd, $R(v) = 0$ (see problem 10.16b, c, d).

8.1.2 *Convolution*

(*a*) *The convolution operation. Convolution* is the time-domain operation of replacing each element of an input function with an output function scaled according to the magnitude of the input element, and then superimposing the outputs. Assume that we feed into a linear system (§10.5) data sampled at regular intervals, Δ, for example, a digital seismic trace. The output of the system can be calculated if we know the *impulse response* of the system, that is, the response of the system when the input is a *unit impulse* (which has zero values everywhere except at the origin where it has the value unity – see §10.33 and (10.102)). We write the unit impulse δ_t or $\delta(t)$ depending on whether we are dealing with sampled data or continuous functions. The impulse response of the system will be zero prior to $t = 0$ and then will have the values $f_0, f_1, f_2, \ldots$ at successive sampling intervals. We represent this process diagrammatically thus:

$$\delta_t \rightarrow \boxed{\text{system}} \rightarrow f_t = [f_0, f_1, f_2, \ldots].$$

Most systems with which we deal are linear and time-invariant (or nearly so). A *linear system* is one in which the output is directly proportional to the input while a *time-invariant system* is one in which the output is independent of the time when the input occurred. Writing δ_{t-n} for a unit impulse which occurs at $t = n$ rather than at $t = 0$, we can illustrate linear and time-invariant systems as follows:

Linear:

$$k\delta_t \rightarrow \boxed{\text{system}} \rightarrow kf_t = [kf_0, kf_1, kf_2, \ldots];$$

Time-invariant:

$$\delta_{t-n} \rightarrow \boxed{\text{system}} \rightarrow f_{t-n} = [\underbrace{0, 0, 0, \ldots, 0}_{n \text{ zeros}}, f_0, f_1, f_2, \ldots]$$

In the last bracket on the right, the first output different from zero is f_0 and occurs at the instant $t = n\Delta$ (which we write as $t = n$ since we count in units of Δ).

Obviously any input which consists of a series of sampled values can be represented by a series of unit impulses multiplied by appropriate amplitude factors. We can then use the above two properties to find the output for each impulse and by superimposing these we get the output for the arbitrary input.

We shall illustrate convolution by considering the output for a filter whose impulse response f_t is $[f_0, f_1, f_2] = [1, -1, \frac{1}{2}]$. When the input g_t is $[g_0, g_1, g_2] = [1, \frac{1}{2}, -\frac{1}{2}]$, we apply to the input the series of impulses $[\delta_t, \frac{1}{2}\delta_{t-1}, -\frac{1}{2}\delta_{t-2}]$ (the last two subscripts meaning that the impulses are delayed by one and two sampling

intervals respectively) and obtain the outputs shown below:

$$\delta_t \rightarrow [1, -1, \tfrac{1}{2}],$$
$$\tfrac{1}{2}\delta_{t-1} \rightarrow [0, \tfrac{1}{2}, -\tfrac{1}{2}, \tfrac{1}{4}],$$
$$-\tfrac{1}{2}\delta_{t-2} \rightarrow [0, 0, -\tfrac{1}{2}, \tfrac{1}{2}, -\tfrac{1}{4}].$$

Summing we find the output

$$[\delta_t + \tfrac{1}{2}\delta_{t-1} - \tfrac{1}{2}\delta_{t-2}] \rightarrow [1, -\tfrac{1}{2}, -\tfrac{1}{2}\tfrac{3}{4}, -\tfrac{1}{4}].$$

Convolution is illustrated in fig. 8.1. This operation is equivalent to replacing each element of the one set by an appropriately scaled version of the other set and then summing elements which occur at the same times. If we call the output h_t and denote the operation of taking the convolution by an asterisk, we may express this as

$$h_t = f_t * g_t$$

$$= \sum_k f_k g_{t-k}$$

$$= [f_0 g_0, f_0 g_1 + f_1 g_0, f_0 g_2 + f_1 g_1 + f_2 g_0, \ldots].$$
$$(8.18, 10.193)$$

(We use the sum sign to mean summation over all appropriate values of the summation index.) Note that we would have obtained the same result if we had input f_t into a filter whose impulse response is g_t; in other words, convolution is commutative:

$$h_t = f_t * g_t = g_t * f_t = \sum_k f_k g_{t-k} = \sum_k g_k f_{t-k}. \quad (8.19)$$

While we have been expressing convolution as an operation on sampled data, one can also convolve a sample set with a continuous function:

$$h(t) = f_t * g(t) = \sum_k f_k g(t - k). \quad (8.20)$$

Each term in the summation represents the function $g(t)$ displaced and scaled (displaced to the right k units and multiplied by f_k). A special case of (8.20) is that of convolving a continuous function $g(t)$ with a unit impulse located at $t = n$:

$$\delta_{t-n} * g(t) = \sum_k \delta_{k-n} g(t - k) = g(t - n)$$

(since δ_{k-n} is zero except for $k = n$, where $\delta_{k-n} = 1$). Hence convolving $g(t)$ with a time-shifted unit impulse displaces the function by the same amount and in the same direction as the unit impulse is displaced.

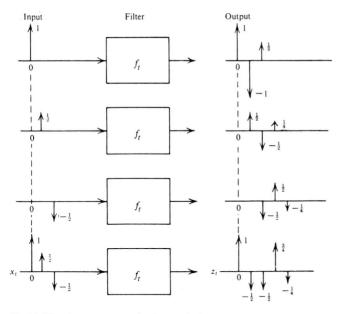

Fig.8.1. Filtering as an example of convolution.

One can also convolve two continuous functions. In this case the summation becomes an integration:

$$h(t) = f(t) * g(t) = \int_{-\infty}^{\infty} f(\tau)g(t - \tau)\,d\tau.$$

$$(8.21, 10.123, 10.174)$$

The *Convolution theorem* states that the Fourier transform of the convolution of two functions is equal to the product of the transforms of the individual functions; we can state the theorem as follows:

$$f_t \leftrightarrow F(v) = |F(v)|\,e^{j\gamma_f(v)},$$
$$g_t \leftrightarrow G(v) = |G(v)|\,e^{j\gamma_g(v)},$$
$$f_t * g_t \leftrightarrow F(v)G(v) = \{|F(v)|\,e^{j\gamma_f(v)}\}\{|G(v)|\,e^{j\gamma_g(v)}\},$$
$$\leftrightarrow |F(v)||G(v)|\,e^{j\{\gamma_f(v) + \gamma_g(v)\}} \qquad (8.22)$$

where $|F(v)|$ and $|G(v)|$ are the amplitude spectra and $\gamma_f(v)$ and $\gamma_g(v)$ the phase spectra. This means that if two sets of data are convolved in the time domain, the effect in the frequency domain is to multiply their amplitude spectra and to add their phase spectra. Equation (8.22) thus provides an alternative way to carry out a convolution operation: (1) transform the function to be convolved, say g_t, and the convolution (or filter) operator f_t into the frequency domain, (2) multiply their amplitude spectra at each frequency value and add their phase spectra, and (3) transform the result back into the time domain.

Because of certain symmetry properties of the Fourier transform, the reciprocal relationship also holds, that is, multiplication in the time domain is equivalent to convolution in the frequency domain:

$$f_t g_t \leftrightarrow F(v) * G(v). \qquad (8.23, 10.124)$$

(b) *Sampling and aliasing.* In analog-to-digital conversion, we replace the continuous signal with a series of values at fixed intervals. It would appear that we are losing information by discarding the data between the sampling instants. The transform relationship in (8.22) and (8.23) can be used to understand sampling and the situations in which information is not lost (see also §10.6.1).

We make use of the *comb* or *sampling function*; this consists of an infinite set of regularly-spaced unit impulses (fig. 8.2b). The transform of a comb is also a comb:

$$\text{comb}(t) \leftrightarrow k_1\,\text{comb}(v), \qquad (8.24, 10.133)$$

where k_1 depends upon the sampling interval (see problem 8.2c). If the comb in the time domain has elements every 4 ms, the transform has elements every $1/0.004 = 250\,\text{Hz}$. We shall also make use of the *boxcar* (fig. 8.2d), a function which has unity value between the values $\pm v_0$ and is zero everywhere else. The transform of a boxcar, box$_{2v_0}(v)$, is a *sinc function*:

$$\text{box}_{2v_0}(v) \leftrightarrow k_2 \, \text{sinc} \, (2\pi v_0 t) = k_2 \frac{\sin 2\pi v_0 t}{2\pi v_0 t}, \quad (8.25, 10.130)$$

where k_2 depends upon the area of the boxcar.

Figure 8.2a shows a continuous function $y(t)$ and its transform $Y(v)$:

$$y(t) \leftrightarrow Y(v).$$

The amplitude spectrum, $|Y(v)|$, is symmetric about zero for real functions, negative frequencies giving the same values as positive frequencies. (Negative frequencies result from the use of Euler's formula when we combine Fourier series in sine-cosine form into the complex exponential form.)

The sampled data which represent $y(t)$ can be found by multiplying the continuous function by the comb (hence the name 'sampling function'). If we are sampling every 4 ms, we use a comb with elements every 4 ms. According to (8.23) and (8.24),

$$\text{comb}(t)y(t) \leftrightarrow k_1 \, \text{comb}(v) * Y(v).$$

Convolution is equivalent to replacing each data element (each impulse in comb (v) in this instance) with the other function, $Y(v)$. This is illustrated in fig. 8.2c. Note that the frequency spectrum of the sampled function differs from the spectrum of the continuous function in this example by the repetition of the spectrum.

We can recover the spectrum of the original function by multiplying the spectrum of the sampled function by a boxcar. The equivalent time-domain operation (see (8.22) and (8.25)) is to convolve the sample data with the sinc function; as shown in fig. 8.2e, this restores the original function in every detail. The sinc function thus provides the precise 'operator' for interpolating between sample values.

In the above instance no information whatsoever was lost in the process of sampling and interpolating. However, if the continuous function had had a spectrum (shown dotted in fig. 8.2a) which included frequency components higher than 125 Hz (in this example), then the time-domain multiplication by the sampling function would have produced an overlap of frequency spectra and no longer would we be able to recover the original spectrum from the spectrum of the sampled data, hence

Fig.8.2. Sampling and reconstituting a waveform.

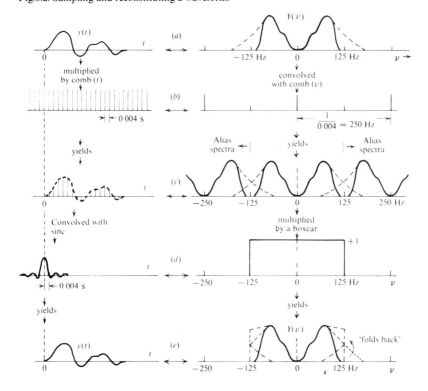

we would not be able to recover the original waveform. Whether or not the original waveform is recoverable depends, therefore, upon whether or not the original waveform contains frequencies higher than half of the sampling frequency.

The relationships demonstrated in the foregoing are summarized by the *sampling theorem*: no information is lost by regular sampling provided that the sampling frequency is greater than twice the highest frequency component in the waveform being sampled (see also §10.6.1). This is equivalent to saying that there must be more than two samples per cycle for the highest frequency. The sampling theorem thus determines the minimum sampling we can use. Since this minimum sampling allows complete recovery of the waveform, we can further conclude that nothing is gained by using a finer sampling. Thus, sampling rates of 2 and 4 ms permit us to record faithfully provided none of the signal spectrum lies above

250 and 125 Hz, respectively. In actual practice the limits are half of the above · frequencies because of analog aliasing filters (see below).

Half the sampling frequency is called the *Nyquist frequency*, v_N, that is,

$$v_N = 1/2\Delta. \tag{8.26}$$

Any frequency present in the signal which is greater than the Nyquist frequency by the amount Δv will be indistinguishable from the lower frequency $v_N - \Delta v$. In fig. 8.3 we see that a sampling rate of 4 ms (that is, 250 samples per second) will allow perfect recording of a 75 Hz signal but 175 Hz and 250 Hz signals will appear as (that is, will *alias* as) 75 Hz and 0 Hz (which is the same as a direct current), respectively. Alias signals which fall within the frequency band in which we are primarily interested will appear to be legitimate signals. To avoid this, *aliasing filters* (see fig. 5.33) are used before sampling to remove frequency

Fig.8.3. Sampling and aliasing. Different frequencies sampled at 4 ms intervals (250 times per second). (*a*) 75 Hz signal; (*b*) 175 Hz signal yields same sample values as 75 Hz; (*c*) 250 Hz signal yields samples of constant value (0 Hz).

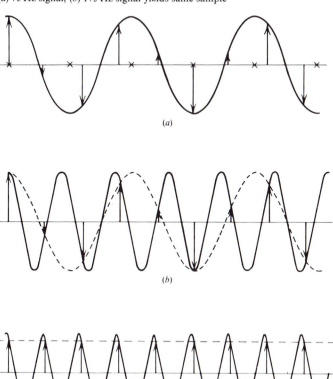

(*a*)

(*b*)

(*c*)

components higher than the Nyquist frequency. This must be done before sampling because afterwards the alias signals cannot be distinguished. Actual alias filters have finite slopes (often 72 dB/octave) and cutoff frequencies an octave below the Nyquist frequency in order to prevent aliasing.

Aliasing is an inherent property of all systems which sample, whether the sampling is done in time, in space or in any other domain such as frequency. Alias filtering must also be done before sampled data are re-sampled at a lower rate, as is often done in processing, so that there are fewer data samples to process.

Spatial sampling occurs when we use geophones to sample a wave at different points in space. Figure 8.3 represents a harmonic wave recorded at a fixed point as a function of time and the sampling is done at fixed time intervals Δt. However, the figure could equally well represent wave motion observed along a line of geophones at a given instant of time, the sampling being done at fixed intervals of distance Δx. In time sampling we take $1/\Delta t$ samples per unit time while in space sampling we take $1/\Delta x$ samples per unit distance. In time sampling we had a Nyquist frequency, ν_N of eq. (8.26), which gave the maximum number of waves per two unit time intervals, and aliasing occurred for components which have frequencies higher than ν_N. In space sampling we have a Nyquist wave number, $\kappa_N/2\pi$:

$$\kappa_N/2\pi = 1/\lambda_N = 1/2\Delta x, \qquad (8.27)$$

which gives the maximum number of waves per two unit lengths and aliasing occurs when a component has more waves per unit length than this, that is, when $\kappa > \kappa_N$ (or $\lambda < \lambda_N$).

(c) *Filtering by the Earth.* We can think of the Earth as a filter of seismic energy. We might consider the wave resulting from an explosion as an impulse $k\delta_t$, that is, the wave motion at the source of the explosion is zero both before and after the explosion and differs from zero only in an extremely short interval (essentially at $t = 0$) and during this infinitesimal interval the motion is very large. Ideally, the signal which we record would be simply $k\delta_t$ convolved with the impulse response of the Earth (assuming the Earth is a linear system – see §10.5). The result would be zero except for sharp pulses corresponding to the arrivals of different reflections. If this were so, we could easily determine from the recorded data the complete solution to the seismic problem. However, in practice we get back not only primary reflections but also multiples, diffractions, surface waves, scattered waves from near-surface irregularities, reflected refractions and

so on, all modified by filtering because of absorption and other causes, and with random noise always superimposed.

The waveform which we finally record as a seismic record is the result of the successive convolutions of the shot impulse with the impulse response of the various portions of the Earth through which the wave travels. We can arrive at an approximate picture by considering the Earth to be divided into three zones: (a) the zone near the shot where stress levels and the absorption of the highest frequencies are very severe – we write s_t for the impulse response of this zone; (b) the reflecting sequence of beds whose impulse response (reflectivity) e_t is the 'message' information which we are seeking to discover by our seismic exploration; and (c) finally, the near-surface zone which exercises considerable filtering action in changing the waveshape – we write n_t for the impulse response of this last zone. Neglecting additional filtering effects, we thus write the seismic trace g_t as the expression

$$g_t = k\delta_t * s_t * e_t * n_t. \qquad (8.28)$$

This equation expresses the *convolutional model*, that is, a seismic trace can be thought of as a series of convolutions; the convolutional model is central to most data processing. Equation (8.28) is also written

$$g_t = w_t * e_t \qquad (8.29)$$

where w_t includes $k\delta_t * s_t * n_t$ and possibly other effects; w_t is called the *equivalent wavelet*. The convolutional model often includes additive noise, r_t, also. The noise is usually (but not necessarily) assumed to be random, in which case the convolutional model of the seismic trace, c_t, is

$$c_t = w_t * e_t + r_t. \qquad (8.30)$$

When we use a Vibroseis™ source, the input to the Earth is a long wavetrain, v_t, and the resulting seismic trace, g_t', is

$$g_t' = v_t * s_t' * e_t * n_t, \qquad (8.31)$$

(where we write s_t' rather than s_t because the filtering processes near the Vibroseis source may be different from those near a shot owing to the different magnitude of the stresses involved).

(d) *Water reverberation and deconvolution.* Let us examine the effect of multiples resulting from reflection at the top and bottom of a water layer (Backus, 1959). We write $n\Delta$ for the round trip traveltime from top to bottom and back, n being an integer. We assume that the reflection coefficients at the surface and bottom of the water layer are such that the ratio of the reflected to incident amplitudes are − 1 and +R respectively, the minus sign denoting phase reversal at the water–air interface. We assume also that the amplitude of a wave returning directly to a hydrophone after reflection at a certain horizon (without a 'bounce' round trip between top and bottom of the water layer) is unity and that its traveltime is t. A wave which is reflected at the same horizon and suffers a bounce either before or after its travel down to the reflector, will arrive at time $t + n\Delta$ with the amplitude − R. Since there are two raypaths with the same traveltime for a single-bounce wave, one which bounced before traveling downward and one which bounced after returning from depth, we have in effect a wave arriving at time $t + n\Delta$ with the amplitude − 2R. There will be three waves which suffer two bounces: one which bounces twice before going downward to the reflector, one which bounces twice upon return to the surface, and one which bounces once before and once after its travel downwards; each of these are of amplitude R^2 so that their sum is a wave of amplitude $3R^2$ arriving at time $t + 2n\Delta$. Continuing thus, we see that a hydrophone will detect successive signals of amplitudes $1, -2R, 3R^2, -4R^3, 5R^4, \ldots$ arriving at intervals of $n\Delta$. We can therefore write the impulse response of the water layer for various water depths, $z = \frac{1}{2} nV\Delta$, where V is the velocity in the water:

$$
\begin{aligned}
f_t &= [1, -2R, 3R^2, -4R^3, 5R^4, \ldots], & (n = 1) \\
&= [1, 0, -2R, 0, 3R^2, 0, -4R^3, \ldots], & (n = 2) \\
&= [1, 0, 0, -2R, 0, 0, 3R^2, \ldots], & (n = 3)
\end{aligned}
\tag{8.32}
$$

etc. Thus the water layer acts as a filter.

If we transform this to the frequency domain we find a large peak (the size of the peak increasing with increasing R) at the frequency $1/2n\Delta$ and at multiples of this frequency. These are the frequencies which are reinforced at this water depth (that is, the frequencies for which interference is constructive). The result of passing a wavetrain through a water layer is the same as multiplying the amplitude spectrum of the waveform without the water layer by the spectrum of the impulse response of the water layer. Whenever the reflection coefficient is large (and hence R is large) and the frequency $(1/2n\Delta)$ (or one of its harmonics) lies within the seismic spectrum, the seismic record will appear very sinusoidal with hardly any variation in amplitude throughout the recording period (see fig. 4.12). Because of the overriding oscillations, it will be difficult to interpret the primary reflections.

A filter i_t which has the property that

$$
f_t * i_t = \delta_t
\tag{8.33}
$$

is called the *inverse filter* of f_t. If we pass the reverberatory output from the hydrophones through the inverse filter (in a data-processing center) we will remove the effect of the water-layer filter. The inverse of the water-layer filter is a simple filter (the Backus filter) with only three non-zero terms:

$$
\begin{aligned}
i_t &= [1, 2R, R^2], & (n = 1) \\
&= [1, 0, 2R, 0, R^2], & (n = 2) \\
&= [1, 0, 0, 2R, 0, 0, R^2], & (n = 3)
\end{aligned}
\tag{8.34}
$$

etc. (see problem 8.3a and §10.6.5). Figure 4.12b shows the result of applying such a filter to the data shown in fig. 4.12a.

The process of convolving with an inverse filter is called *deconvolution* and is one of the most important operations in seismic data-processing (Middleton and Whittlesey, 1968). While we have illustrated deconvolution as removing the singing effect of a water layer, we could also deconvolve for other filters whose effects we wish to remove if we know enough about the filters and the signal (see §8.2.1 and §10.8).

(e) Multidimensional convolution. In the foregoing sections we have assumed only one independent variable, time. We have assumed that each seismic trace is being processed by itself and that the only data available are the succession of samples of that trace, the actual situation for many processes. There is, however, no need to restrict ourselves to a succession of time samples for only one trace and some processes involve convolution with two or even more variables. The convolution operation, (8.18), can be written for two variables, t and w, as

$$
h_{t,w} = f_{t,w} * g_{t,w} = \sum_k \sum_m f_{k,m} g_{t-k, w-m}
\tag{8.35, 10.142}
$$

This equation expresses that the convolution result is the superposition of values at nearby times and locations (if t and w respectively indicate time and location) after each has been weighted by the filter $f_{t,w}$. As with one-dimensional convolution, the operation is commutative. It is also possible to carry out the operation by transforming $f_{t,w}$ and $g_{t,w}$ using a two-dimensional Fourier transform (see §10.3.2), multiplying the two-dimensional amplitude spectra and adding the phase spectra, and then transforming back to give $h_{t,w}$. If t is time and w is distance, then the transformed domain is the frequency-wavenumber domain.

8.1.3   *Correlation*

(a) Cross-correlation. The cross-correlation function is a measure of the similarity between two data sets. One data set is displaced varying amounts relative to the

other and corresponding values of the two sets are multiplied together and the products summed to give the value of cross-correlation. Wherever the two sets are nearly the same, the products will usually be positive and hence the cross-correlation is large; wherever the sets are unlike, some of the products will be positive and some negative and hence the sum will be small. If the cross-correlation function should have a large negative value, it means that the two data sets would be similar if one were inverted (that is, they are similar except that they are out-of-phase). The two data sets might be dissimilar when lined up in one fashion and yet be similar when one set is shifted with respect to the other; thus the cross-correlation is a function of the relative shift between the sets. By convention we call a shift positive if it involves moving the second function to the left with respect to the first function.

We express the cross-correlation of two data sets x_t and y_t as

$$\phi_{xy}(\tau) = \sum_k x_k y_{k+\tau}, \qquad (8.36, 10.194)$$

where τ is the displacement of y_t relative to x_t. (Note that $\phi_{xy}(\tau)$ is a data set rather than a continuous function, because x and y are data sets.) Let us illustrate cross-correlation by correlating the two functions, $x_t = [1, -1, \frac{1}{2}]$ and $y_t = [1, \frac{1}{2}, -\frac{1}{2}]$, shown in fig. 8.4. Diagram (c) shows the two functions in their normal positions. Diagram (a) shows y_t shifted two units to the right; corresponding coordinates are multiplied and summed as shown below the diagram to give $\phi_{xy}(-2\Delta)$. Diagrams (b) to (e) show y_t shifted varying amounts while (f) shows the graph of $\phi_{xy}(\tau)$. The cross-correlation has its maximum value (the functions are most similar) when y_t is shifted one unit to the left ($\tau = +\Delta$). Obviously we get the same results if we shift x_t one space to the right. In other words,

$$\phi_{xy}(\tau) = \phi_{yx}(-\tau). \qquad (8.37, 10.135)$$

The similarity between (8.36) and the convolution (8.18) should be noted. We may rewrite (8.36) in the form

$$\phi_{xy}(\tau) = \phi_{yx}(-\tau) = \sum_k y_k x_{k-\tau} = \sum_k y_k x_{-(\tau-k)}$$

$$= y_\tau * x_{-\tau} = x_{-\tau} * y_\tau. \qquad (8.38, 10.136)$$

Hence cross-correlation can be performed by reversing the first data set and convolving.

If two data sets are cross-correlated in the time domain, the effect in the frequency domain is the same as multiplying the complex spectrum of the second data set by the conjugate of the complex spectrum of the first set. Since forming the complex conjugate involves only rever-

sing the sign of the phase, cross-correlation is equivalent to multiplying the amplitude spectra and subtracting the phase spectra. In mathematical terms,

$$
\begin{aligned}
x_t &\leftrightarrow X(v) &&= |X(v)|\, e^{j\gamma_x(v)}, \\
y_t &\leftrightarrow Y(v) &&= |Y(v)|\, e^{j\gamma_y(v)}, \\
x_{-t} &\leftrightarrow \overline{X}(v) &&= |X(v)|\, e^{-j\gamma_x(v)}, \\
\phi_{xy}(\tau) &\leftrightarrow \overline{X}(v)\, Y(v) &&= |X(v)||Y(v)|\, e^{-j\{\gamma_x(v)-\gamma_y(v)\}}.
\end{aligned}
$$
$$(8.39, 10.125)$$

We note that changing the sign of a phase spectrum is equivalent to reversing the trace in the time domain. Anstey (1964) gives a particularly clear explanation of correlation.

(b) *Autocorrelation.* The special case where a data set is being correlated with itself is called *autocorrelation*. In this case (8.36) becomes:

$$\phi_{xx}(\tau) = \sum_k x_k x_{k+\tau}. \qquad (8.40)$$

Autocorrelation functions are symmetrical because a time shift to the right is the same as a shift to the left; from (8.37),

$$\phi_{xx}(\tau) = \phi_{xx}(-\tau). \qquad (8.41)$$

The autocorrelation has its peak value at zero time shift (that is, a data set is most like itself before it is time-shifted). If the autocorrelation should have a large value at some time shift $\Delta t \neq 0$, it indicates that the set tends to be periodic with the period Δt. Hence the autocorrelation function may be thought of as a measure of the repetitiveness of a function.

We can express the preceding concepts in integral form applicable to continuous functions. Equations (8.36) and (8.40) now take the forms

$$\phi_{xy}(\tau) = \int_{-\infty}^{\infty} x(t)\, y(t+\tau)\, dt, \qquad (8.42, 10.125)$$

$$\phi_{xx}(\tau) = \int_{-\infty}^{\infty} x(t)\, x(t+\tau)\, dt. \qquad (8.43, 10.139)$$

(c) *Normalized correlation functions.* The autocorrelation value at zero shift is called the *energy* of the trace:

$$\phi_{xx}(0) = \sum_k x_k^2. \qquad (8.44, 10.140)$$

(This terminology is justified on the basis that x_t is usually a voltage, current or velocity and hence x_t^2 is proportional to energy.) For the autocorrelation function, (8.39) becomes

$$\phi_{xx}(\tau) \leftrightarrow |X(v)|^2. \qquad (8.45, 10.139)$$

In continuous-function notation,

$$\phi_{xx}(0) = \int_{-\infty}^{\infty} |x(t)|^2 \, dt = \int_{-\infty}^{\infty} |X(v)|^2 \, dv. \qquad (8.46, 10.140)$$

Since the zero-shift value of the autocorrelation function is the energy of the trace, $|x(t)|^2$ is the energy per unit of time or the *power* of the trace and $|X(v)|^2$ is the energy per increment of frequency, usually called the *energy density* or *spectral density*.

We often normalize the autocorrelation function by dividing by the energy:

$$\phi_{xx}(\tau)_{norm} = \frac{\phi_{xx}(\tau)}{\phi_{xx}(0)}. \qquad (8.47)$$

The cross-correlation function is normalized in a similar manner by dividing by the geometric mean of the energy of the two traces:

$$\phi_{xy}(\tau)_{norm} = \frac{\phi_{xy}(\tau)}{\{\phi_{xx}(0)\phi_{yy}(0)\}^{\frac{1}{2}}}. \qquad (8.48)$$

Normalized correlation values must lie between ± 1. A value of $+1$ indicates perfect copy, a value of -1 indicates perfect copy if one of the traces is inverted.

Fig.8.4. Calculating the cross-correlation of two functions.

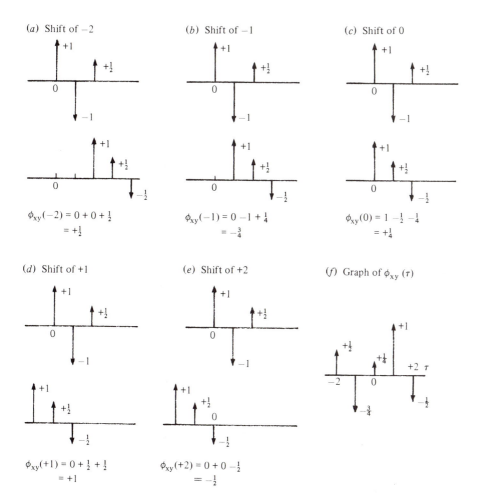

(a) Shift of -2

$$\phi_{xy}(-2) = 0 + 0 + \tfrac{1}{2}$$
$$= +\tfrac{1}{2}$$

(b) Shift of -1

$$\phi_{xy}(-1) = 0 - 1 + \tfrac{1}{4}$$
$$= -\tfrac{3}{4}$$

(c) Shift of 0

$$\phi_{xy}(0) = 1 - \tfrac{1}{2} - \tfrac{1}{4}$$
$$= +\tfrac{1}{4}$$

(d) Shift of $+1$

$$\phi_{xy}(+1) = 0 + \tfrac{1}{2} + \tfrac{1}{2}$$
$$= +1$$

(e) Shift of $+2$

$$\phi_{xy}(+2) = 0 + 0 - \tfrac{1}{2}$$
$$= -\tfrac{1}{2}$$

(f) Graph of $\phi_{xy}(\tau)$

(d) *Vibroseis analysis.* The signal g_t' which our geophones record when we use a Vibroseis source (fig. 8.5a) bears little resemblance to e_t, the impulse response of the Earth. To obtain a meaningful record, the data are correlated with the Vibroseis sweep (control) signal v_t. The recorded signal g_t' is

$$g_t' = v_t * e_t',$$

where we let $e_t' = s_t' * e_t * n_t$ in (8.31). Using (8.38) we find for the cross-correlation of the sweep and the recorded signal

$$\begin{aligned}\phi_{vg}(t) = g_t' * v_{-t} &= (v_t * e_t') * v_{-t} \\ &= e_t' * (v_t * v_{-t}) \\ &= e_t' * \phi_{vv}(t).\end{aligned} \quad (8.49)$$

(The next to the last step is possible because convolution is commutative.) Hence the overall effect is that of convolving the Earth function with the autocorrelation of the Vibroseis sweep signal. The autocorrelation function, $\phi_{vv}(t)$, is quite sharp and has sizeable values only over a very narrow range of time shifts. Therefore, the overlap produced by the passage of a long sweep through the Earth has been eliminated almost entirely. This is shown in fig. 8.5, where (a) and (b) are the same Vibroseis record before and after cross-correlation.

(e) *Sign-bit recording.* In a conventional Vibroseis field recording such as shown in fig. 8.5, the information about any individual reflection is distributed over the time duration of the sweep. The information content at any time is small compared to the noise and can be thought of as a bias superimposed on stationary random noise. The probability that any individual sample will be positive or negative will depend on the magnitude of this information bias compared to the noise level. If we record only the sign of each sample (ignoring completely the magnitude of each sample) and then sum many samples, we will in effect measure the time variations of this probability and thus the magnitude of the signal bias.

Figure 8.6 is a synthetic example to show that the sign-bit method gives results comparable to conventional Vibroseis. Traces 1–10 are Vibroseis traces with different amplitudes and arrival times to represent signals from 10 reflectors; trace 11 is the sum of these, i.e., the trace which would be recorded in the absence of noise. Random noise three times larger than the rms amplitude of trace 11 has been added to it to give trace 12. Trace 13 shows the sign-bit record equivalent to trace 12; it flips back and forth between the only two possible states, + and −. Trace 14 shows the result of correlating trace 12 with the sweep trace and thus is a conventionally correlated Vibroseis

trace as would result from a single sweep. Trace 15 shows the result of correlating trace 13 with a sign-bit sweep trace. Trace 16 is the sum of 20 traces manufactured like trace 12 except that each has different random noise; it thus represents the field Vibroseis record which would result from vertical stacking of 20 sweeps. Trace 17 is the sum of 20 traces manufactured like trace 13. Trace 18 results from correlating trace 16 with a sweep trace and trace 19 results from correlating trace 17 with a sign-bit record of the sweep. Thus traces 18 and 19 represent respectively a conventional correlated Vibroseis trace and a correlated sign-bit trace.

With sign-bit recording, only one bit is recorded per sample so that the volume of data is greatly reduced and instrumentation can be simplified. Instrumentation fidelity requirements are also greatly relaxed; geophone nonlinearity, for example, becomes less important because only the sign and not the magnitude of the output is measured. Geocor™ uses sampling boxes in the field with 16 geophone channels connected to each box. The sampling box determines whether each channel output is positive or negative at the sampling instant and records this information as one of the bits in a 16-bit word which is then relayed to the recording truck. A 16-to-1 saving in the number of cable channels is thus achieved. Geocor field systems employing 1024 channels were in use in 1982.

(f) *Multichannel coherence.* The cross-correlation function can only be used as a measure of the coherence between two traces. As a coherence measure for a large number of traces we could make use of the fact that when we stack several channels together the resulting amplitude is generally large where the individual channels are similar (coherent) so that they stack in-phase, and small where they are unlike (incoherent). The ratio of the energy of the stack compared to the sum of the energies of the individual components would therefore be a measure of the degree of coherence.

If we let g_{ti} be the amplitude of the individual channel i at the time t, then the amplitude of the stack at time t will be $\sum_i g_{ti}$ and the square of this will be the energy. If we call E_t the ratio of the output energy to the sum of the energies of the input traces, we may write

$$E_t = \frac{\left(\sum_i g_{ti}\right)^2}{\sum_i (g_{ti}^2)}. \quad (8.50)$$

We expect a coherent event to extend over a time interval; hence a more meaningful quantity than E_t is the

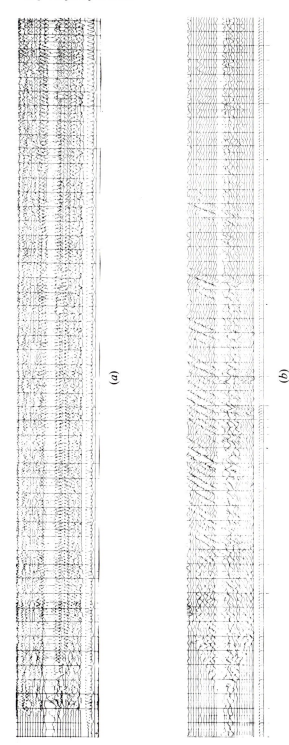

(a)

(b)

Fig. 8.5. Vibroseis record. (Courtesy United Geophysical.) (*a*) Before correlation (the uncorrelated record is very long, hence only the first portion is shown); (*b*) after correlation.

Fig. 8.6. Comparison of conventional Vibroseis and sign-bit recordings. (Courtesy Geophysical Systems.)

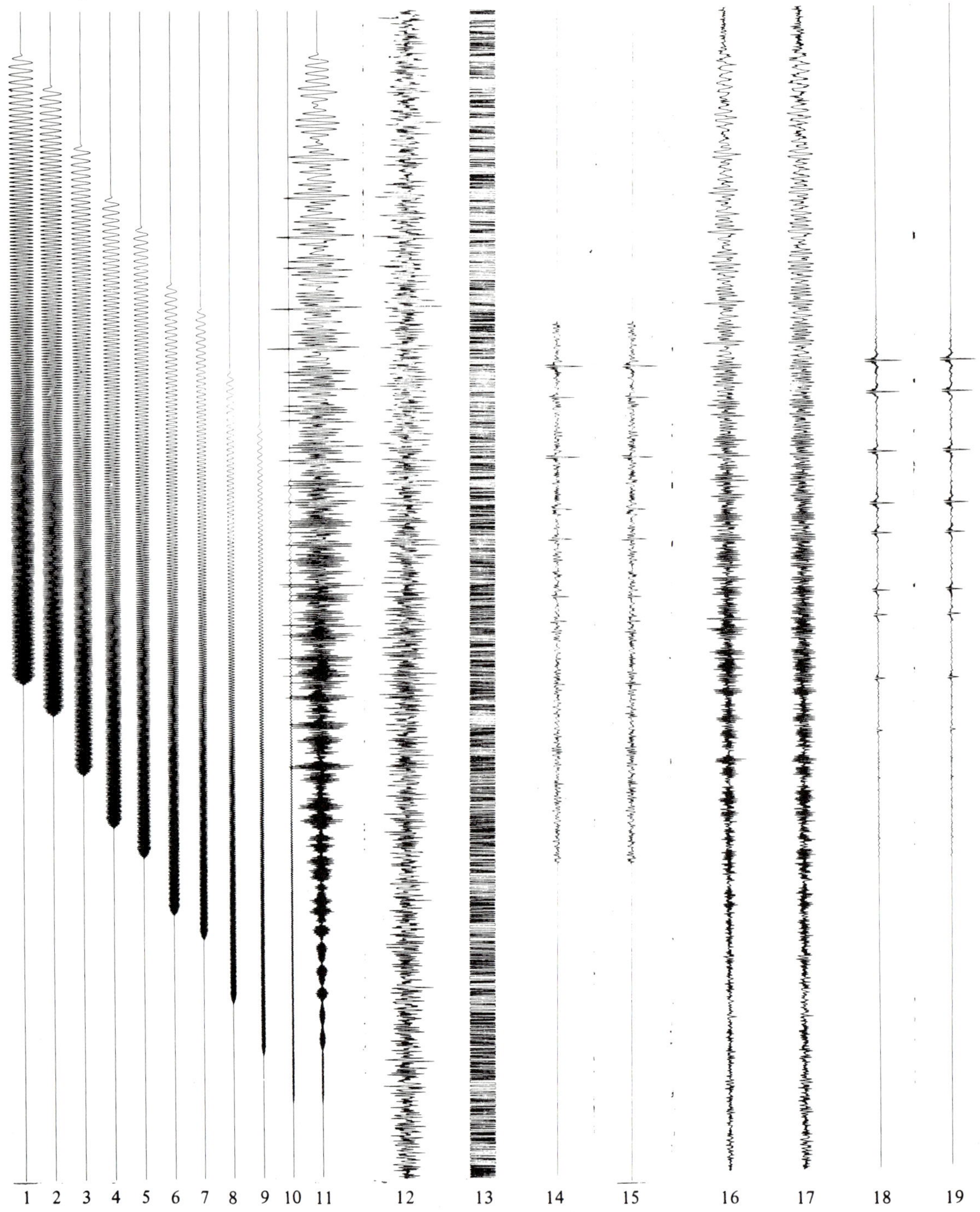

semblance, S_t (Neidell and Taner, 1971), which denotes the ratio of the total energy of the stack, within a gate of length Δt, to the sum of the energy of the component traces within the same time gate. Using the same terminology as before we can write

$$S_t = \frac{\sum\limits_{t=t}^{t+\Delta t}\left(\sum\limits_{i} g_{ti}\right)^2}{\sum\limits_{t=t}^{t+\Delta t}\sum\limits_{i}(g_{ti}^2)}. \qquad (8.51)$$

The semblance will not only tend to be large when a coherent event is present but the magnitude of the semblance will also be sensitive to the amplitude of the event. Thus strong events will exhibit large semblance and weak events will exhibit moderate values of semblance while incoherent data will have very low semblance.

Semblance and other coherence measures are used to determine the values of parameters which will 'optimize' a stack. The semblance is calculated for various combinations of time shifts between the component channels and the optimum time shifts are taken to be those which maximize the semblance. Semblance therefore can be used to determine static corrections or normal-moveout corrections.

8.1.4   *Phase considerations*

The Fourier synthesis of wavetrains according to (8.2) involves adding together cosine waves of different frequencies and different phases. If the same components are added together with different phase relations, different waveforms will result. Changing the waveform changes the location of a particular peak or trough and hence measurements of arrival times will be affected by variations in the phase spectra. Because seismic exploration involves primarily determining the arrival times of events, preservation of proper phase relationships during data-processing is essential.

Out of all possible wavelets with the same amplitude spectrum, that wavelet whose energy builds up fastest is called the *minimum-delay* wavelet; its phase is always less than the other wavelets with the same amplitude spectrum and hence it is also called *minimum-phase*. The simplest wavelet (except for an impulse) is a data set which contains only two elements, the set (a, b). The amplitude spectrum of this data set is identical with that of the set $(b, a,)$ but no other data set has the same spectrum. If $a > b$, energy is concentrated earlier in the wavelet in the set (a, b) than in the set (b, a) and hence (a, b) is minimum-phase (or minimum-delay). Larger wavelets

can be expressed as the successive convolution of two-element wavelets (see §10.6.6a); a large wavelet is minimum-phase if all of its component wavelets are minimum-phase. Minimum-phase can also be defined in other ways, for example, by the location of roots in the z-domain (§10.6.6a). Most seismic sources generate waves which are nearly minimum-phase and the impulse response of many of the natural filtering processes in the Earth are minimum-phase.

Minimum-phase does not necessarily mean that the first half-cycle is the largest, however. In the presence of interfering events and noise, it is often difficult to tell whether a reflected wavelet has the same or opposite sign as the downgoing wavelet, that is, whether the reflection coefficient is positive or negative. It is also difficult to tell the onset time of a reflection and it is this time which is needed in determining reflector depth.

The equivalent wavelet can be changed in processing to a zero-phase wavelet (see §10.6.6d) to facilitate interpretation. A *zero-phase wavelet* has its phase spectrum identically zero, i.e., $\gamma(v) = 0$ for all v. Such a wavelet is symmetrical about a central peak (or trough), which has higher amplitude than any other peaks or troughs. (An autocorrelation function is zero-phase.) We shift the time scale so that the amplitude maximum gives the arrival time. Such a wavelet is anticipatory because half of the wavelet precedes the arrival time.

Some filtering processes require that assumption be made about the phase of the signal; generally minimum-phase is assumed (Sherwood and Trorey, 1965). Thus deconvolution based upon autocorrelation information has to assume the phase because the phase information of the waveform was lost when its autocorrelation was formed. This can be seen from (8.45) where we note that the autocorrelation function, $\phi_{xx}(t)$, has the transform, $|X(v)|^2$ with zero-phase for all values of frequency. Thus all of the phase information present in $X(v)$ has been lost in the autocorrelation. However, most signals and natural filtering processes are represented by real, causal functions (see §10.6.6a). The Hilbert transform technique (§10.3.11) can be used to determine the phase information if the function is real and causal.

Causal wavelets which are not minimum-phase can be made minimum-phase by applying an exponential gain (taper) which cuts down the size of the latter part of the wavelet. Many actual effective wavelets contain only a few non-minimum-phase roots and can be made minimum-phase by using an exponential multiplier with a very gentle slope, perhaps 0.995^t where t is the time in milliseconds (see problem 8.17). A criterion for determining how much taper to apply is discussed in §10.7.

8.2　Processes to improve signal-to-noise
8.2.1　*Deconvolution and frequency filtering*

(a) *General*. In §8.1.2*d* we defined deconvolution as convolving with an inverse filter. Equations (8.28) to (8.30) expressed the seismic trace as a convolution of an Earth reflectivity function e_t and a series of distorting filters, whose combined expression is the equivalent wavelet w_t. The ultimate objective of deconvolution is to extract the reflectivity function from the seismic trace but often it has the more limited objectives of only undoing the effects of some prior filtering.

Deconvolution to extract the reflectivity function e_t is non-unique unless additional information is available or additional assumptions are made. Equation (8.30) contains three unknowns, w_t, e_t and r_t, but only one known, c_t. The most common additional assumptions are that w_t is minimum-phase and/or e_t has a flat (white) spectrum (at least over some limited bandpass). Additional constraints sometimes used include imposing a maximum length to w_t or use of a multichannel procedure.

Deconvolution operations are sometimes cascaded, a deconvolution to remove one type of distortion being followed by a different type to remove another type of distortion. Some of the types of deconvolution are described in the following sections. Webster (1978) includes most of the literature on deconvolution.

(b) *Deterministic inverse-filtering*. Where the nature of the distorting filter is known, the inverse can sometimes be found in direct deterministic manner. In §8.1.2*d* we gave an example of water reverberation wherein we used a model of a water layer to derive the distorting filter, f_t. We also expressed the inverse filter, i_t by (8.33) but this equation cannot be used to find the sequence i_t (except on a trial-and-error basis). However, we can transform (8.33) into the z-domain (see §10.6.3), carry out the division,

$$I(z) = 1/F(z) = \sum_k i_k z^k \qquad (8.52)$$

(provided the division doesn't 'blow up'), and then transform back to get i_t. Even where we know the nature of the filter, we may have to determine part of the solution by trial-and-error or other techniques; in any case, we still have to determine the value of R in (8.34).

Deterministic (or semi-deterministic) solutions are also used to remove the filtering effects of recording and processing systems. The source waveshape is sometimes recorded (though not always with proper ghost effects) and used in a deterministic source-signature correction. In marine work the source waveform is often assumed to be constant; the waveshape may be recorded in deep water with a hydrophone suspended below the source. Another procedure is to monitor each energy release using hydrophones near the source and calculate changes in the signature (see §5.5.3*f*) from such monitor records; this procedure corrects for shot-to-shot variations before stacking. Still another procedure uses the seafloor reflection to determine the source waveshape, assuming that the seafloor is a simple, sharp reflector.

(c) *Deghosting and recursive filtering*. With a shot below the base of the weathering (fig. 8.7), ghost energy from the shot is reflected at the base of the weathering (see §4.2.2*b*) where the coefficient (approaching from below) is $-R$ (ignoring ghost energy reflected at the free surface). An impulse followed by its ghost constitutes a filtering action; the transform can be written as

$$F(z) = 1 - Rz^n \qquad (8.53)$$

where z^n represents the delay associated with the two-way traveltime from the shot to the reflector producing the ghost. The inverse filter is an infinite series,

$$F^{-1}(z) = 1/(1 - Rz^n)$$
$$= 1 + Rz^n + (Rz^n)^2 + (Rz^n)^3 + \dots \qquad (8.54)$$

Because R is less than unity, this series converges and so could be used as a satisfactory deghosting filter. An input which includes ghost effects, g_t, can be convolved with this inverse filter to give a deghosted output h_t; in z-transform form,

$$H(z) = G(z) F^{-1}(z) = G(z)/(1 - Rz^n),$$
$$H(z)(1 - Rz^n) = G(z),$$
$$H(z) = G(z) + Rz^n H(z). \qquad (8.55)$$

Fig.8.7. Reflection plus ghost.

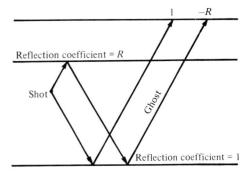

The last term on the right represents a delay of $H(z)$ by n units; this last equation can be written in time notation as

$$h_t = g_t + Rh_{t-n}. \qquad (8.56)$$

Thus we can determine an output value by adding to the input a proportionate amount of a previous output value. If the ghost delay is the sample interval, $n = 1$ and

$$h_0 = g_0,$$

because there was no output prior to zero;

$$h_1 = g_1 + Rh_0, \quad h_2 = g_2 + Rh_1, \quad h_3 = g_3 + Rh_2, \ldots$$

Filtering which involves feeding back part of the output is called *feedback* or *recursive filtering* (see §10.8.2). Recursive filtering allows us to carry out complete deghosting without using many terms and thus is economical in computing. In more general terms, we can express a filter $F(z)$ as a quotient of polynomials,

$$H(z) = G(z)F(z) = G(z)\{N(z)/D(z)\},$$

and write

$$G(z)N(z) = H(z)D(z).$$

The right side is the 'recursive' part where previous output values are used in deriving future output values.

(d) Least-squares (Wiener) filtering. Sometimes we wish to determine the filter which will do the best job of converting an input into a desired output. The filter which most nearly accomplishes this objective in the least-squares error sense is called the *least-squares filter* or the *Wiener filter*, occasionally the *optimum filter* (Robinson and Treitel, 1967).

Let the input data set be g_t, the filter which we have to determine be f_t, and the desired output set be h_t. The actual result of passing g_t through this filter is $g_t * f_t$ and the 'error' or difference between the actual and the desired outputs is $(h_t - g_t * f_t)$. With the least-squares method (§10.1.5) we add together the squares of the errors, find the partial derivatives of the sum with respect to the variables f_i (the elements of f_t) and set these derivatives equal to zero. This gives the following simultaneous equations where g_t and h_t are known:

$$\frac{\partial}{\partial f_i} \sum_t (h_t - g_t * f_t)^2 = 0, \quad i = 0, 1, 2, \ldots n \qquad (8.57)$$

or

$$\sum_t (h_t - g_t * f_t)\frac{\partial}{\partial f_i}(g_t * f_t) = 0, \quad i = 0, 1, \ldots, n;$$

one such equation is obtained for each of the $n + 1$ elements in f_t. Writing the convolution as a sum using (8.19) gives

$$\sum_t \left(h_t - \sum_k g_k f_{t-k} \right) \frac{\partial}{\partial f_i} \left(\sum_k g_k f_{t-k} \right) = 0.$$

The only terms in the convolution which involve f_i are those containing g_{t-i}. Hence

$$\sum_t \left(h_t - \sum_k g_k f_{t-k} \right) g_{t-i} = 0,$$

$$\sum_t h_t g_{t-i} = \sum_t \sum_k g_k g_{t-i} f_{t-k}.$$

The left side is $\phi_{gh}(i)$ according to (8.36) and (8.37). We let $j = t - k$ and sum over j instead of over k:

$$\phi_{gh}(i) = \sum_t \sum_j g_{t-j} g_{t-i} f_j = \sum_j f_j \sum_t g_{t-j} g_{t-i}$$

upon interchanging the order of summation. The last term is $\phi_{gg}(i - j)$ according to (8.40) (see problem 8.12). Hence we arrive at the *normal equations*,

$$\sum_{j=0}^{n} \phi_{gg}(i - j)f_j = \phi_{gh}(i), \quad i = 0, 1, 2, \ldots n \qquad (8.58, 10.27)$$

(see also problem 10.10). The normal equations for least-squares filtering also have an integral expression for continuous functions,

$$\int_{-\infty}^{\infty} \phi_{gg}(\tau - t)f(t)\,dt = \phi_{gh}(\tau). \qquad (8.59, 10.254)$$

These equations can be used to *cross-equalize* traces, that is, to make traces as nearly alike as possible. Suppose we have a group of traces to be stacked, such as the components of a common-depth-point stack. After the normal-moveout corrections have been made, the traces may still differ from each other because they have passed through different portions of the near-surface. The normal equations can be used to find the filters which will make all the traces as nearly as possible like some *pilot trace*, such as the sum of the traces. This procedure will improve the trace-to-trace coherence before the stack and hence improve the quality of the stacked result.

(e) Whitening, predictive deconvolution and wavelet shaping. The normal equations are used to accomplish spiking deconvolution. The Earth's impulse response, e_t, is assumed to be random, that is, knowledge of the shallow reflections does not help in predicting the deeper reflections. Consequently the autocorrelation of e_t is

negligibly small except for zero shift and we can write

$$\phi_{ee}(\tau) \approx k\delta_t. \tag{8.60}$$

The geophone input g_t is regarded (see (8.28) and (8.29)) as the convolution of e_t with various filters (the most important of which results from near-surface effects), the overall effect being represented by the single-equivalent filter w_t:

$$g_t = e_t * w_t.$$

The desired output h_t is the Earth's impulse function e_t (which is assumed to be minimum-phase); hence using (8.38), we can write

$$\begin{aligned}
\phi_{gh}(t) &= h_t * g_{-t} \\
&= e_t * (e_{-t} * w_{-t}) \\
&= (e_t * e_{-t}) * w_{-t} \\
&= k\delta_t * w_{-t} \\
&= kw_{-t}. \tag{8.61}
\end{aligned}$$

There can be no output from the filter w_t until after there has been an input to the filter; this is equivalent to saying that g_t is causal. Hence $w_t = 0$ for $t < 0$. Thus

$$\phi_{gh}(t) = 0 \quad \text{for} \quad t < 0. \tag{8.62}$$

Therefore if we concern ourselves only with positive values of i, we have the values required to solve (8.58) for the spiking deconvolution filter.

The ability to transform into the frequency or other domains not only provides alternative computing methods but also provides insights as to what deconvolution methods imply. Thus, thinking of the Earth's response, e_t, as random implies equal probabilities for the amplitudes of all frequencies and reminds us of white light, and we call spiking deconvolution *whitening*. It is equivalent to finding an inverse filter whose transform is $I(v)$ where

$$I(v) = 1/G(v) \tag{8.63}$$

(see fig. 8.8a), $G(v)$ being the transform of the input, so that the product $I(v) G(v)$ is constant (the constant is set equal to 1 since we are only concerned with relative values of I: such ignoring of scale factors is proper in many processing methods). In applying the inverse filter, the phase has to be known. Often we start with ϕ_{gg}, assume that $G(v)$ is minimum-phase so that it can be determined uniquely from Φ_{gg} (see §10.6.6c and problem 10.34), then invert $G(v)$ to obtain $I(v)$ (note that $I(v)$ is also minimum-phase; see problem 10.31a).

Equation (8.63) applies to any frequency value, for example, $I(v_1) = 1/G(v_1)$. If $G(v_1)$ should be small, then $I(v_1)$ will be large. Thus a whitening filter emphasizes the weak frequency components, resulting in improvement to the extent that they are attenuated signals. Above some frequency v_u noise dominates, so whitening is performed only over a limited bandpass. If, as in fig. 8.8a, the signal $G(v)$ should be especially weak over a narrow portion of

Fig.8.8. Spectra of signal before deconvolution and of inverse filter used to achieve whitening. (a) Without addition of white noise; the bandwidth is usually specified without knowledge of the exact spectrum; (b) with white noise added for inverse-filter design purposes.

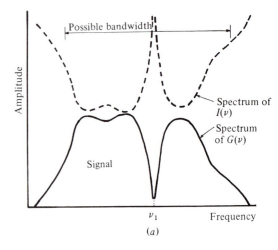

(a)

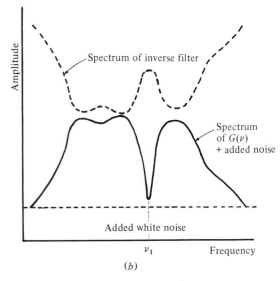

(b)

the passband (sometimes referred to as a notch in the signal spectrum), as near the frequency v_1, the inverse filter will magnify noise at this frequency, sometimes with disastrous results. To prevent excessive magnification of noise, white noise, A_w, is sometimes added when the filter is being designed, the magnitude of A_w being small compared to the average of $G(v)$. This does not substantially change the filter (fig. 8.8b) at most frequencies but makes it smaller at the notch frequency v_1 (that is, $1/\{G(v) + A_w\} \approx 1/G(v)$ except when $G(v)$ is very small). The white noise is added only for filter-design purposes and a 'white noise added' comment on a seismic section thus indicates less noise generated by the whitening deconvolution.

The least-squares inverse filter is designed from autocorrelation values (8.58). The central peak of the autocorrelation function represents shifts of less than half the dominant period; it contains most of the information about average wavelet shape (fig. 8.9a). On the other hand, peaks and troughs for greater lags represent repetition of information, such as produced by multiples. If our desired wavelet h_t is to be essentially the same as the early part of the wavelet involved in g_t but is to die out rapidly but smoothly (since sharp changes induce ringing – see §10.3.5), we can take for ϕ_{gh} the early part of ϕ_{gg} multiplied by a suitable taper which truncates it after perhaps one cycle (fig. 8.9b), giving us everything needed to solve (8.58).

Predictive deconvolution (Peacock and Treitel, 1969) attempts to remove multiple effects, which can be predicted from knowledge of the arrival time of the primaries involving the same reflectors. Predictive deconvolution operators often do not begin to exert an effect until after some time L (called the *prediction lag*), which is usually the two-way traveltime to the first multiple-generating reflector. We use the portion of ϕ_{gg} after the time L as ϕ_{gh} in (8.58), so that the filter predicts the multiples (fig. 8.9c), that is, we write

$$\sum_{j=0}^{n} \phi_{gg}(i - j)f_j = \phi_{gh}(L + i); \qquad (8.64)$$

then we subtract the predicted trace from the observed trace to give the prediction error, which is the trace with the predicted multiples removed:

$$h_t = g_t - g_{t-L} * f_t. \qquad (8.65)$$

Where the first multiple-generating reflector is deep, as with marine data in deep water, the deconvolution operator may be set to zero over portions of its length (corresponding to the zeros in the inverse filters in (8.34))

to make the computation more economical; this is called *gapped deconvolution* (Kunetz and Fourmann, 1968).

Corrections may be made for variations in the seismic wavelet shape from shot-to-shot. One procedure is to record the initial waveform, as mentioned in §5.5.3f. Another is to sum the autocorrelations of all the traces from each shot and estimate the waveshape from this autocorrelation sum, assuming that the waveform is minimum-phase, then apply a Wiener filter to convert this wavelet into a desired constant waveshape. This sort of procedure may be used to correct for variations in wavelet shape produced by various factors, such as part of a line being recorded with land geophones and part with hydrophones, or parts recorded with different recording instruments or involving different near-surface conditions, etc.

Fig.8.9. Determining autocorrelation values to be used in deconvolution. (*a*) Autocorrelation of a trace; (*b*) tapering of ϕ_{gg} to use for ϕ_{gh} in (8.58); (*c*) prediction of multiple effects for predictive deconvolution filter design.

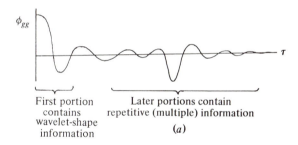

(*a*)

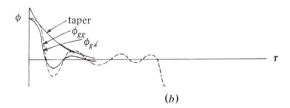

(*b*)

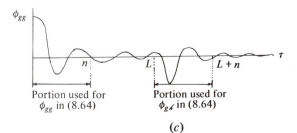

(*c*)

(*f*) *Other types of deconvolution*. Homomorphic (or cepstral) deconvolution, Kalman filtering and other techniques are occasionally employed.

Homomorphic deconvolution involves a transformation from a space where functions are convolved (the time domain) to one where they are added (the cepstrum domain; see §10.7). The transformation of a time-domain function g_t to a cepstrum-domain function $\hat{g}(\zeta)$ is accomplished in three steps,

$$
\left.
\begin{aligned}
g_t &\leftrightarrow G(z), \\
\ln\{G(z)\} &= \hat{G}(z), \\
\hat{G}(z) &\leftrightarrow \hat{g}(\zeta).
\end{aligned}
\right\}
\qquad (8.66, 10.214)
$$

The cepstrum-domain equivalent of (8.29) is

$$
\hat{g}(\zeta) = \hat{w}(\zeta) + \hat{e}(\zeta), \qquad (8.67)
$$

thus the contributions of the wavelet, $\hat{w}(\zeta)$, and of the reflectivity, $\hat{e}(\zeta)$, add. The wavelet is usually slowly varying and its cepstrum lies mainly at low ζ-values whereas that of the reflection coefficients is mostly spread out over larger values. Thus low-pass and high-pass filtering in the cepstrum domain (called *liftering*) achieves a large measure of separation. An inverse transformation of the high-pass (reflectivity) portion back to the time domain then completes the *homomorphic deconvolution*.

The cepstrum of a minimum-phase function is one-sided, that is, $\hat{g}(\zeta) = 0$ for $\zeta < 0$ if $G(z)$ is minimum-phase; this fact is sometimes used in separating minimum- and maximum-phase elements. To take advantage of the one-sided aspect in the cepstral domain one sometimes forces $G(z)$ to be minimum-phase by applying exponential gain (§8.1.4), that is, by using $G'(z) = G(z)k^t$ where $k \lesssim 1$ (Stoffa *et al.*, 1974). The reflectivity $E(z)$ is then found by liftering the high-pass portion of $\hat{g}'(\zeta)$, transforming back to find $E'(z)$, and then applying the inverse exponential weighting $E(z) = E'(z)\,k^{-t}$.

Otis and Smith (1977) use spatial averaging in the cepstrum domain as a way of determining source wavelet shape. They assumed that the source wavelet is stationary and the Earth's response is spatially non-stationary, so that phase contributions of the Earth's response at different locations disappear in the averaging.

Entropy is a measure of the chaos or lack of order in a system. Primary reflections are non-predictable from preceding data and thus lack order. *Maximum-entropy deconvolution* attempts to extract such reflections by separating orderly (for example, equivalent wavelet) from disorderly (for example, signal) elements. Maximum-entropy deconvolution is discussed further in §10.8.6*d*.

Most of the foregoing techniques assume stationarity, that is, that the statistics of the waveshape do not change with time. However, we know that higher frequencies are attenuated more rapidly than lower frequencies and that peg-leg multiples and other factors cause the downgoing wavetrain to lengthen with time. One should deconvolve more effectively if the change in waveshape with time were accounted for. Kalman filtering (Crump, 1974) and other types of *adaptive filtering* attempt to take changes with time into account by continuously updating the statistics on which the filters are based. The most common mode of *time-variant deconvolution* (often abbreviated TV decon; see Clarke, 1968) involves designing one operator based on an autocorrelation of the early portion of the data and another based on an autocorrelation of the late portion of the data, each data window being 1 s or longer so as to give adequate statistics. The early and late operators are then applied at any given time in inverse proportion to the difference in time to the centers of the design windows, that is, the early operator is gradually *ramped* out as the late operator is ramped in. Sometimes the design windows overlap and sometimes more than two design windows are used.

(*g*) *Wavelet processing*. A variety of different processes which involve determining, assuming, or operating on the effective wavelet shape go under the name *wavelet processing*. Some of these (1) attempt to make the wavelet shape everywhere the same, (2) some change the effective wavelet to some 'more desirable' shape, and (3) some endeavor to separate the Earth's reflectivity from wavelet shape effects.

Wavelet processing, which attempts to make the wavelet shape everywhere the same, should be done as a pre-stack process so that all the component traces to be stacked have the same effective wavelet shape. Low-frequency components are more likely to be stacked in-phase than high-frequency components so that stacking often acts as a filter attenuating higher frequencies; this type of wavelet processing decreases this filtering action. Sometimes the source wavelet is actually recorded for every shot in marine recording and then used in a deterministic wavelet processing. More commonly the wavelet is determined from the autocorrelation function by summing the autocorrelations of all traces recorded from the same source and assuming that the only common element is the source wavelet, so that the autocorrelation sum is simply the autocorrelation of the source wavelet. An example of this type of wavelet processing is shown in fig. 8.10.

The second type of wavelet processing, changing to some more desirable waveform, is used to correct for filtering actions (especially phase shifts) associated with instrumentation, such as to change hydrophone-recorded data to look more like geophone-recorded data or to produce a better match between lines shot with different recording instruments. Sometimes the effective wavelets associated with certain source types have been measured and 'catalogued' and the catalog wavelet is used in this type of wavelet processing. Sometimes the effective wavelet is determined from the seafloor reflection. Usually we desire the output for the final display to be zero-phase.

The third type of wavelet processing attempts to remove wavelet shape effects and leave the Earth's reflectivity function, that is, to separate w_t and e_t in (8.29), such as in the liftering example above. Such processing is usually a final deconvolution applied after other processing has removed as much of the noise as possible. Most wavelet processing techniques are trade secrets as of the date of this writing. They usually improve the high-frequency response and, consequently, the resolution. They often precede trace inversion (§9.4.5).

(*h*) *Frequency filtering.* Reflection signals often dominate over noise only within a limited frequency band. The filter should pass frequencies where the signal dominates and not pass those where the noise dominates in order to optimize the signal-to-noise ratio. Filter panels, such as shown in fig. 8.11, which display a portion of record section filtered by a succession of narrow bandpass filters are often used to determine the optimum bandpass limits.

The frequency spectrum of seismic reflections usually becomes lower with increasing arrival time as the higher-frequency components are attenuated faster by absorption, peg-leg multiple and other natural filtering processes (see §2.3.2c, §4.2.2b). Hence we often wish to shift the passband towards lower frequencies for later portions of the records, that is, we wish to accomplish *time-variant* (*TV*) *filtering*. Decisions as to the time-variant filter parameters are often based on filter panels such as shown in fig. 8.11, the deepest coherent energy in any passband being taken as the point where noise begins to predominate over signal.

Any discontinuous change, such as an abrupt change in bandpass parameters, will produce undesirable

Fig.8.10. Wavelet processing. (Courtesy Seiscom Delta.) (*a*) Portion of migrated seismic section; (*b*) after processing to broaden bandwidth of equivalent wavelet and make it zero-phase.

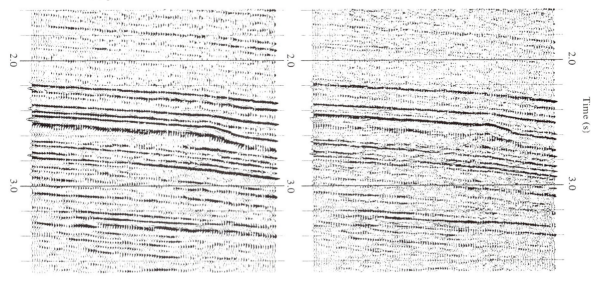

(*a*)   (*b*)

effects on the seismic section, however, including Gibbs' phenomena (§10.3.5). Changes are therefore usually distributed over a *merge zone*. For example, filter A might be used down to a time t_A and filter B below time t_B ($t_A < t_B$); the merge zone then is between times t_A and t_B. A *linear ramp* may be used in the merge zone; the data in this zone may be filtered with both filters A and B and the data at $t_A + \Delta t$ within the merge zone will be the sum of the results of applying these two filters where the results are weighted according to the position within the zone, that is, the weights would be $\Delta t/(t_B - t_A)$ and $(t_B - t_A - \Delta t)/(t_B - t_A)$, respectively. More than two filters and hence more than one merge zone may be used.

In addition to filtering, other processes such as deconvolution, statics correction, etc. are sometimes applied in *time-variant* mode following similar procedures. Changes in filter parameters or in the parameters

in other processes should not be made in the region where mapping is to be done lest the effects of changing parameters be misinterpreted as having structural or stratigraphic significance.

(i) Multichannel deconvolution. Most of the foregoing discussions imply that the design of the deconvolution operator is based on data from the same trace as that to which it is to be applied. One of the wavelet-processing methods described was based on the sum of a number of autocorrelations and then the application was to all of the components of the sum. Occasionally other multichannel schemes are utilized.

The radial multiple-suppression method given by Taner (1980) involves using data from one trace as the basis for designing the operator to be applied to another trace. For flat reflectors, the angle of incidence is the same

Fig.8.11. Filter panel. (Courtesy Seiscom Delta.)

FIELD DATA
FILTER PANEL

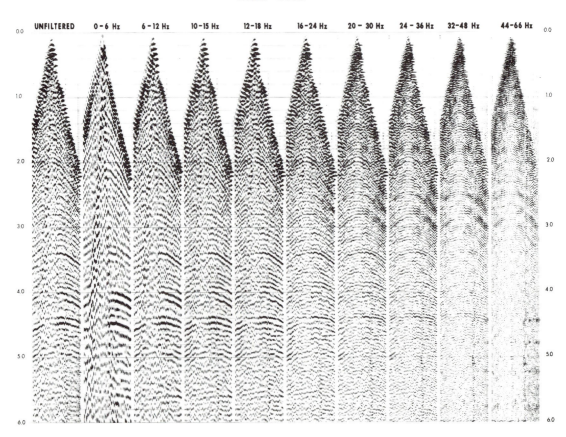

for the first multiple as it is for a primary at half the offset distance (see fig. 4.15) and for the second multiple as for the primary at a third the offset distance, etc. The reflectivity for the seafloor reflection changes so much because of the changing angle of incidence that predictive deconvolution applied to the same trace as that used in the operator design does not work well. Taner achieved much better multiple attenuation by designing his operators on traces where the angle of incidence for primaries was more nearly the same as for the multiples to be attenuated.

8.2.2   *Automatic statics determination*

(a)   *Interrelation of statics and normal-moveout corrections.* Statics corrections can be determined most easily after normal-moveout corrections have been optimized, but (as will be seen) normal-moveout determination is best when statics corrections are optimum. Since one of these determinations must precede the other, the calculations are often repeated with more refined inputs. Corrections for elevation differences or corrections based on uphole or first-break information and estimated velocity are usually made before the first statics determination. This is then followed by normal-moveout determination using these statics values. The values determined from the first statics and normal-moveout determinations are applied and then a second statics determination is made. The cycle of refining parameter values may be repeated several times to obtain an optimum solution.

(b)   *The surface-consistent model.* Automatic statics determination is often based on a *surface-consistent model* which associates a delay R_i with geophone group i and a delay S_j with source j. All data received by geophone group i will be delayed by R_i, possibly because the geophone group is at a higher elevation or because there is a thicker or slower low-velocity layer underneath it. All data from source j will be delayed by S_j, possibly because there was a delay between the source firing signal and the actual energy release, from the source being in (or on) a medium with lower velocity than other sources, from a higher source elevation or shallower shothole, etc. Following the method of Taner *et al.* (1974), we refer the subscripts i, j to a common origin and make the station increments equal (as in the surface stacking chart, fig. 5.5b); hence offset distance is proportional to $(j - i)$. If there is structure along the line, a delay L_k may be associated with the location k (that is, L_k is some sort of average of time shifts because of structure at different depths below k). For flat reflectors, $k = \frac{1}{2}(i + j)$ and, if the

dip is gentle, k is nearly constant for common reflecting point traces. If the normal-moveout correction is only approximate, some residual normal moveout M_k will remain and this residual normal moveout will vary as the square of the offset distance. Since the residual normal moveout varies with arrival time, the delay associated with M_k will be some sort of average, as was L_k. For the surface-consistent model, the total time shift for a trace, t_{ij}, will be given by

$$t_{ij} = R_i + S_j + L_k + M_k(j - i)^2. \qquad (8.68)$$

While we may not know the amount of time shift to be associated with any trace, cross-correlation affords a means of determining $(t_{ij} - t_{mn})$, the time shift of one trace relative to another which produces the optimum alignment of the two traces:

$$t_{ij} - t_{mn} = R_i - R_m + S_j - S_n + L_{i+j} - L_{m+n}$$
$$+ M_{i+j}(j - i)^2 - M_{m+n}(n - m)^2. \qquad (8.69)$$

(We use the subscript $i + j$ for L rather than $k = \frac{1}{2}(i + j)$ to assure that the subscripts are integers; the subscript magnitudes are not important since they are merely index values.) The shift which maximizes the cross-correlation produces the optimum alignment (match) of the two traces, and the magnitude of the cross-correlation indicates quantitatively how much improvement such a shift produces. With CDP data we have many combinations of traces which have some of the unknowns R_i, S_j, L_{i+j} or M_{i+j} in common. Since we can cross-correlate any two traces, we have more relative shift data than we have unknowns, that is, we have an 'overdetermined' set of equations to be satisfied. However, we also have uncertainty in our measurements, because of which opposite sides of each of (8.68) differ by some 'error'. The solution for R_i, S_j, L_{i+j}, M_{i+j} is usually by the least-squares method, sometimes in an iterative manner.

The least-squares problem is to minimize the sum of the squares of the errors:

$$E = \sum e_p^2 = \sum \{t_{ij} - t_{mn} - R_i + R_m - S_j + S_n$$
$$- L_{i+j} + L_{m+n} - M_{i+j}(j - i)^2$$
$$+ M_{m+n}(n - m)^2\}^2$$
$$= \text{minimum}. \qquad (8.70)$$

We wish to solve (8.70) for the best set of $R_i, S_j, L_{i+j}, M_{i+j}$. The least-squares solution is found by setting

$$\partial E/\partial R_i = 0, \quad \partial E/\partial S_j = 0,$$
$$\partial E/\partial L_{i+j} = 0, \quad \partial E/\partial M_{i+j} = 0. \qquad (8.71)$$

This results in many equations since there are as many R_i as there are geophone group locations, as many S_j as there are sources, etc.

We do not want to produce an overall time shift. One way is to require that $\sum R_i = 0$ and $\sum S_j = 0$. Taner *et al.* (1974) achieved this by adding the extra equations and modifying (8.70), writing

$$\sum e_p^2 + \lambda \left\{ \sum_i R_i^2 + \sum_j S_j^2 + \sum_{i+j} L_{i+j}^2 + \sum_{i+j} M_{i+j}^2 \right\}$$
$$= \text{minimum}, \quad \lambda > 0, \qquad (8.72)$$

λ being a weighting factor expressing the relative emphasis to be given to the latter part of the equation (e.g., see Claerbout, 1976, pp. 112–4). Equation (8.72) has a unique solution for any λ. It is often solved in an iterative manner to achieve any desired degree of accuracy.

Additional constraints are often applied to some of the variables in (8.70). One can remove part of the ambiguity between L_{i+j} and R_i or S_j by limiting L_{i+j} to small values. One may postulate a relation between R_i and S_j, for example, that the receiver and source statics for the same location should be similar, especially when using surface sources.

Taner *et al.* (1974) showed that solutions to (8.72) have five arbitrary constants which represent intrinsic indeterminacies, some of which correspond to (*a*) overall time shift of the section, that is, all events are too shallow or too deep, (*b*) overall tilt of the section, which may create fictitious structure, and (*c*) masking of real structure, that is, making structure show up as a statics correction or *vice versa*.

Cross-correlation may result in indeterminacy by one (or more) cycles, that is, the match between two traces may be nearly as good if the traces are displaced by a cycle. Usually the minimum trace shift is the preferred one but occasionally this will cause a cycle jump.

The cross-correlation concept implies that the best match of traces results in the best match of primary reflections. Sometimes other types of energy are so strong that the corrections determined optimize non-reflection rather than reflection alignments. Since cross-correlation is performed over a window, the best solution may be to narrow the window so that obvious noise is excluded. However, if the window is made too narrow, an alignment will be indicated even if there are no reflections. The window should include as many primary reflections as possible while excluding non-primary energy.

Equations (8.69), (8.70) and (8.72) plus other equations which express additional constraints can be written in matrix form (§10.1.5) as

$$\mathscr{X} \mathscr{A} - \mathscr{Y} = \boldsymbol{\varepsilon} \qquad (8.73)$$

where

$\mathscr{X}$ = matrix of coefficients and weightings, λ,

$\mathscr{A}$ = matrix of unknowns, R_i, S_j, L_{i+j}, M_{i+j},

$\mathscr{Y}$ = matrix of time shifts, $t_{ij} - t_{mn}$, plus other values for auxiliary equations,

$\boldsymbol{\varepsilon}$ = matrix of error terms.

The solution is

$$\mathscr{A} = (\mathscr{X}^{\mathrm{T}} \mathscr{X})^{-1} \mathscr{X}^{\mathrm{T}} \mathscr{Y} \qquad (8.74, 10.44)$$

(*c*) *Maximizing the power of the stacked trace.* Another approach assumes that the optimum static corrections are those which maximize the power of the stacked trace. A time-shift relation similar to (8.69) provides the starting point, with the R_i, S_j, L_k, M_k quantities being regarded as independent variables, x_n. Appropriate traces are stacked and the square of the amplitude (proportional to the power P) determined. The amount by which the power changes for changes in each variable, that is, $(\partial P / \partial x_n) \Delta x_n$ is determined for each variable and Δx_n is selected so that P increases. This is the method of *steepest ascent* and similar methods are used in many data-processing methods. In practice two problems are encountered: (1) how is one to find the correct maximum if there are several maxima, and (2) how to get to the maximum with the fewest calculations.

To solve the first problem, one assumes that the first estimate is on the slope of the correct maximum (seismic data are semiperiodic and adjacent maxima usually represent cycle jumps). Sometimes a search is made for other maxima so that one can determine which is the largest. Another technique is to make a first solution after filtering out higher frequencies so that the maxima are broader and fewer; the first solution is then used as the starting point for solving the problem with the unfiltered data.

The ideal solution to the second problem is to climb toward the maximum in relatively few steps without overshooting the top by very much. The step size is often related to $\partial P / \partial x_n$. Another technique is to calculate the curvature (or second derivative) to estimate how far away the maximum is. To minimize calculations, problems are often subdivided, limiting the number of variables being considered at one time.

Figure 8.12 illustrates the improvement in data quality which can result from application of automatic statics. Marked improvement is often achieved.

8.2.3 *Velocity analysis*

(*a*) *Conventional velocity analysis.* The variation of normal moveout with velocity and arrival time has

Fig.8.12. Improvement resulting from use of automatic statics. (Courtesy Seiscom Delta.) (*a*) Section using field-determined statics; (*b*) section using statics determined by a surface-consistent program; (*c*) velocity analysis using field statics; (*d*) velocity analysis after application of surface-consistent statics.

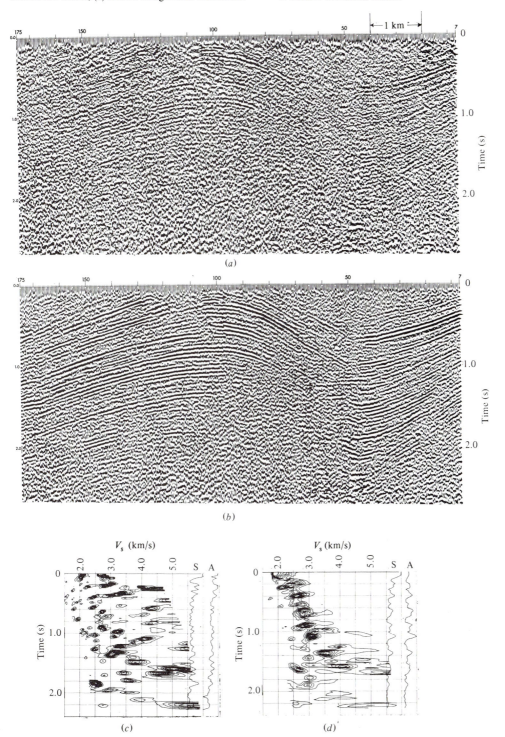

already been discussed in connection with (3.7). Several techniques utilize the variation of normal moveout with record time to find velocity (Garotta and Michon, 1967; Cook and Taner, 1969; Schneider and Backus, 1968; Taner and Koehler, 1969). Most assume a stacking velocity (V_s) as discussed in §7.3.3*a* and apply the normal moveouts appropriate for the offsets of the traces being examined as a function of arrival time, and then measure the coherence (degree of match) among all the traces available to be stacked. Several measures of coherence may be used; some of these were discussed in §8.1.3*f* (see (8.50) and (8.51)). Another stacking velocity is then assumed and the calculation repeated, and so on, until the coherence has been determined as a function of both stacking velocity and arrival time. (Sometimes normal moveout is the variable rather than stacking velocity.)

A velocity-analysis display is shown in fig. 8.13. This is a good analysis because the data involved in fig. 8.13*a* are good. Peaks on the peak amplitude trace (fig. 8.13*d*) correspond to events. The locations of the highs on fig. 8.13*b* yield the velocities (or normal moveouts) which have to be assumed to optimize the stack (hence the name stacking velocity). Multiples as well as primaries will give rise to peaks and hence the results have to be interpreted to determine the best values to be used to stack the data. In many areas where the velocity increases more-or-less monotonically with depth, the peaks associated with the highest reasonable stacking velocities are assumed to represent primary reflections and peaks associated with lower velocities are attributed to multiples of various sorts. In other areas the relationships are not as obvious and even where the velocity relationships are generally regular, difficulties may be encountered.

A compromise has to be made between using the small amount of data appropriate to a specific spot, in which case the velocity analysis is apt to be non-definitive, and using more data but distributed over a larger area, in which case velocity may be defined better but the velocity measurements are then averages over a sizable region. The compromise is often to use data for 5 – 7 adjacent midpoints. Measurements are also usually based on all the data within a window, which is often 50 – 100 ms long, in order to increase the amount of data and hence improve the velocity definition.

Velocity analysis involves an appreciable number of calculations and hence is fairly expensive to execute; therefore, analyses are often run only every 1 – 5 km along the line. Velocity analyses are often plotted at the same time scale as the seismic section, so that the times and velocities picked on them may be more easily related to events on the seismic section by overlaying.

The precision of reading values from velocity analyses is usually ± 0.010 s in arrival time and ± 50 m/s in velocity but the accuracy is often less than this. Velocity values have to be interpolated for intermediate points and intervening locations. Occasional velocity picks which are poor because of undue noise influence or non-systematic picking of successive analyses (not always picking the same events) can introduce irregularities in stacking velocities, especially in doubly-interpolated values (often the velocities for times between picks are interpolated linearly and then the velocities for traces between analyses are interpolated from these). A plot showing the interpolated values (fig. 8.14) provides valuable control by making the consequences of velocity assumptions clear.

(*b*) *Velocity panels.* Velocity panels (fig. 8.15) provide another display from which stacking velocity can be determined. A set of data is plotted several times, each plot being based on a different stacking velocity. The central two panels, fig. 8.15*d, e*, utilize an approximate velocity function (such as derived from an analysis of the type shown in fig. 8.13); the panels to the left use velocities successively lower by some velocity increment and those to the right utilize higher velocities. Such a set of velocity panels shows whether increasing or decreasing the velocity will enhance individual events. Since stacking velocity is not necessarily single-valued (see fig. 8.16), different events might require different velocities to be optimized. A velocity panel is often run as a check on the interpretation of velocity analyses of the type shown in fig. 8.13.

The major objective of velocity analysis is to ascertain the amount of normal moveout which should be removed to maximize the stacking of events which are considered to be primaries. This does not necessarily optimize the primary-to-multiple energy ratio and better stacks can be achieved with respect to identifiable multiples. This is not often done, however. An auxiliary objective of velocity analysis is identifying lithology.

8.2.4 *Preservation of amplitude information*
The amplitude of a reflection depends on the acoustic impedance contrast at the reflecting interface. However, other factors, such as those listed in fig. 8.17, often obscure the acoustic-impedance-contrast information. The effects of spherical divergence and raypath curvature can be calculated and corrected for. The gain of the recording instruments is usually known. Array directivity usually will not significantly affect the amplitude of events which do not dip appreciably, and the variation of reflection coefficient with angle will be small except for large offsets. Migration (§8.3) can correct for reflector curvature effects.

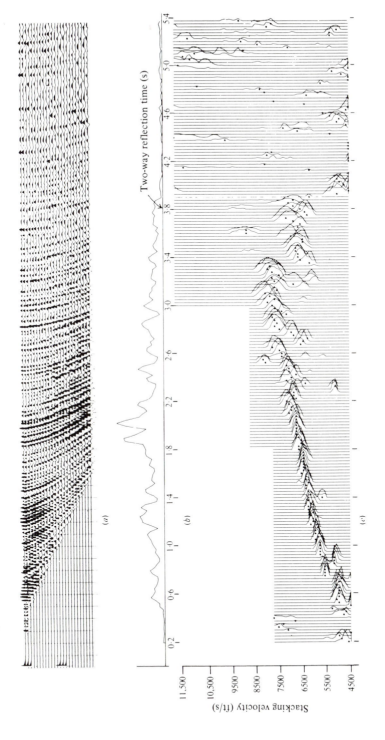

Fig.8.13. Velocity analysis. (Courtesy Petty-Ray Geophysical.) (*a*) Common-depth-point gather showing the data involved in the analysis; (*b*) maximum amplitude achievable in stacked traces. Peaks indicate the stronger events; (*c*) amplitude of stacked trace as a function of stacking velocity at 100-ms intervals of t_0. The dots indicate the stacking velocity giving maximum amplitude. The stacking velocity increases with depth from about 5000 ft/s at 0.6 s to about 7700 ft/s at 3.3 s. The data between 2.7 and 3.9 s having velocities around 6600 ft/s are undoubtedly multiples involving the strong reflections at 1.9–2.1 s. There are probably few primary reflections below 3.3 s.

Fig.8.14. Stacking velocity along a seismic line. Values
are interpolated by the computer from the events shown
by dashes. (Courtesy Seiscom Delta.)

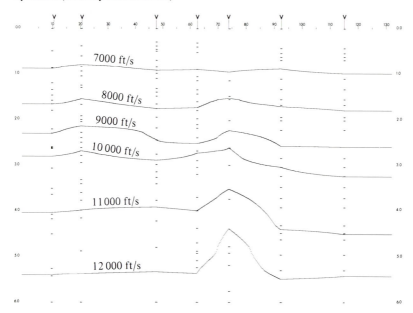

Fig.8.15 Velocity panels. Panels (*e*) and (*f*) show a CDP
gather for the applied stacking velocity, $V_s(t)$; a mute has
been applied in (*f*). Panels (*a*), (*b*), (*c*) and (*d*) show the
results where the stacking velocity has been decreased by
$n\Delta V_s$, $n = 4, 3, 2, 1$; ΔV_s is often 200–500 ft/s. Panels (*g*), (*h*),
(*i*) and (*j*) show results for increasing velocity by $n\Delta V_s$.
(Courtesy Seiscom Delta.)

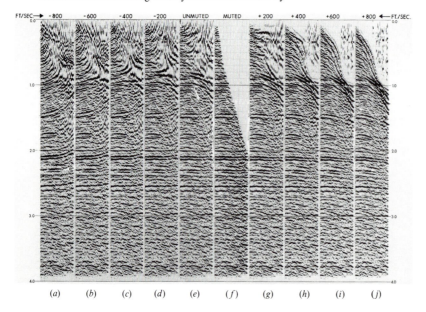

Remaining effects are mostly of two kinds: (1) those associated with energy losses because of absorption, scattering, transmissivity losses and peg-leg multiples, and (2) those which vary with source strength and source coupling, geophone sensitivity and geophone coupling, and offset. The effects in the first group are difficult to determine but they usually do not vary appreciably along a line and so may not obscure lateral variations. The high multiplicity of CDP data permits determining the second group of effects in a manner similar to that employed for time delays in automatic-statics-correction programs (§8.2.2 and Taner and Koehler, 1981).

One processing routine adjusts amplitude in several steps. After first correcting for amplitude adjustments made in recording, a time-dependent spherical

divergence correction is applied; such a correction makes the range of amplitude values smaller and therefore easier to handle but does not allow for the effect of velocity changes in spherical divergence. These corrections constitute the 'preliminary gain adjustment' shown in 'editing' phase of fig. 8.36. Surface-consistent amplitude analysis and/or correction is then done during one or more of the processing passes in the 'main processing' phase. After velocity has been determined, the spherical divergence correction is changed so as to depend on distance rather than travel time. Sometimes an additional arbitrary exponential gain is applied to make the range of amplitude values smaller for display purposes. Sometimes the above step-by-step amplitude adjustment is simply replaced by an arbitrary gain function.

8.2.5 *Common-depth-point stacking*

(*a*) *Gathers.* Common-depth-point (CDP) stacking is probably the most important application of data-processing in improving data quality. The principles involved have already been discussed along with the field procedures used to acquire the data. The component data are sometimes displayed as gathers; a *common-depth-point gather* (see fig. 8.15) has the components for the same midpoint arranged side-by-side, and a *common-offset gather* has the components for which the shotpoint-to-geophone distance is the same arranged side-by-side. For dipping events the reflecting points of the traces of a common-depth-point gather will not be the same (fig. 8.18) even though the midpoints between shotpoints and

Fig.8.16. Illustrating different velocity values for the same arrival time; reflections from *B* and *C* might arrive at the same time at point *A*, but velocities along the two paths might differ.

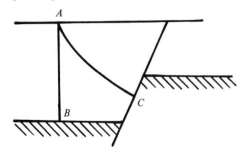

Fig.8.17. Factors which affect amplitude. (After Sheriff, 1975.)

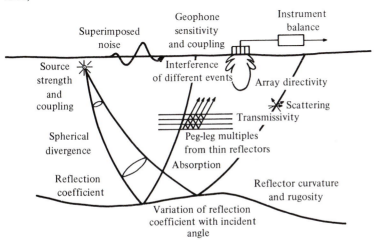

geophones are the same (also shown in fig. 3.21). CDP stacking of dipping reflections thus involves smearing of reflecting points. Moreover, dip causes peg-leg multiples to divide into two sets, one with apparent stacking velocity higher than the zero-dip stacking velocity, the other lower (Levin and Shah, 1977); hence stacking will alter the character of events which include appreciable peg-leg energy.

(b) *Muting.* First-break data and the refraction wavetrains which follow the first-breaks are usually so strong that they have to be excluded from the stack to avoid degrading the quality of shallow reflections. This is done by *muting*, which involves arbitrarily assigning zero values to traces during a desired interval. Figure 8.15*f* shows the result of muting fig. 8.15*e* during the arrival of first-breaks and following strong wavetrains. Thus the multiplicity of a stack increases by steps, the shallowest data often being a two-fold stack, slightly deeper data being a four-fold stack, and so on until the full multiplicity of the stack is achieved after the muted events have passed beyond the most distant geophones. To avoid amplitude discontinuities associated with changes in the multiplicity, the amplitude is usually divided by the number of non-zero traces which have been added.

Sometimes traces near the shotpoint become very noisy as time after the shot increases, perhaps because of *hole noise* (noise produced by the oscillation and venting of gases generated by the shot and/or ejection of material from the borehole) or airwave or ground roll; a *tail mute* may be applied to such traces after a certain time. Occasionally a wedge of data across the gather (such as a portion dominated by ground roll) will also be muted, although it is more common to use apparent-velocity filtering (§8.2.7) in such situations.

(c) *Diversity stacking.* Much data-processing is far less exotic than is suggested by the mathematical relationships expressed in the foregoing pages. Some of these processes involve merely excluding certain elements of the data, such as the muting operation which has already been discussed. It is almost always better to throw away noisy data than to include it on the theory that its adverse effects will be averaged out. A very powerful processing technique which is not used as much as it should be is to simply look at the data and delete the portions which appear to be mainly noise.

Diversity stacking is another technique used to achieve improvement by excluding noise. Records in high-noise areas, such as in cities, often show bursts of large-amplitude noise while other portions of the records are relatively little distorted by noise. Under such circumstances amplitude can be used as a discriminant to determine which portions are to be excluded. This can take the form of merely excluding all data where the amplitude exceeds some threshold or perhaps some form of inverse weighting might be used. Such noise bursts are often randomly located on repeated recordings so that sufficient vertical stacking after the weighting tends to produce records free from the high-amplitude noise.

Fig.8.18. Ray-trace model showing the updip migration of reflecting points as offset increases. (From Taner *et al.*, 1970.)

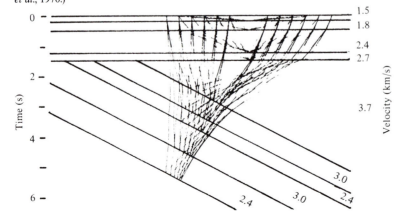

8.2.6 *Simplan stacking*

Although occasionally large source arrays are used, most sources are effectively points; hence seismic waves are spherical or nearly so. An alternative to CDP stacking of component spherical-wave records is sometimes used to simulate sections which would have been generated by plane or cylindrical waves; such sections are called 'Simplan' sections (Taner, 1976).

Fig.8.19. Synthetic common-source gather and Simplan trace. (Courtesy Seiscom Delta.) (*a*) Gather showing reflections symmetrical about the trace $x = -2h \sin \xi$ where ξ is the dip and h the distance to the reflector, as in fig. 3.2; (*b*) the Simplan trace which results from summing all the traces; in effect only the first Fresnel zone contributes.

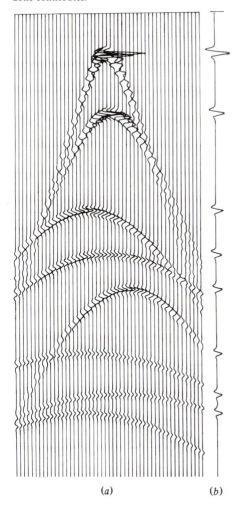

(*a*)   (*b*)

The Simplan process utilizes reciprocity (§3.3.3) and superposition (§10.5.1). The sum of the outputs of a geophone for a number of in-line point sources simulates the output from a line source, that is, a cylindrical wave. Figure 8.19 shows a split-spread record and the Simplan trace which results from simple stacking without making any time shifts for normal-moveout correction. Note that dipping reflectors produce full-amplitude response on the Simplan trace even though the incident wavefront is not parallel to them. Only those traces of the gather which lie within the first Fresnel zone effectively contribute to the Simplan trace, so dipping events suffer no attenuation unless their dip is so extreme that their first Fresnel zone extends beyond the traces included in the gather. The first Fresnel zone also includes more traces as arrival time increases so that the rate of amplitude decay on the Simplan trace is less than on the traces of the gather (this must be the case since the Simplan trace undergoes cylindrical divergence rather than the spherical divergence of the component traces). The traces from geophones closely spaced can be used in the same way as the traces from sources closely spaced. Customary group spacing and range of offset distances are usually sufficient to avoid undesirable end-effects.

Split-spread and Simplan records can be simulated from end-on records. Note (fig. 8.20) that the trace at (r_{k+i}, s_k) on the surface diagram is the same as the trace at (r_k, s_{k+i}) by reciprocity. Thus end-on records can be used to produce a split-spread record for twice the number of channels, using the common-source and common-receiver traces respectively, for the two halves of the split.

Fig.8.20. Reciprocal relations between traces on a surface stacking chart. Traces on one side of the zero-offset line are identical to traces symmetrically disposed on the opposite side of the line.

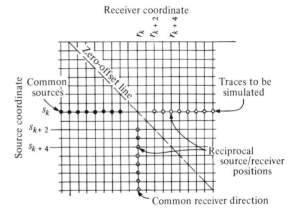

Figure 8.21 shows a 96-trace split-spread record simulated from 48-trace end-on records. The stack of these 96 traces yields one Simplan trace.

Simplan sections contain all primaries, multiples and diffractions without amplitude bias or waveform distortion, whereas CDP stacking with proper NMO corrections emphasizes primary reflections compared to multiples and diffractions. Simplan sections before deconvolution usually have poorer signal-to-noise ratios than CDP sections because multiples are not attenuated by the stacking. We call a Simplan section 'physically realizable' because it would be possible (but very costly) to actually fire a sufficiently long array of shots to generate a cylindrical wave and produce directly a Simplan trace. On the other hand, no possible configuration of shots and geophones can be used to generate a CDP trace without the intervening non-linear processing step of NMO correction. Predictive deconvolution (and most processing) implies physical realizability and thus performs better on Simplan than on CDP sections, that is, the amplitudes and waveshapes of primaries and their associated multiples more nearly conform to the assumptions of predictive deconvolution.

By shifting traces by an amount proportional to their location, one can effectively 'beam' the wave in an arbitrary direction (*beam steering*); the wavefront from a line source in this instance is conical in shape. The apparent dip is changed by beaming the wavefront, but the Simplan traces can be time-shifted to compensate. The positions of events after such compensation do not depend on the direction in which wavefronts are beamed, although the amplitudes do. Preparing several Simplan sections by beaming in different directions, migrating each and then stacking, results in improved signal-to-noise ratio without reflecting-point smearing.

8.2.7 *Apparent-velocity (2-D) filtering*

This method, also called *fan, moveout* or *pie-slice filtering* (Fail and Grau, 1963; Treital *et al.*, 1967) for reasons which will become obvious, depends upon the apparent velocity (defined by (3.13)) of a signal as it approaches an array of geophones. Using (2.48) and (3.13), we can define an *apparent wavenumber*, κ_a, which is related to the apparent velocity V_a by the equation,

$$V_a = \omega/\kappa_a = 2\pi v/\kappa_a. \qquad (8.75)$$

For a fixed V_a, the plot of frequency versus wavenumber is a straight line. For a seismic spread along the x-axis, κ_a is plus or minus according as V_a is in the plus or minus direction; for a vertically traveling signal, $V_a = \infty$ and the

(v, κ_a) curve is the v-axis. For most reflection signals, $V_a > V_m$, some minimum apparent velocity, and hence the reflections lie within a relatively narrow wedge containing the v-axis, as shown in fig. 8.22a. Coherent noise generally has a lower V_a than the signals and therefore separates from them in the (v, κ_a) plot.

We can use two-dimensional transforms (§10.3.2) to define an apparent velocity filter,

$$\left. \begin{aligned} F(v, \kappa_a) &= 1, \quad |\kappa_a| < 2\pi v/V_m \\ &= 0, \quad |\kappa_a| > 2\pi v/V_m \end{aligned} \right\} \qquad (8.76)$$

which will pass the signal but reject the noise (as shown on fig. 8.22a). Such a filter which passes a narrow slice in the (v, κ_a) domain is a 'pie-slice' filter. Of course, neither signal, noise, nor filter need be symmetric about the v-axis. For example, there are hardly any coherent alignments dipping to the left in fig. 5.14 and so fig. 5.15, if extended to the left of the v-axis, would be essentially blank. Apparent-velocity filters can also be designed to remove a noise wedge rather than pass a signal wedge; such a filter is called a 'butterfly' filter.

Just as frequencies above the Nyquist frequency may alias back into the passband unless excluded by alias filters before the sampling, so spatial sampling involves the aliasing of data for wavenumber values exceeding the Nyquist wavenumber (see (8.27)). The only way to prevent aliasing is to filter before sampling, which is not possible with respect to spatial sampling, or to move the Nyquist points farther out by sampling more closely. Figure 8.22b shows how noise can alias into the signal region because of spatial aliasing.

The filter in the space–time domain (x, t) equivalent to the filter given by (8.76) is obtained by taking the two-dimensional inverse Fourier transform (see (10.95)):

$$f(x, t) = (1/2\pi) \int_{-\kappa_N}^{+\kappa_N} \int_{-v_N}^{+v_N} F(v, \kappa_a) e^{j(\kappa_a x + 2\pi vt)} d\kappa_a dv$$

$$= (1/2\pi) \int_{-\kappa_N}^{+\kappa_N} \int_{-v_N}^{+v_N} \cos(\kappa_a x + 2\pi vt) d\kappa_a dv, \qquad (8.77)$$

since $f(x, t)$ must be real. The convolution of $f(x, t)$ with the input (signal + noise), $g(x, t)$, gives the output $h(x, t)$,

$$h(x, t) = g(x, t) * f(x, t)$$

$$= \int_{-\infty}^{\infty} \int_{-\infty}^{\infty} g(\sigma, \tau) f(x - \sigma, t - \tau) d\sigma d\tau. \qquad (8.78, 10.142)$$

This equation can also be written in digital form:

$$h_{x,t} = \sum_m \sum_n g_{m,n} f_{x-m, t-n} \qquad (8.79)$$

Fig.8.21. Split-spread record with 96 traces simulated
from 48-trace end-on records. (Courtesy Seiscom Delta.)

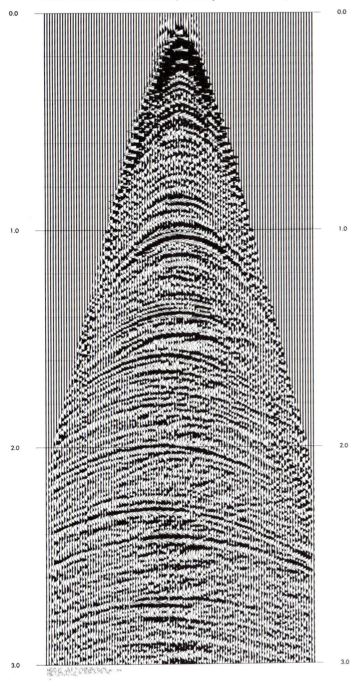

Fig.8.22. The frequency–wavenumber domain. (*a*) A
pie-slice filter passing the region between $\kappa_a = \pm 2\pi \nu / V_m$
will pass the signal but exclude the coherent noise.
(*b*) Aliasing will fold some of the coherent noise back
into the signal passband.

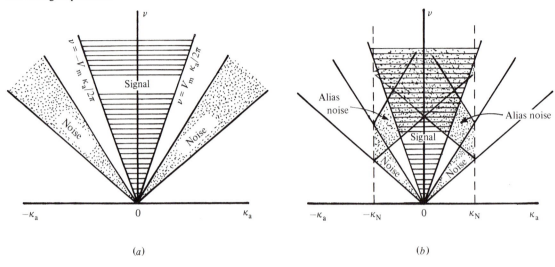

(*a*)   (*b*)

Fig.8.23. An example of apparent-velocity filtering.
(Courtesy Seiscom Delta.) (*a*) Two seismic records
showing severe horizontally traveling energy; (*b*) same
after apparent-velocity filtering; reflections stand out
much more clearly and can be followed to longer offset
distances, permitting better stacking-velocity
determination.

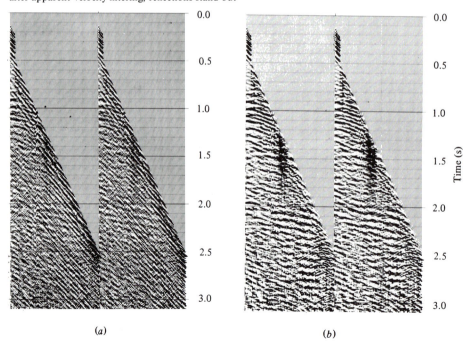

(*a*)   (*b*)

where the space-sample interval is usually the trace spacing in the *x*-direction and the time-sample interval in the *t*-direction.

Instead of transforming the 2-D filter to the time-domain and calculating $g(x,t) * f(x,t)$ as we did in (8.35), we can transform $g(x,t)$ to the (v,κ_a) domain, multiply $G(v,\kappa_a)$ by $F(v,\kappa_a)$ and use the two-dimensional convolution theorem (10.143) to obtain $h(x,t)$.

The use of 2-D filtering to attenuate severe ground-roll on common-source gathers is illustrated in fig. 8.23. Using 2-D filtering reduces the amount of muting required so that more reflection data can be used in velocity analysis and in stacking, providing better stacking velocity definition and better attenuation of multiples in stacking.

8.2.8   *Polarization filtering*

Where data showing ground motion in different directions are available, as when 3-component geophones are used, phase relationships between the components can be used to attenuate modes of motion. The technique, called *polarization filtering*, involves phase-shifting and amplitude-adjusting of the components and then combining (stacking) them. Since different propagation modes and directions of wave travel involve systematic relationships among the components, polarization filtering can be used to preferentially select or reject any particular mode, such as groundroll. However, multicomponent geophones are not used very much as of 1982. An example of the selectivity of polarization filtering for in-seam studies (§5.3.9*d*) is given in fig. 8.24.

8.3   **Processes to reposition data**
8.3.1   *Introduction*

Seismic data prior to migration are oriented with respect to the observation point. Migration (see §5.6.3) involves repositioning data elements to make their locations appropriate to the locations of the associated reflectors or diffracting points. The need to migrate seismic data to obtain a structural picture was recognized at the beginning of seismic exploration and the very first seismic reflection data in 1921 were migrated (fig. 1.3*b*).

Migration generally is based on the premise that all data elements represent either primary reflections or diffractions. The migration of noise, including energy which does not travel along simple reflection paths, produces meaningless results. Migration requires a knowledge of the velocity distribution; changes in velocity bend raypaths and thus affect migration. Although migration can be extended to three dimensions, we usually assume that the cross-dip is zero which results in two-dimensional migration. Ignoring cross-dip sometimes results in under-migration, but an undermigrated section is at least easier

Fig.8.24. Use of polarization filtering to separate P- and S-headwaves on transmission in-seam data. (From Millahn, 1980.) (*a*) Records of components perpendicular and parallel to the gallery in which measurements were made; (*b*) after polarization filtering and rotation to emphasize components perpendicular and parallel to the direction to the source.

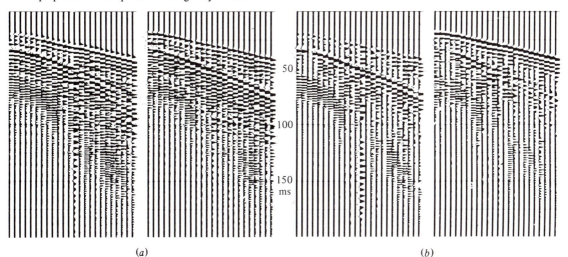

(*a*)   (*b*)

to interpret than one not migrated at all. Moreover, cross-dip information is often not available, two-dimensional migration is appreciably more economical, and the results are often adequate.

The simplest approach to migration is to determine the direction of approach of energy and track the raypath backwards to the reflecting point at half the traveltime, or to find the common tangent to the wavefronts for half the traveltime; these methods were extensively used in hand-migrating data. Computer methods generally involve solutions of the wave equation. We replace the time with half the traveltime, that is, in effect we start with the energy originating at each reflector, as if each reflector were covered by elementary point sources as postulated by Huygens' principle, all actuated at the instant $t = 0$ (the 'exploding-reflector' model). We regard $\psi(x, z, c)$ as a vertical section showing the wave motion at the point (x, z) at time $t = c$, that is, an unmigrated seismic section corresponds to $\psi(x, 0, t)$ whereas a migrated seismic section corresponds to $\psi(x, z, 0)$. There are various ways of solving for $\psi(x, z, 0)$, including (a) integral methods based on Kirchhoff's equation (§8.3.2) where the integration is over those elements in unmigrated space which contribute to an element in migrated space, (b) methods based on a solution in the frequency–wavenumber domain (§8.3.3), and (c) finite-difference solutions of the scalar wave equation (§8.3.4), which accomplish backward-tracing of seismic waves in a downward-continuation manner.

The methods discussed in the next sections accomplish full-waveform migration; they involve large numbers of calculations and so are generally restricted to computer implementation.

8.3.2 *Diffraction-stack migration*

Diffraction-stack migration is based on a concept of Hagedoorn (1954). We assume constant velocity V and convert arrival times to distances by multiplying by $\frac{1}{2}V$. Figure 8.25a shows a series of source points and coincident receivers S_0, S_1, etc. A reflector with dip ξ passes through P, at a depth z_0, where $S_0 P$ is perpendicular to the reflector. If we swing arcs with centers S_0, S_1, etc., with radii equal to the distances to the reflector, we define a straight line MN, the reflection on an unmigrated section with the vertical scale in depth. If we assume that P is a diffracting point, the diffraction arrival times will correspond to the distances PS_0, PS_1, etc. Hagedoorn called the unmigrated diffraction curve PMR a *curve of maximum convexity*, since no other event from the depth z_0 can have greater curvature (see fig. 4.18a). The diffraction curve is a

hyperbola with apex at P and the unmigrated reflection is tangent at M (see problem 8.19).

The concept for carrying out migration is to plot a diffraction curve for each depth and slide it along the unmigrated section (keeping the top lined up with zero depth) until a segment of a reflection is tangent to one of the curves; on the corresponding migrated section the reflector is located at the crest of the diffraction curve tangent to the wavefront which passes through the point of tangency of the reflection to the diffraction curve (fig. 8.25b). The principle is the same if the velocity is not constant and if the sections, wavefronts and diffraction curves are plotted in time rather than in depth.

To carry out diffraction-stack migration, the diffraction curve is calculated for each point on the migrated section. The data at each of the points on the unmigrated section where the diffraction curve intersects

Fig.8.25. Wavefront and diffraction curves intersecting at the unmigrated and migrated positions.
(a) Unmigrated reflection MN migrates onto reflector PQ; (b) relation between wavefront and diffraction curves (from Hagedoorn, 1954).

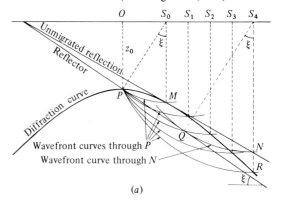

(a)

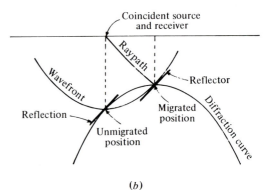

(b)

traces are summed together to give the amplitude at the point on the migrated section. If there is energy coming from the point at the crest of the diffraction curve, then addition along the diffraction curve will produce the value appropriate to the energy involving that point; if only noise is present, positive and negative values will be equally probable along the diffraction curve so the sum will be very small. An example of migration is shown in fig. 8.26.

In effect, diffraction-stack migration treats each element of an unmigrated reflection as a portion of a diffraction, that is, a reflector is thought of as a sequence of closely-spaced diffracting points. The relationship between points shown in fig. 8.25*b* suggests that the data at each point could be distributed along the wavefront through that point (wavefront smearing) and when the smears for all points are superimposed they will reinforce where reflectors exist but otherwise positive and negative

Fig.8.26. Diffraction-stack migration. (Courtesy AGIP and Western Geophysical.) (*a*) Before migration; (*b*) after migration. Multiple-branch reflection events from sharply-folded basement migrate so as to show a relatively continuous basement reflector. Deeper data, which are probably multiples, out-of-the-plane diffractions and other types of events, are not migrated properly.

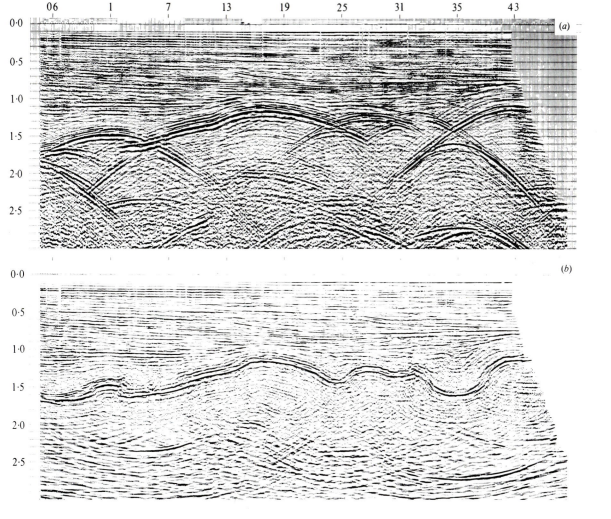

values will be equally probable so the sum will be small. Migration by the method of wavefront smearing produces results identical to diffraction-stack migration, the only difference being in that operations are performed in a different sequence. The 'common-tangent' method of migration (Sheriff, 1978) is in effect wavefront smearing.

A more elegant formulation of diffraction-stack migration is based on the Kirchhoff integral (see (2.34) or (2.35); Schneider, 1978). This approach makes it clear that this technique is an integral solution to the wave equation, as opposed to a finite-difference solution (usually called 'wave-equation migration') or a Fourier-transform solution of the wave equation (usually called 'frequency-domain migration').

8.3.3 *Migration in the frequency–wavenumber domain*

This method starts with (2.23) in two dimensions, the x-axis being along the profile direction and the z-axis positive vertically downward. Thus,

$$\frac{\partial^2 \psi}{\partial t^2} = V^2 \left(\frac{\partial^2 \psi}{\partial x^2} + \frac{\partial^2 \psi}{\partial z^2} \right). \tag{8.80}$$

We use (10.94) to take the three-dimensional transform of $\psi(x, z, t)$ and obtain

$$\psi(x, z, t) \leftrightarrow \Psi(\kappa_x, \kappa_z, \omega) = \int_{-\infty}^{\infty} \int_{-\infty}^{\infty} \int_{-\infty}^{\infty} \psi(x, z, t)$$

$$\exp\{-j(\kappa_x x + \kappa_z z + \omega t)\} \, dx \, dz \, dt.$$

Equation (10.119) now gives

$$\frac{\partial^2 \psi}{\partial t^2} \leftrightarrow (j\omega)^2 \Psi(\kappa_x, \kappa_z, \omega),$$

$$\frac{\partial^2 \psi}{\partial x^2} \leftrightarrow (jk_x)^2 \Psi(\kappa_x, \kappa_z, \omega),$$

$$\frac{\partial^2 \psi}{\partial z^2} \leftrightarrow (jk_z)^2 \Psi(\kappa_x, \kappa_z, \omega).$$

Substituting in (8.80), we get

$$\omega^2 - V^2(\kappa_x^2 + \kappa_z^2) = 0. \tag{8.81}$$

Returning to (8.80), we take the two-dimensional transform of $\psi(x, z, t)$ with respect to x and z, obtaining

$$\Psi_{xz}(\kappa_x, \kappa_z, t) = \int_{-\infty}^{\infty} \int_{-\infty}^{\infty} \psi(x, z, t)$$

$$\exp\{-j(\kappa_x x + \kappa_z z)\} \, dx \, dz. \tag{8.82}$$

If we restrict the solution to harmonic waves, we can write

$$\Psi_{xz}(\kappa_x, \kappa_z, t) = \Psi_{xz}(\kappa_x, \kappa_z, 0) e^{-j\omega t} \tag{8.83}$$

(to verify this relation, substitute $t = 0$; the first factor on the right is the required solution to our problem.

To find $\Psi_{xz}(\kappa_x, \kappa_z, 0)$ we start by calculating the transform of the recorded data with respect to x and t:

$$\Psi_{xt}(\kappa_x, \omega) = \int_{-\infty}^{\infty} \int_{-\infty}^{\infty} \psi(x, 0, t)$$

$$\exp\{-j(\kappa_x x + \omega t)\} \, dx \, dt; \tag{8.84}$$

inverting the transform we have

$$\psi(x, 0, t) = (1/2\pi)^2 \int_{-\infty}^{\infty} \int_{-\infty}^{\infty} \Psi_{xt}(\kappa_x, \omega)$$

$$\exp\{j(\kappa_x x + \omega t)\} \, d\kappa_x \, d\omega. \tag{8.85}$$

From (8.82) and (8.83) we obtain

$$\psi(x, z, t) = (1/2\pi)^2 \int_{-\infty}^{\infty} \int_{-\infty}^{\infty} \{\Psi_{xz}(\kappa_x, \kappa_z, 0) e^{-j\omega t}\}$$

$$\exp\{j(\kappa_x x + \kappa_z z)\} \, d\kappa_x \, d\kappa_z;$$

hence,

$$\psi(x, 0, t) = (1/2\pi)^2 \int_{-\infty}^{\infty} \int_{-\infty}^{\infty} \Psi_{xz}(\kappa_x, \kappa_z, 0)$$

$$\exp\{j(\kappa_x x + \omega t)\} \, d\kappa_x \, d\kappa_z. \tag{8.86}$$

Comparing (8.85) with (8.86) we see that

$$\Psi_{xt}(\kappa_x, \omega) \, d\omega = \Psi_{xz}(\kappa_x, \kappa_z, 0) \, d\kappa_z,$$

or

$$\Psi_{xz}(\kappa_x, \kappa_z, 0) = \Psi_{xt}(\kappa_x, \omega) \frac{\partial \omega}{\partial \kappa_z}$$

$$= V \Psi_{xt}(\kappa_x, \omega)\{1 + (\kappa_x/\kappa_z)^2\}^{-\frac{1}{2}}$$

from (8.81), ω being equal to $V(\kappa_x^2 + \kappa_z^2)^{\frac{1}{2}}$. Since $\Psi_{xt}(\kappa_x, \omega)$ is known from (8.84), we can calculate $\Psi_{xz}(\kappa_x, \kappa_z, 0)$, then invert it to get the solution, $\psi(x, z, 0)$. Chun and Jacewitz (1981) discuss frequency-domain migration from a geometric viewpoint.

The migrated solution clearly depends on the velocity, but the above derivation assumes a constant velocity. However, velocity is usually a function of z. To get a section with effectively constant velocity, we stretch the time scale, even though the velocity as a function of time differs for reflections with different amounts of dip. Where velocity variations with depth involve serious problems, migration can be carried out by transforming with respect to t only and then downward-continuation can be carried out in the space – frequency (x, ω) domain. One could also transform into the wavenumber – time (κ_x, t) domain.

8.3.4   Finite-difference method of wave-equation migration

The concept underlying time-domain migration is downward-continuation of the seismic wavefield. Continuation is a familiar process with gravity and magnetic fields. It utilizes the continuity property of fields, one expression of which is that we can determine the field over any arbitrary surface if we know the field completely over one surface provided the field satisfies Laplace's equation, $\nabla^2 \phi = 0$. Evidences of subsurface features spread out as distance from the features increases, or, conversely, the evidences converge on the location of the features as they are approached. If we let $t' = jVt$ and assume harmonic waves, we can write the scalar wave equation (8.80) as

$$\frac{\partial^2 \psi}{\partial x^2} + \frac{\partial^2 \psi}{\partial z^2} - \frac{1}{V^2}\frac{\partial^2 \psi}{\partial t^2} = 0 = \frac{\partial^2 \psi}{\partial x^2} + \frac{\partial^2 \psi}{\partial z^2} + \frac{\partial^2 \psi}{\partial t'^2}$$

$$= \nabla^2 \psi(x, z, t') = 0,$$

thus expressing the wave equation in the form of Laplace's equation. We know the wavefield at the surface of the Earth, $z = 0$, so our problem is to continue this field downward to determine what geophones would see if they were buried at arbitrary depths. We downward-continue in a series of steps, effectively lowering the geophones gradually downward through the Earth. At each geophone depth we expect to get a clear picture of reflectors which lie immediately below the geophones, so we retain this portion of the downward-continued record section from each continuation step and combine the upper portion of these to give our complete migrated section.

We let t be the one-way traveltime (half the arrival time for coincident source-detector data). A plane wave approaching the surface at the angle θ is given by

$$\Psi(x, z, t) = A \exp\left[j\omega\{t - (x/V)\sin\theta - (z/V)\cos\theta\}\right]. \tag{8.87}$$

If we restrict ourselves to small angles of θ, we can approximate $\sin\theta \approx \theta$ and $\cos\theta \approx 1 - \tfrac{1}{2}\theta^2$, so that (8.87) becomes

$$\psi(x, z, t) = A \exp\{j\omega(t - x\theta/V - z/V + z\theta^2/2V)\}. \tag{8.88}$$

We now define a new time scale, $t^* = t - z/V$. This change means that our coordinate system effectively rides along on an upcoming wavefront (see problem 8.20). We now have

$$\psi^*(x, z, t^*) = A \exp\{j\omega(t^* - x\theta/V + z\theta^2/2V)\}, \tag{8.89}$$

and

$$\frac{\partial\psi}{\partial t} = \frac{\partial\psi^*}{\partial t^*}\frac{\partial t^*}{\partial t} = \frac{\partial\psi^*}{\partial t^*}; \quad \frac{\partial^2\psi}{\partial t^2} = \frac{\partial^2\psi^*}{\partial t^{*2}}$$

$$\frac{\partial\psi}{\partial x} = \frac{\partial\psi^*}{\partial x}; \quad \frac{\partial^2\psi}{\partial x^2} = \frac{\partial^2\psi^*}{\partial x^2}$$

$$\frac{\partial\psi}{\partial z} = \frac{\partial\psi^*}{\partial z} + \frac{\partial\psi^*}{\partial t^*}\frac{\partial t^*}{\partial z} = \frac{\partial\psi^*}{\partial z} - \frac{1}{V}\frac{\partial\psi^*}{\partial t^*}$$

$$\frac{\partial^2\psi}{\partial z^2} = \frac{\partial^2\psi^*}{\partial z^2} - \frac{2}{V}\frac{\partial^2\psi^*}{\partial z\partial t^*} + \frac{1}{V^2}\frac{\partial^2\psi^*}{\partial t^{*2}}$$

Substituting into the wave equation (8.80) gives a new wave equation,

$$\frac{\partial^2\psi^*}{\partial x^2} + \frac{\partial^2\psi^*}{\partial z^2} - (2/V)\frac{\partial^2\psi^*}{\partial z\partial t^*} = 0 \tag{8.90}$$

For waves which are traveling nearly vertically, the change in ψ^* with respect to z is small in our moving coordinate system, so we neglect $\partial^2\psi^*/\partial z^2$. This is called the '15° approximation' and neglecting this term means that we shall not be able to migrate steep dips well. We thus get

$$\frac{\partial^2\psi^*}{\partial x^2} - (2/V)\frac{\partial^2\psi^*}{\partial z\partial t^*} = 0. \tag{8.91}$$

(A '45° approximation' gives the equation

$$\frac{\partial^3\psi^*}{\partial x^2\partial t^*} - \tfrac{1}{2}V\frac{\partial^3\psi^*}{\partial x^2\partial z} - (2/V)\frac{\partial^3\psi^*}{\partial x\partial t^{*2}} = 0;$$

see Claerbout, 1976, pp. 196–202.)

If we think of ψ^* as a three-dimensional array (fig. 8.27a), that is, having values at discrete intervals of $\Delta x, \Delta z$ and Δt^*, the plane $z = 0$ represents the unmigrated time section (which is our starting point) and the diagonal plane $t = t^* - z/V$ represents the migrated time section. (The projection of this diagonal plane onto the $t^* = 0$ plane would also give a migrated depth section.) We approximate derivatives by finite differences:

$$\frac{\partial^2 \psi^*}{\partial x^2}$$

$$\approx \frac{\psi^*(x,z,t^*) - 2\psi^*(x - \Delta x, z, t^*) + \psi^*(x - 2\Delta x, z, t^*)}{\Delta x^2}$$

$$\frac{\partial^2 \psi^*}{\partial z \partial t^*} \approx \{\psi^*(x,z,t^*) - \psi^*(x,z - \Delta z, t^*)$$

$$- \psi^*(x,z,t^* - \Delta t^*)$$

$$+ \psi^*(x, z - \Delta z, t^* - \Delta t^*)\}/\Delta z \Delta t^*.$$

Equation (8.91) now becomes

$$\psi^*(x,z,t^*) = \frac{\Delta z \, \Delta t^* \, \Delta x^2}{2\Delta z^2 - V\Delta z \, \Delta t^*} \left\{ \frac{\psi^*(x, z - \Delta z, t^*)}{\Delta z \, \Delta t^*} \right.$$

$$+ \frac{\psi^*(x, z, t^* - \Delta t^*)}{\Delta z \, \Delta t^*} - \frac{V\psi^*(x - \Delta x, z, t^*)}{\Delta x^2}$$

$$- \frac{\psi^*(x, z - \Delta z, t^* - \Delta t^*)}{\Delta z \, \Delta t^*}$$

$$+ \left. \frac{V\psi^*(x - 2\Delta x, z, t^*)}{2\Delta x^2} \right\}. \tag{8.92}$$

Fig.8.27. Relationship of elements in (x, z, t^*) space. (a) Seismic traces on the top surface $z = 0$ show the unmigrated section, those at successive layers show what would be recorded by geophones buried at the depth z; (b) elements entering into the time-domain calculation of $\psi^*(x, z, t^*)$; (c) table of values of $\Psi_x^*(z, t^*)$.

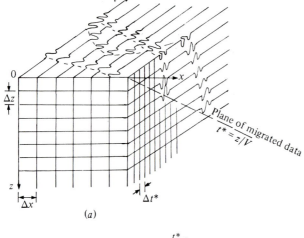

(a)

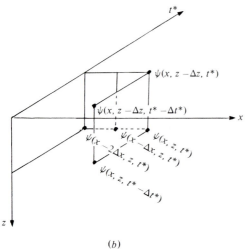

(b)

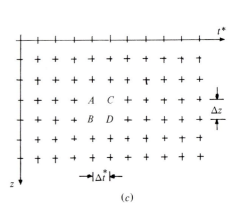

(c)

This is a relation among six elements of the array, as shown in fig. 8.27b:

$$\psi^*(x, z, t^*) = a_1\psi^*(x, z - \Delta z, t^*)$$
$$+ a_2\psi^*(x, z, t^* - \Delta t^*)$$
$$+ a_3\psi^*(x, z - \Delta z, t^* - \Delta t^*)$$
$$+ V\{a_4\psi^*(x - \Delta x, z, t^*)$$
$$+ a_5\psi^*(x - 2\Delta x, z, t^*)\}.$$

This relation can be used to extend the three-dimensional array in the $+x$, $+z$, $+t^*$ directions. Our problem is getting enough of the array to begin with so that we can extend it. For negative z, $\psi^* = 0$ because this refers to wave values in the air above the surface of the ground. If our data begin at $x = 0$, we need values of ψ^* for $x = -\Delta x$ and $x = -2\Delta x$, which are not available to us; we guess these starting values and it turns out that the solution is not affected very much if these are in error, so that a stable solution can be achieved.

Various alternatives to the foregoing method of approximating the derivatives are available, some of which lead to highly stable algorithms and permit calculations using a fairly coarse grid, which of course makes the calculation much more economical. One approach is to take the Fourier transform with respect to x:

$$\psi^*(x, z, t^*) \leftrightarrow \Psi_x^*(\kappa_x, z, t^*),$$

so that (8.91) becomes

$$\kappa_x^2\Psi_x^* - (2/V)\frac{\partial^2\Psi_x^*}{\partial z\partial t^*} = 0. \tag{8.93}$$

We assume that we have a table of values of Ψ_x^* for discrete values of z, t^*, as illustrated in fig. 8.27c and we consider the portion of the table centered between the values A, B, C, D. The approximate value of Ψ_x^* at this point is $\frac{1}{4}(A + B + C + D)$. We can approximate $\partial^2\Psi_x^*/\partial z\partial t^*$ by the expression

$$\left(\frac{D - C}{\Delta z} - \frac{B - A}{\Delta z}\right)\frac{1}{\Delta t^*} = \frac{A - B - C + D}{\Delta z\,\Delta t^*},$$

that is, as the difference of differences. We now write (8.93) in the form

$$\varepsilon\begin{Vmatrix} 1 & 1 \\ 1 & 1 \end{Vmatrix} - \begin{Vmatrix} 1 & -1 \\ -1 & 1 \end{Vmatrix} = \begin{Vmatrix} \varepsilon - 1 & \varepsilon + 1 \\ \varepsilon + 1 & \varepsilon - 1 \end{Vmatrix} = 0$$

where $\varepsilon = \frac{1}{8}V\kappa_x^2\,\Delta z\,\Delta t^*$. This box of four compartments is laid over the table of Ψ_x^*-values, and each is multiplied by the overlying factor $(\varepsilon \pm 1)$, and the sum is set equal to

zero. If we know three of the four values, we can calculate the fourth. Again, some values have to be guessed to get started.

The upper surface in fig. 8.27a, the unmigrated time section observed at the surface of the Earth, thus provides the basis for calculating the time section which would be observed at $z = \Delta z$, etc. In effect we are calculating the output of geophones buried at $z = z_1$ as the superposition of filtered outputs of the geophones in the layer $z = z_1 - \Delta z$ and previously-calculated values in the layer z_1. The filters are mainly phase-shifting to allow for the difference in traveltime from a geophone directly above the buried geophone to those at adjacent locations.

In continuing the wave field from $z = z_1$ to $z = z_1 + \Delta z$, we use the velocity of the layer between z_1 and $z_1 + \Delta z$, so that accommodating vertical variations of velocity is fairly easy and straightforward in time-domain migration. The velocity V can also be made a function of x, but often this is not done. A common way of accounting for lateral velocity variations is to migrate with different velocities and then merge the two sections, that is, $\psi = k\psi_1 + (1 - k)\psi_2$ where k varies linearly over the merge region, the same type of procedure as is used to accomplish 'time-variant' filtering.

An example of wave-equation migration is shown in fig. 8.28.

8.3.5   *Other migration considerations*

(a) *Relative merits of different migration methods.* In the preceding three sections we discussed three migration methods (and mentioned that still others are sometimes used). In practical implementation each method involves approximations and limitations which affect data with different characteristics in different ways, so that one method may migrate better for one data set but another method might be superior for a different data set, or one implementation of a method may yield better results than another implementation of the same method. Among the characteristics of the methods are the following:

Diffraction-stack migration:
   Migrates steep dips
   Allows weighting and muting according to dip or
      coherency
   Aperture can be varied explicitly
   Usually not adapted to accommodate lateral velocity variation

Frequency–wavenumber migration:
   Migrates dips up to spatial-aliasing limitations
   Difficult to accommodate lateral velocity variations
   Can be applied to specific limited areas
   Is often the most economical method

Finite-difference migration:

> Migrates dips up to 45° (or 15° with one version)
>
> Produces less migration noise
>
> Is effective in low signal-to-noise areas
>
> Can accommodate lateral velocity variations

Frequency–space migration:

> Migrates dips up to spatial-aliasing limitations
>
> Often the easiest method for depth migration (§8.3.5c) because velocity as a function of space is explicit

One important point should be made: migration (by almost any method) almost always produces a result which is closer to the correct picture than failure to migrate, even where the fundamental assumptions are grossly violated.

(b) Resolution of migrated sections. In §4.3.2c it was pointed out that the horizontal resolution of seismic data is limited by noise and migration considerations rather than Fresnel-zone size, as with unmigrated data. One of the basic assumptions underlying migration is that the data show only the reflected wavefield; data of any other type (including multiples) will be migrated as if they were primary reflected energy and will be rearranged and superimposed on the migrated images of reflectors. Noise of limited extent which might be quite recognizable as noise in the unmigrated data may smear out and degrade the sharpness of reflector images. For example, a noise burst on a single trace will become a complete wavefront (sometimes called a 'smile') on a migrated section. Data with a dip component perpendicular to the line will not be

Fig.8.28. Migration by finite-difference method. (Courtesy Prakla-Seismos.) (a) Unmigrated section; (b) migrated section.

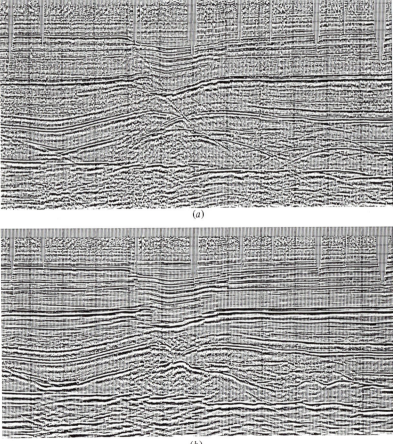

(a)

(b)

migrated correctly, thus degrading horizontal resolution. Spatial-aliasing considerations limit the amount of dip which can be migrated. The use of a limited migration aperture and the approximations involved in the migration algorithms further restrict the resolution. With good data and good migration algorithms, features are sometimes defined very sharply, for example faults can sometimes be located to within the trace spacing distance. Usually, however, the net effect of the various limiting factors limits the precision to much less than this.

(c) Depth migration. Migrated sections are often needed with the vertical scale linear in depth rather than in time. Most often such sections are produced by simply stretching migrated time sections according to the vertical velocity function (fig. 8.29b). However, where velocity varies appreciably in the horizontal direction, raypath bending introduces additional complications. Depth migration (Judson *et al.*, 1980; Schultz and Sherwood, 1980; Larner *et al.*, 1981) attempts to accommodate such situations.

Depth migration is sometimes accomplished by stretching time-migrated data into depth along raypaths calculated according to a velocity model, as in fig. 8.29c, d. Depth migration can also be accomplished directly in frequency–space domain migration. Obviously depth migration requires a detailed accurate knowledge of the velocity distribution. Unfortunately, such knowledge is often not available in the structurally-complex areas where depth migration is most needed.

(d) Migration before stack. Most seismic data are acquired by the common-depth-point method. Since migration is often a fairly expensive process to carry out, stacking usually precedes migration in order to reduce the amount of data to be migrated (by approximately the fold of the stack). The stacking of dipping data, however, introduces a smear of the reflecting points (fig. 8.18) which degrades the data and attenuates diffraction tails which are required for complete migration. Migrating before stacking almost always produces significantly better results but is usually not done because of the increased costs.

Various schemes (Sattlegger and Stiller, 1974; Sattlegger *et al.*, 1980; Schultz and Sherwood, 1980; Jain and Wren, 1980) are sometimes employed to obtain much of the benefit of migration before stack without incurring excessive cost. Most commonly, several stacks are made, each with data involving only a very limited range of offsets, so that the smear produced by stacking is minor for each of them. These partial stacks are then individually migrated and the migrated results are stacked. Thus

the full stack and full noise attenuation are achieved without also producing significant reflection-point smear and without excessive cost.

8.3.6   *Three-dimensional migration*

In concept at least, the various techniques used for two-dimensional migration can be extended to accommodate three-dimensional migration. In two-dimensional diffraction stacking we sum along all possible diffraction hyperbolae; in three dimensions we sum over all possible hyperboloids. In frequency–wavenumber migration we transform $\psi(x, z, t) \leftrightarrow \Psi(\kappa_x, \kappa_z, \omega)$; in three dimensions, $\psi(x, y, z, t) \leftrightarrow \Psi(\kappa_x, \kappa_y, \kappa_z, \omega)$. In finite-difference migration we calculate a new section for each continuation step; in three dimensions we calculate a new cube of data for each step. The problem in achieving three-dimensional migration is the large number of computations required. These can be handled by very large computers, but at significantly higher cost.

Most commonly, 3-D migration is approximated by first doing 2-D migration on lines running in one

Fig.8.30. Illustrating three-dimensional migration. The data to be migrated to t_0 lie on a hyperboloid appropriate to a stacking velocity V_1; the aperture should include a circular area centered under t_0. The migration is often done as a two-stage operation to economize on computing time, that is, the data are two-dimensionally migrated along lines in the *y*-direction and then these intermediate results are migrated in the *x*-direction. The effect is to take the triangular data element and migrate it to the square element in the *y*-direction migration, and then to move it to the t_0 location in the *x*-direction migration. However, the stacking velocity V_2 would be used for the *y*-direction migration rather than the correct velocity V_1. The aperture in the *y*-direction migration would also be different than an aperture symmetrical about t_0 as should be used for proper three-dimensional migration. (*a*) Isometric sketch of (x, y, t) space; (*b*) stacking velocity versus time.

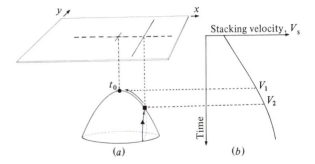

(a)   (b)

Fig.8.29. Depth migration. (From Hatton *et al.*, 1981.)
(*a*) Unmigrated time section; (*b*) finite-difference
migration vertically stretched to a depth scale;

(*c*) ray-tracing through velocity layers; (*d*) depth
migration stretched along raypaths in (*c*).

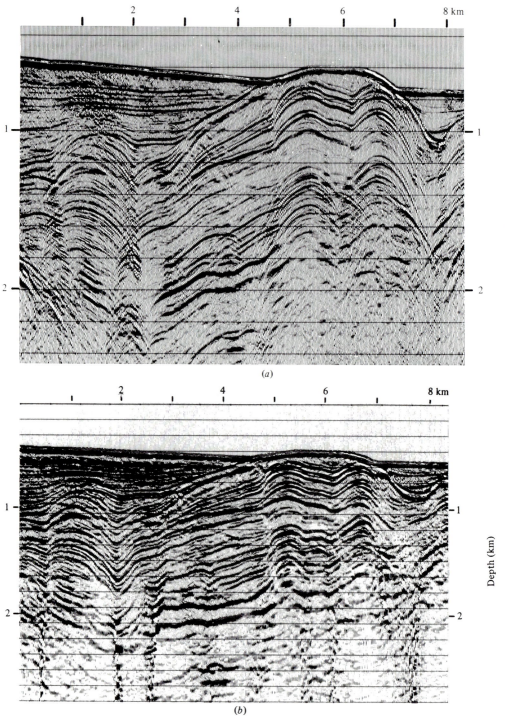

(*a*)

(*b*)

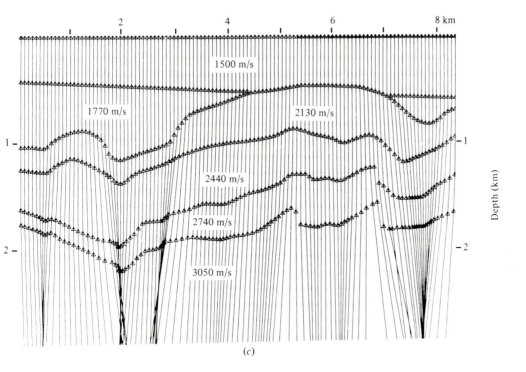

1500 m/s

1770 m/s   2130 m/s

2440 m/s

2740 m/s

3050 m/s

Depth (km)

(*c*)

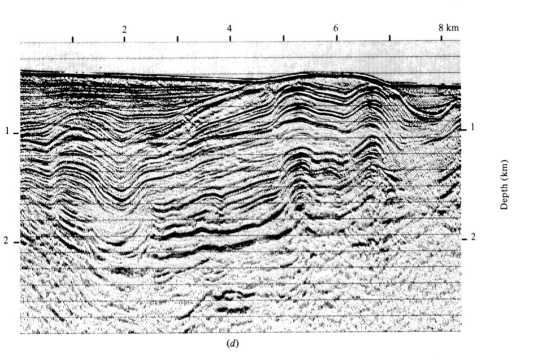

Depth (km)

(*d*)

direction and then doing 2-D migration on these migrated data in the perpendicular direction. Figure 8.30 shows that a diffraction from a point to the side of a seismic line can be migrated to the crest of the diffraction hyperboloid by two-step (dual 2-D) migration. Such schemes migrate data to the correct location if the velocity is constant, but they may not preserve amplitudes. Usually the range of data to be included in the migration (the 'migration aperture') is limited and the two-step approach results in

the inclusion of different amounts of data than would a pattern of spherical symmetry about the diffraction point. A 'splitting' approach is also sometimes used; the 3-D operator is approximated by 2-D operators acting in perpendicular directions, the two being applied successively at each downward-continuation step.

If the velocity is not constant, then the wrong velocity will be used in migration by the two-step methods

Fig.8.31. Two- versus three-dimensional migration. (Courtesy Western Geophysical.) (*a*), (*b*) Lines N1 and E17 migrated by two-dimensional migration; (*c*), (*d*)

migrated by three-dimensional migration. The area of data was 6.4 by 3.2 km and the lines are at right-angles in the principal directions.

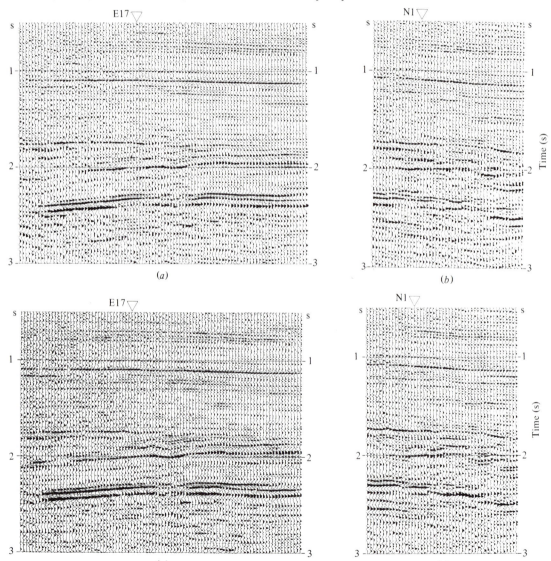

(a)

(b)

(c)

(d)

(see fig. 8.30). The shape of the diffraction hyperboloid is defined by the velocity at its crest; the hyperbola on a line to the side of the crest will have a larger apex time and hence be associated with too-large a velocity (in the usual case of increase of velocity with depth) and thus will not be as sharp.

The inaccuracies resulting from failure to handle properly 3-D considerations may not be much greater than those that result from approximations in the migration methods and uncertainty in the definition of velocity. Areas where 3-D migration is especially needed are apt to also be the same complex areas where velocity uncertainties and lateral velocity gradients are of most concern. Three-dimensional migration comparisons are shown in fig. 8.31.

The migration of unmigrated reflection-time maps (Kleyn, 1977) often provides an excellent way of achieving correct depth-contour maps. Usually the unmigrated map surface (the consequence of picking a reflection event, posting and contouring) and also intermediate velocity surfaces are approximated by a grid of small planar elements (fig. 8.32). Rays are then traced, the starting direction at the surface being given by the attitude of the elements on the unmigrated map (which give the angle of approach at the surface), the rays being bent in accordance with Snell's law whenever they encounter an intermediate velocity interface. The consequent depth map (fig. 8.33) then has data posted in an irregular manner, so these are usually regridded and contoured by the computer.

8.4   Other processing techniques
8.4.1   *Automatic picking*
Events can be picked and graded automatically using as criteria coherence measures such as semblance (Paulson and Merdler, 1968; Bois and la Porte, 1970; Garotta, 1971). At any point on the record where the coherence exceeds a certain threshold value, an event can be marked; the arrival time can be determined by varying the position of the time gate to maximize the coherence. The normal moveout, and hence the stacking velocity, can be found by varying the normal moveout to maximize the coherence. Dip moveout can be measured likewise. Grades can be assigned according to some arbitrary scale. The subsurface distance over which coherence can be maintained can be included as one of the factors in such measurements. The picks can be automatically migrated and plotted on a depth scale, as shown in fig. 8.34. Amplitudes and various attributes (§8.4.2) can also be measured and posted or contoured on sections or maps as aids in interpretation.

Automatic picking can be expanded to incorporate intersecting seismic lines. The picks can also be posted on a map automatically and the resulting data contoured automatically. Thus, in principle the output of processing can be contoured depth maps of reflecting horizons and so some of the work usually thought of as 'interpretation' can be automated. However, in the processing many decisions must be made. Criteria have to be specified for determining which events are primary reflections and which multiples, for deciding what to do when events interfere or when events terminate, and so on. The programmer has to define the criteria for each of these decisions in advance and the process breaks down if he has not anticipated the need for a certain decision or has not defined the criteria adequately. The value of the final result depends largely on the soundness of the decisions which were made.

Fig.8.32. Map migration. The unmigrated map (top) is subdivided into elements and a raypath is traced downward, the direction of the raypath being determined by the values at the corners of the elements. Isovelocity surfaces (center) are likewise subdivided into elements and when the raypath strikes one of the isovelocity elements, it is bent according to Snell's law. The (x, y, z) coordinates when the traveltime is satisfied are contoured to give the migrated map.

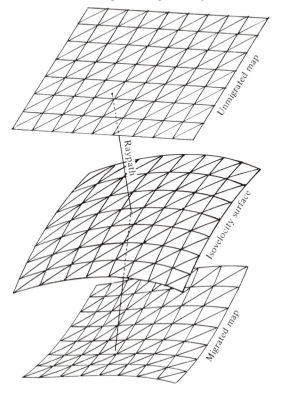

Unmigrated map

Raypath

Isovelocity surface

Migrated map

Fig.8.33. Map migration. (Courtesy Prakla-Seismos.)
(*a*) Unmigrated map, subdivided into square elements;
(*b*) migrated map showing values for each element with
the values contoured.

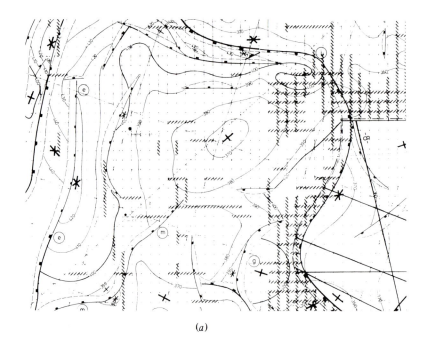

(*a*)

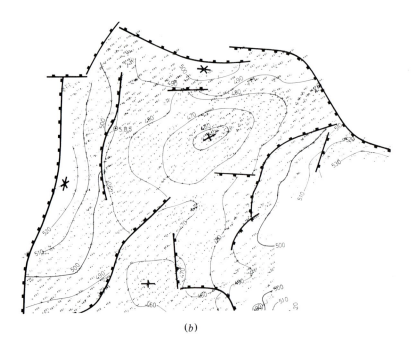

(*b*)

8.4.2   *Complex-trace and attribute analysis*

Let us assume a seismic trace of the form

$$g(t) = A(t)\cos 2\pi v t \qquad (8.94)$$

where $A(t)$ varies slowly with respect to $\cos 2\pi v t$; $A(t)$ is the envelope of $g(t)$. For $A(t)$ constant, the Hilbert transform (§10.3.11) of $g(t)$ is given by

$$g_\perp(t) = -A(t)\sin 2\pi v t \qquad (8.95)$$

(see problem 10.22*a*). Thus, we can form a complex signal, $h(t)$ where

$$h(t) = g(t) + jg_\perp(t) = A(t)\mathrm{e}^{-j2\pi v t}, \qquad (8.96)$$

$h(t)$ being known as the *analytical* or *complex trace*, (Bracewell, 1965), $g_\perp(t)$ as the *quadrature trace* of $g(t)$ (see fig. 8.35). If v is not constant but varies slowly, we define the *instantaneous frequency*, $v_i(t)$, as the time derivative of

Fig.8.34. Automatically-picked migrated section. (From Paturet, 1971.)

Fig.8.35. The complex trace, shown as a helix of variable amplitude in the direction of the time-axis. Projection onto the real plane gives the actual seismic trace and onto the imaginary plane the quadrature trace.

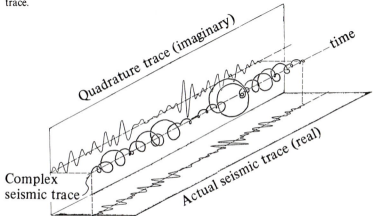

the phase, $\gamma(t)$; thus,

$$2\pi v_i(t) = \frac{d\gamma(t)}{dt} = \frac{d}{dt}(2\pi vt) \qquad (8.97)$$

The quantities $A(t)$, $\gamma(t)$, $v_i(t)$ and other measurements derived from the seismic data are called *attributes*.

To find $A(t)$, $\gamma(t)$ and $v_i(t)$, we obtain $h(t)$, either by (10.154), that is,

$$g_\perp(t) = g(t)*(1/\pi t) = (1/\pi) \sum_{n=-\infty}^{\infty} g_{t-n}(e^{jn\pi} - 1)/n$$

$$(8.98, 10.154)$$

for digital functions (see problem 10.22c), or by using (10.155), that is, we calculate the transform of $g(t)$, set the result equal to zero for negative frequencies, multiply by two, and then inverse-transform to get $h(t)$. Since $A(t)$ is real and $|e^{j2\pi vt}| = 1$, we see that

$$A(t) = |h(t)|,$$

$$\gamma(t) = 2\pi vt = \tan^{-1}\{g_\perp(t)/g(t)\},$$

$$v_i(t) = (1/2\pi)\frac{d}{dt}\{\gamma(t)\}.$$

Complex-trace analysis can be used in convolution, correlation, semblance, and other types of calculations (Taner *et al.*, 1979), sometimes facilitating the calculations.

Attributes sometimes reveal features which are not as obvious otherwise, especially lateral changes along the bedding such as those associated with stratigraphic changes or hydrocarbon accumulations (§9.7 and §9.8); see Taner and Sheriff (1977). Phase plots facilitate picking weak coherent events, and lateral discontinuities in phase facilitate picking reflection terminations as at faults, pinchouts, etc. Instantaneous frequency patterns tend to characterize the interference patterns resulting from closely-spaced reflectors and thus aid in correlating from line to line or across faults.

8.5 Data-processing procedures

8.5.1 *Typical processing sequence*

(a) *Initial steps.* Seismic data-processing often follows a basic sequence (fig. 8.36) which is varied to tailor to specific needs of data.

A small group of people is usually responsible for processing. They prepare specific instructions, choose processing parameters, and monitor the quality of results. This group needs to know the objectives of the processing in order to make optimal decisions. Processing which is optimal for one set of objectives may not be optimal for another set.

The first step after field tapes are received at a processing center is to verify the arrangement of data on the magnetic tape. This involves *dumping* (displaying the tape's magnetic pattern on a printout) the first few (possibly 10) records and comparing with what is expected. With Vibroseis data, format verification includes a check on the Vibroseis sweep length and spectrum.

(b) *Editing.* Editing follows format verification. The data are rearranged or *demultiplexed*; field data are time-sequential, that is, the first sample for each channel is recorded before the second sample from any channel, whereas most processing requires trace-sequential data, that is, all the data for the first channel before the data for the second channel.

If the input waveform has been recorded for each shot, a deterministic source-signature correction (§8.2.1e) may be applied to make the effective waveform the same for all records. If the source is Vibroseis or some other source which continues for some time, the equivalent of source-signature correction is accomplished by correlation with the source waveform.

Field data encompass a very wide range of amplitude values which have been recorded in encoded form, so the field gain is decoded and a first approximation of a spherical-divergence correction is applied to give a smaller range of values for input to subsequent processes. Sometimes data are also vertically stacked or resampled to make less data for future processing. Resampling may have to be preceded by alias filtering.

Editing may involve detecting and changing dead or exceptionally noisy traces. 'Bad' data may be replaced with interpolated values. Anomalously high amplitudes, which are probably noise, may be reduced to the level of the surrounding data or possibly to zero.

Outputs after editing usually include: (1) a plot of each file, so that one can see what data need further editing and what types of noise attenuation are required; (2) a plot of the shortest-offset trace from each record (near-trace plot), to give a quick look at the geologic structure and for use in making decisions as to where velocity analyses are to be made; (3) near-trace autocorrelation plots, which indicate multiple problems and aid in making deconvolution decisions; (4) a trace-sequential tape which will be used for subsequent processing. The field tape is then returned to a tape library.

(c) *Parameter determination.* The object of this sequence of processes is to determine processing parameters which are data-dependent, such as static time shifts, amplitude

Fig.8.36. Typical processing flow chart.

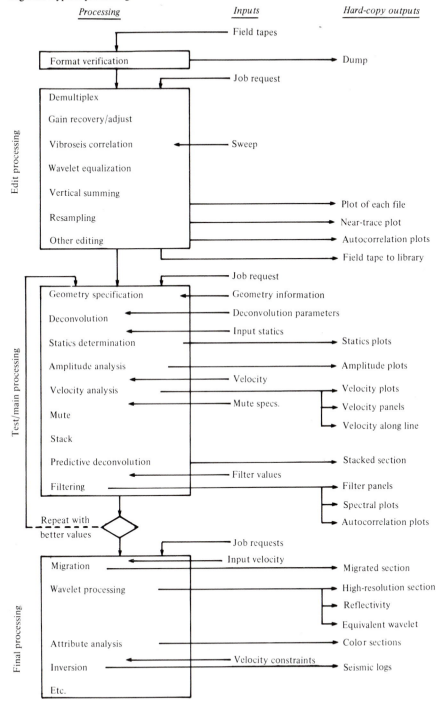

adjustments, normal-moveout values and frequency content. Field-geometry information is input so the computer can determine which data involve common geophones or have other factors in common, and so the offset for each trace is known for normal-moveout determinations.

If the near-trace plot shows strong coherent noise trains from surface waves, shallow refractions or other horizontally traveling energy, apparent-velocity filtering may be applied to remove them. Statistical wavelet contraction may be applied. The analysis of the statistical characteristics of the data for wavelet contraction and for statics and velocity analysis should be based on data which exclude known bad traces and zones of non-reflection energy, such as the first-breaks region. A wide window is usually chosen so as to provide appreciable statistics, but the window should exclude zones where primary reflections are not expected, such as below the basement.

The statics-analysis program looks for systematic variations such as would be expected if time shifts were associated with particular shots, particular geophones, etc. Preliminary statics, as determined in the field office from first-break information and from the elevation of geophone stations, is usually input before the statics analysis so that the statics analysis determines residual statics errors. The results of this analysis are output on a control plot.

An analysis similar to the time-shift analysis of the statics program is carried out for amplitude, to determine systematic amplitude effects such as might be associated with a weak shot, poor geophone plant, etc., and a control plot is output.

Velocity analyses are usually run 1–2 km apart at locations selected because they are relatively free of structural complications. The locations for analyses are based on the near-trace output from the editing pass. A first guess as to velocity is input prior to velocity analysis so that the velocity analysis determines residual normal-moveout. If the near-trace plots show dip, this information is input because velocity determination depends on dip. Compromises have to be made as to how much data should be included in an analysis; more data yield better statistics but then they do not apply at specific points, so the compromise is between determining a more accurate average and a less accurate value which applies to a specific location. Outputs from velocity analysis may include (1) a velocity spectral plot, such as in fig. 8.13, which shows the coherence achieved when various stacking velocities are assumed; (2) velocity panels, as shown in fig. 8.15, which show the data stacked according to the input velocity information and also according to vel-

ocities slightly smaller and larger than the input velocity; such panels allow one to see if certain events require different stacking velocity than other events, since stacking velocity is not necessarily a single-valued function, and also how sensitive the stack is to velocity assumptions; (3) a graph such as in fig. 8.14, showing how velocity determinations at different locations along the lines relate to each other.

The data may also be filtered by a sequence of narrow bandpass filters yielding a filter panel (fig. 8.11) which is used to determine subsequent filtering parameters. Autocorrelations and spectral plots of various sorts may also be output. Preliminary stack sections may be made to show the effectiveness of processing parameters and as an aid in diagnosing additional problems.

(d) *Principal processing pass*. The main processing pass usually begins with the tape from the editing pass and the sequence of operations is almost the same as in the parameter determination pass, the differences being in the values that are input for statics, normal moveout, etc. The correct spherical divergence based on the actual velocities replaces preliminary gain assumptions. The main processing pass or portions of it may be repeated using more refined values, especially statics and normal-moveout values.

The final output from the main processing pass will be one or more stacked sections which may differ in the processes applied or in the choices of parameters, especially display parameters such as amplitude, polarity, and filtering choices.

(e) *Migration and other processing*. The stacked data may then be used as input to various other processes, such as migration and attribute analysis. Velocity information is required for migration; the velocity may be based on stacking velocity values where dips are not extreme, but in general the optimum velocity for migration differs from the optimum velocity for stacking.

Either the stacked or the migrated data may be further analyzed for amplitude, frequency content, apparent polarity, etc., and displayed in various ways. The data may also be used in iterative modeling or other types of processing.

8.5.2 *Interactive processing*

Many decisions have to be made during seismic processing and criteria for making these in optimum fashion may not be available. Often one would like to see the consequences of decisions without having to wait very long. The time required for many processes is so great that an operator becomes bored waiting at a terminal for response to his decisions and the cost of tying up computers or computer memories while waiting for an operator to make decisions has militated against interactive processing. However, 'intelligent' terminals with enough memory and computing power to execute many programs, time-sharing arrangements with large computers so that one only has to pay for actual processing time, higher-quality graphic displays, simpler hard-copy printers and other computer advances are rapidly changing the situation.

One of the most useful applications of intelligent terminals is simply checking input parameters and instructions to make sure that these are complete and consistent before involving an expensive computer in the execution of the instructions. Intelligent terminals can prompt an inputter regarding overlooked decisions and calculate checks for consistency of parameters. Once the data are entered into the terminal in adequate fashion, the terminal can then communicate them directly to a larger computer for the actual processing.

Velocity analysis is an especially critical process where many decisions have to be made. In the past these were often made individually without any easy way of checking the consistency of picks, either internally or with respect to adjacent analyses. The ability to pick an analysis plot on a video screen while also seeing adjacent plots and having the corresponding interval velocities calculated and displayed promises significant improvement in data-processing. Other types of processing also lend themselves to interactive decision making.

Access to data banks from which information can be retrieved as required is another area where interactive processing works well. Seismic line locations, picked horizons, faults and other features, well locations and information about and derived from wells, geographical and ownership information, etc., can be stored in data banks for arbitrary retrieval. The data bank includes tables showing what information is in the bank, so that an operator may know what information can be called up. Processes are available for combining and manipulating data, posting them on sections or maps, contouring, calculating isopachs between data sets, converting from traveltime to depth, drawing graphs, etc.

Problems

8.1. Derive (8.6) from (8.3). [Hint: multiply both sides of (8.3) by $\exp(-j2\pi v_n t)$ and integrate over one cycle, e.g. from $-\frac{1}{2}T$ to $+\frac{1}{2}T$. Before substituting the values at the limits, use Euler's theorem, $e^{\pm jx} = \cos x \pm j \sin x$, to express the complex exponentials in terms of sines and cosines.]

8.2. (a) Since $\delta(t)$ is zero except for $t = 0$ where it equals $+1$ (see §10.3.3), we can apply (8.12) and find that

$$\delta(t) \leftrightarrow +1;$$

show that

$$\delta(t - t_0) \leftrightarrow e^{-j2\pi v t_0}.$$

(b) Show that $\delta(t) * g(t) = g(t)$ and $\delta_t * g_t = g_t$. (c) The comb can be written

$$\text{comb}(t) = \sum_{n=-\infty}^{+\infty} \delta(t - n\Delta);$$

the transform of this expression is clearly

$$\text{comb}(t) \leftrightarrow S(v) = \sum_{n=-\infty}^{+\infty} e^{-j2\pi n v \Delta}.$$

Show (see Papoulis, 1962, p. 44) that this represents an infinite series of impulses of height $2\pi/\Delta$ spaced $2\pi/\Delta$ apart, that is,

$$\sum_{n=-\infty}^{+\infty} \delta(t - n\Delta) \leftrightarrow \left(\frac{2\pi}{\Delta}\right) \sum_{m=-\infty}^{+\infty} \delta(2\pi v - 2\pi m/\Delta),$$

or

$$\text{comb}(t) \leftrightarrow \omega_0 \text{comb}(\omega), \quad \omega_0 = 2\pi/\Delta.$$

(d) Show that a boxcar of height h and extending from $-v_0$ to $+v_0$ in the frequency domain has the transform

$$h\,\text{box}_{2v_0}(v) \leftrightarrow A\,\text{sinc}(2\pi v_0 t) = A\frac{\sin 2\pi v_0 t}{2\pi v_0 t},$$

where $A = 2hv_0 = $ area of the boxcar.

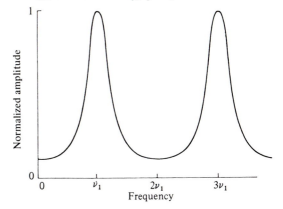

Fig.8.37. Spectrum of water-layer filter for water-bottom reflection coefficient of 0.5. If z = water depth and V = water velocity, $v_1 = V/4z$.

8.3. (a) Verify that (8.34) is the inverse filter for water reverberation by convolving (8.34) with (8.32), that is, by substituting the expressions given by (8.32) and (8.34) in (8.33). (b) The spectrum of the water-layer filter is shown in fig. 8.37 for $n = 1$; the large peaks occur at the 'singing frequency'; sketch the amplitude spectrum of the inverse filter. [Hint: time-domain convolution such as shown in (8.33) corresponds to frequency-domain multiplication (see (8.22)); the frequency spectrum of the unit impulse is $+1$, that is flat.] (c) Verify your sketch of the water-reverberation inverse filter by transforming

$$[1, 2R, R^2] \leftrightarrow 1 + 2R e^{-j2\pi v\Delta} + R^2 e^{-j4\pi v\Delta}.$$

and calculating the value of the spectrum for $v = 0$, v_1 and $2v_1$.

8.4. (a) Convolve $[2, 5, -2, 1]$ with $[6, -1, -1]$; (b) Cross-correlate $[2, 5, -2, 1]$ with $[6, -1, -1]$; for what shift are these functions most nearly alike? (c) Convolve $[2, 5, -2, 1]$ with $[-1, -1, 6]$; compare with the answer in (b) and explain. (d) Autocorrelate $[6, -1, -1]$ and $[3, -5, -2]$; the autocorrelation of a function is not unique to that function (see §10.6.6a), for example other wavelets having the same autocorrelations as the preceding are $[-1, -1, 6]$ and $[-2, -5, 3]$; which of the four is the minimum-delay wavelet? (e) What is the normalized autocorrelation of $[6, -1, -1]$? What is the normalized cross-correlation in (b)? What do you conclude from the magnitude of the largest value of this normalized cross-correlation?

8.5. (a) The operator $f_t = [-1, +1]$ is called the 'derivative operator'; explain why. (b) What is the integral operator?

8.6. Four causal wavelets are given by $a_t = [2, -1]$; $b_t = [4, 1]$; $c_t = [6, -7, 2]$; $d_t = [4, 9, 2]$. (a) Which are minimum-phase? (b) Find $a_t * b_t$ and $a_t * c_t$ by calculations in the time-domain. (c) Repeat (b) except using z-transforms. (d) Find $a_t * b_t * c_t$. (e) Does the maximum value of a minimum-phase function have to come at $t = 0$? (f) Can a minimum-phase wavelet be zero at $t = 0$?

8.7. Using the data in problem 8.6, calculate ϕ_{ab}, ϕ_{ac}, ϕ_{ca}, ϕ_{aa} and ϕ_{cc} by calculating in the time-domain and in the frequency domain.

8.8. Fill in the values in table 8.1.

8.9. Using the wavelets $W_1(z) = (2 - z)^2(3 - z)^2$, $W_2(z) = (4 - z^2)(9 - z^2)$, calculate the composite wavelets: $W_1(z) + W_2(z)$, $W_1(z) + zW_2(z)$, $zW_1(z) + W_2(z)$ and $z^{-1}W_1(z) + W_2(z)$. Plot the composite wavelets in the time-domain. The results illustrate the effects of phase shifts (note that all of the composite wavelets have the same frequency spectrum but different phases since multiplication by z^n shifts the phase – see problem 10.27).

Table 8.1

	$t = -3$	$t = -2$	$t = -1$	$t = 0$	$t = +1$	$t = +2$	$t = +3$	$t = +4$	$t = +5$
$a_t = [\overset{\downarrow}{2}, 1, -2, 1]$									
$b_t = -2a_t$									
$c_t = 3a_{t-2}$									
$d_t = \frac{1}{2}a_{-t}$									
$e_t = \pi a_{3-t}$									
$f_t = [-\overset{\downarrow}{1}, 1]$									
$g_t = a_t * f_t$									
δ_{t+2}									
δ_{2-t}									
$\phi_{ff}(t)$									
$\phi_{fa}(t)$									

8.10. A shot is located 7 m below the base of the LVL. Given that $V_H = 2.0$ km/s, $V_W = 0.3$ km/s, $\rho_H = 2.3$ g/cm³, $\rho_W = 1.8$ g/cm³, $\Delta = 4$ ms and that the reflected signal is $[6, -7, -2.8, 5.6, -1.6]$, find the original wavelet using: (a) the inverse filter of (8.54); (b) (8.56).

8.11. The following wavelet is approximately minimum-phase: $[0, 11, 14, 5, -10, -12, -6, 3, 5, 2, 0, -1, -1, 0]$ (fig. 8.38a), the sampling interval being 2 ms. Use $V_{sd} = 2.0$ km/s for the velocity in sand and the following reflection coefficients (these have been scaled up and rounded off to simplify the calculations): shale-to-sand $= +0.1 =$ sand-to-limestone. (a) Determine the reflection waveshape for a sand 0, 2, 4, 6, 8 and 10 m thick encased in shale. (The thickness of 6 m is approximately a quarter-wavelength.) (b) Repeat for the sand overlain by shale and underlain by limestone. (c) Determine the waveshape for two sands each 6 m thick and separated by 4.5 m of shale, the sequence encased in shale (assume $V_{sh} = 1.5$ km/s); this illustrates a 'tuned' situation. (d) Repeat (a) and (b) with the following wavelet: $[0, 6, 11, 14, 14, 10, 5, -2, -10, -11, -12, -10, -6, 0, 3, 4, 5, 4, 3, 1, 0]$ (fig. 8.38b, a minimum-phase wavelet stretched out so that it has about half the former dominant frequency); comparison of the results with those of (a) and (b) illustrates the effect of frequency on the resolvable limit. (e) Repeat parts (a) and (b) using the zero-phase wavelet, $[1, 1, -1, -4, -6, -4, 10, 17, 10, -4, -6, -4, -1, 1, 1]$ (fig. 8.38c, having the same frequency spectrum as the wavelet of fig. 8.38a).

Fig.8.38. Determining a composite reflection.
(a) Minimum-phase wavelet; (b) minimum-phase wavelet of lower frequency; (c) zero-phase wavelet corresponding to (a).

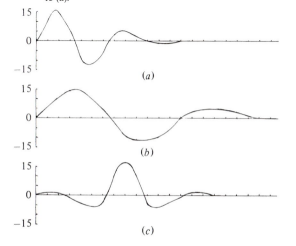

(a)

(b)

(c)

8.12. Show that

$$\sum_t g_{t-j}g_{t-i} = \phi_{gg}(i-j)$$

(see the derivation of (8.58)).

8.13. The techniques and concepts of convolution, aliasing, z-transforms, etc. can be applied to other than the time–frequency domain. Express the source and group patterns of fig. 5.13a as functions of x (horizontal coordinate) and convolve the two to verify the effective pattern shown in fig. 5.13b.

8.14. Assuming the signature of an airgun array is a unit impulse, find the inverse filter for the recorded wavelet: $[-12, -4, +3, +1]$. How many terms should the filter include to get 1% accuracy?

8.15. In §8.2.1 several deconvolution methods were described. List the assumptions of the different methods, such as invariant wavelet, randomness of the reflectivity or of the noise, that a source wavelet is the same as a wavelet recorded near the source or measured from a seafloor reflection by a group offset a few hundred meters, etc.

8.16. In figure 8.14 the hump in the 8–12,000 ft/s contours might cause some concern. If it is known that this is the same section as that shown in figure 9.29, what would the hump imply?

8.17. The wavelet $[-0.9505, -0.0120, 0.9915]$ is not minimum-phase. How would you make it minimum-phase, at the same time changing it as little as possible? Give two methods.

8.18. Figure 8.39 shows three reflections before and after NMO removal. Explain (a) the broadening of the wavelets by the NMO corrections; (b) why the reflections do not have straight alignments after NMO correction.

8.19. Show that: (a) the equation of a diffraction curve (curve of maximum convexity; see fig. 8.25) is

$$z^2 - x^2 = z_0^2$$

where O is the origin and $z_0 = OP$; (b) the unmigrated reflection is tangent to the diffraction curve; (c) the coordinates of P and the slope of the wavefront at P (hence the dip also) can be obtained from the recorded data.

8.20. Show that the coordinate system (x, z, t^*) in eq. (8.89) in effect 'rides along on an upcoming wavefront.'

8.21. The standard alias filter such as shown in fig. 5.33 has its 3 dB point at about half the Nyquist frequency and a very steep slope so that noise above the Nyquist frequency is highly attenuated relative to the passband of the system. (a) Assuming an initially flat spectrum, alias filtering with a 125 Hz 72 dB/octave filter and subsequent resampling from 2 to 4 ms (without additional alias filtering), graph the resulting alias noise versus frequency. (b) Some believe standard alias filters are unnecessarily severe; assume a 90 Hz 72 dB/octave filter and 4 ms sampling and graph the alias noise versus frequency.

8.22. On a N – S line the noise arriving from the south is confined to the band $V_a \leqslant 6$ km/s while the noise arriving from the north is in the band $V_a \leqslant 3$ km/s. (a) Given that

$\Delta x = 50$ m, redraw fig. 8.22b. (b) Repeat (a) for $\Delta x = 25$ m. (c) Calculate $f(x, t)$ (see (8.76), (8.77)) for parts (a) and (b).

8.23. Given the wavelet $[\overset{\downarrow}{10}, -8, 0, 9, -11, 6, 0, -7, 12, -5, 0, 0]$, calculate: (a) the quadrature function, $g_\perp(t)$, (b) $\gamma(t)$ (add multiples of π to obtain a monotonically increasing function), (c) $h(t)$, $A(t)$ and $v(t)$ at $n = 4$. Take $\Delta = 4$ ms.

8.24. Derive (8.92).

Fig.8.39. End-on records from a model with four horizontal layers of velocities 1490, 1895, 2215 and 2440 km/s. (a) Before NMO correction; (b) after.

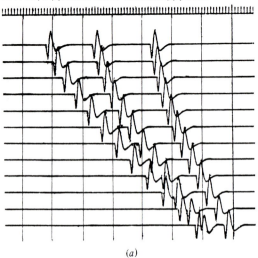

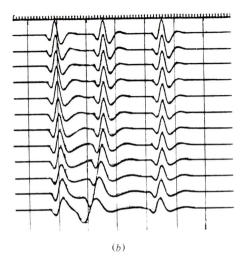

(a)   (b)

9
Geologic interpretation of reflection seismic data

Overview

Interpretation, as we use the word in this chapter, involves determining the geologic significance of seismic data. This necessarily involves geologic terminology and we follow the usage in Bates and Jackson (1980). Interpretation also sometimes includes data reduction, selecting events believed to be primary reflections, and locating the reflectors with which they are associated. Indeed, a number of decisions have to be made in data-processing, acquisition, and even in the initial planning of a survey which prejudice the geologic conclusions and thus could be legitimately included as part of interpretation.

There are few books on the geologic interpretation of seismic data; we cite Fitch (1976), Anstey (1977), Sheriff (1978), and McQuillin *et al.* (1979).

It is rare that the correctness or incorrectness of an interpretation can be ascertained because the actual geology is rarely ever known in adequate detail. The test of a good interpretation is consistency rather than correctness (Anstey, 1973). Not only must a good interpretation be consistent with all the seismic data, it also must be consistent with all that is known about the area, including gravity and magnetic data, well information and surface geology, as well as geologic and physical concepts.

One can usually be consistent and still have a choice of interpretations, the more so when data are sparse. The interpreter should explore various possibilities, but usually only one interpretation is wanted, that which offers the greatest possibilities for significant profitable hydrocarbon accumulation (assuming this is the objective). An interpreter must be optimistic, that is, he must find the good possibilities. The optimistic interpretation is usually preferable to the 'most probable,' because the former will probably cause additional work to be done to test (and perhaps modify) the interpretation, whereas a non-optimistic interpretation may result in abandoning the area. Management is usually tolerant of optimistic interpretations which are disproven by subsequent work, but failing to recognize a possibility is an 'unforgivable sin.' It should be noted that 'success' or 'failure,' that is, finding or failing to find hydrocarbons in commercial quantity, is often a poor test of an interpretation because many factors critical to commercial accumulations cannot be predicted from seismic data.

Seismic data are usually interpreted by geophysicists or geologists. The ideal interpreter combines, in one person, training in both fields. He is fully aware of the processes involved in the generation and transmission of seismic waves, the effects upon the data of the recording equipment and data-processing, and the physical significance of the seismic data. At the same time his geologic experience helps him assimilate the mass of data, much of it conflicting, and arrive at the most plausible geologic picture. Unfortunately, not all interpreters have the requisite knowledge and experience in both geology and geophysics, and often the next best alternative is to have a geophysicist–geologist team working in close cooperation.

Deducing geologic significance from the aggregate of many minor observations not only tests the ingenuity of an interpreter, it also tests his in-depth understanding of physical principles. For example, downdip thinning of reflection intervals might result from a normal increase of velocity with depth as well as from thinning of the sediments, and flow of salt or shale may cause illusory structure on deeper horizons. Geometric focusing produced by reflector curvature can produce various effects, especially if not migrated correctly, and energy which comes from a source located off to one side of the line can interfere with the patterns of other reflection events to produce effects which might be interpreted erroneously, unless their true nature is recognized. Improper processing likewise can create opportunities for misinterpreting data (Tucker and Yorsten, 1973).

Inasmuch as our interpretation objective is usually locating hydrocarbon accumulations, this chapter begins with a summary of concepts about the generation and migration of hydrocarbons and the types of situations which trap hydrocarbon accumulations.

A section on interpretation procedures includes both discussions of the philosophy of seismic interpretation, questions to be answered by the interpretation, and cautions to help avoid erroneous conclusions. Deducing the geologic history of the area is one interpretation objective. Well data have to be related to seismic data so that the interpretation is consistent with both.

Structural maps are commonly the foremost interpretation objective. The tectonic setting usually governs which types of structures are present and how structural features relate to each other, so a review of structural styles precedes discussions of the evidences of various geologic features. Among the features examined are faults, folded and flow structures, reefs, unconformities, channels and stratigraphic traps.

Modeling provides a major tool in interpretation. Direct modeling, the making of a synthetic seismogram to show what should be expected from a geologic model, helps in understanding what seismic features should be looked for as evidences of sought-for geologic anomalies. Inverse modeling, the making of synthetic acoustic-impedance or sonic logs from seismic data, aids in seeing the geologic significance of seismic waveshape variations near well control, especially in locating nearby stratigraphic changes suggested by well data.

Lateral variations in velocity can produce the illusion of unreal structural features. Unrecognized three-dimensional aspects can also lead to misinterpretation.

Stratigraphic interpretation involves delineating seismic sequences which represent different depositional units, recognizing seismic facies characteristics which suggest the depositional environment, and analyzing reflection character variations to locate both stratigraphic changes and hydrocarbon accumulations. Hydrocarbon accumulations are sometimes evidenced by amplitude, velocity, frequency or waveshape changes.

Seismic data are also proving useful in crustal studies, as aids in indicating and delineating deeper features than those usually associated with sedimentary basin exploration.

9.1 Basic geologic concepts
9.1.1 *Generation and migration of hydrocarbons*

The interpretation of seismic data in geologic terms is the objective and end product of seismic work. However, before discussing this most important and critical phase of interpretation, we shall review briefly some basic geologic concepts which are fundamental in petroleum exploration.

Petroleum is a result of the decomposition of plant or animal matter in areas which are slowly subsiding.

These areas are usually in the sea or along its margins in coastal lagoons or marshes, occasionally in lakes or inland swamps. Sediments are deposited along with the organic matter and the rate of deposition of the sediments must be sufficiently rapid that at least part of the organic matter is preserved by burial before being destroyed by decay. Restricted circulation and reducing (rather than oxidizing) conditions favorable for hydrocarbon preservation are found in the deeper portions of both marine and lacustrine waters. As time goes on and an area continues to sink slowly (because of the weight of sediments deposited or because of regional tectonic forces), the organic material is buried deeper and hence is exposed to higher temperatures and pressures. Eventually chemical changes result in the generation of petroleum, a complex, highly variable mixture of hydrocarbons, including both liquids and gases (part of the gas being in solution because of the high pressure). Temperature in the Earth generally increases at a rate of 20–55°C/km, in some places (e.g., Sumatra) by as much as 100°C/km. The habitat of liquid petroleum generation is generally 65–150°C, which is usually in the 1.5–3 km range. At depths of 3–6 km reservoirs predominantly contain gas rather than oil, and at still greater depths the temperature is apt to be so high as to cause gas to decompose.

Sedimentary rocks are porous and, as originally deposited, generally have about 45% porosity. As sediments are piled on top, the weight of the overburden compacts the rocks and the porosity becomes less (fig. 7.6). Some of the water which filled the interstices in the rock (*interstitial water*) usually escapes during the compaction process until the pressure of the water equals that of the hydrostatic head corresponding to its depth of burial. If the formation water cannot escape, it becomes overpressured (see §7.2.4).

Petroleum collects in the pore spaces in the source rock or in a rock adjacent to the source rock, intermingled with the remaining water which was buried with the sediments. When a significant fraction of the pores is interconnected so that fluids can pass through the rock, the·rock is *permeable*. Permeability permits the gas, oil and water to separate partially because of their different densities. The oil and gas tend to rise and they will eventually reach the surface of the Earth and be dissipated unless they encounter a barrier which stops the upward migration. Such a barrier produces a *trap*.

9.1.2   *Types of traps*

The essential characteristic of a trap is a porous, permeable bed, (*B* in fig. 9.1*a*) overlain by an impermeable bed (*A*) which prevents fluid from escaping. Oil and gas

can collect in the reservoir of an anticline until the anticline is filled to the *spill point*. While fig. 9.1*a* is two-dimensional, similar conditions must hold for the third dimension, the structure forming an inverted bowl. The spill point is the highest point at which oil or gas can escape from the anticline, the contour through the spill point is the *closing contour* and the vertical distance between the spill point and the highest point on the anticline is the *amount of closure*. In fig. 9.1*b* the closing contour is the −2085 m contour and the closure is 30 m. The quantity of oil which can be trapped in the structure depends upon the amount of closure, the area within the closing contour, and the thickness and porosity of the reservoir beds.

Figure 9.1*a* might be the cross-section of an anticline or a dome, and the trap is called an *anticlinal trap*. Other structural situations can also provide traps. Figure 9.1*c* shows *fault traps* in which permeable beds, overlain by impermeable beds, are faulted against impermeable beds. A trap exists if there is also closure in the direction parallel to the fault, for example, because of folding, as shown by the contours in fig. 9.1*d*. Figure 9.1*e* shows possible traps associated with thrust faulting.

Figure 9.1*f* shows a *stratigraphic trap* in which a permeable bed grades into an impermeable bed, as might result where a sand grades into a shale. Sometimes permeable beds gradually thin and eventually pinch out to form *pinchout traps*. (Stratigraphic traps of various types are also shown in fig. 9.38.) Closure must also exist at right angles to the diagram, possibly because of folding or faulting. Many traps involve both stratigraphic and structural aspects.

Figure 9.1*g* shows *unconformity traps*, which may result from permeable beds onlapping an unconformity or beds truncated by erosion at an unconformity (see also fig. 9.38). If the permeable beds are overlain by impermeable ones and if there is closure at right angles to the diagram, hydrocarbons can be trapped at the unconformity.

Figure 9.1*h* shows a limestone *reef* which grew upwards on a slowly subsiding platform. The reef was originally composed of coral or other marine animals with calcareous shells which grow prolifically under the proper conditions of water temperature and depth. As the reef subsides, sediments are deposited around it. Eventually the reef stops growing, perhaps because of a change in the water temperature or the rate of subsidence, and the reef may be buried. The reef material is often highly porous and covered by impermeable sediments. Reefs sometime produce arching in overlying sediments because of differential-compaction effects, the reef being generally less compactable than the sediments on either side of it.

Fig.9.1. Sedimentary structures which produce hydrocarbon traps. Permeable beds are dotted in the cross-sections, hydrocarbon accumulations are in black. (a) Vertical section through anticline along line *MN* in (b); (b) map of top of the permeable bed in (a) with the spill-point contour dashed; (c) vertical section through fault traps; (d) map of the middle permeable bed in (c); (e) possible traps associated with thrust faulting; (f) stratigraphic traps produced by lithology change and pinchout; (g) unconformity traps; (h) a trap in a reef and in draping over the reef; (i) possible traps associated with a saltdome.

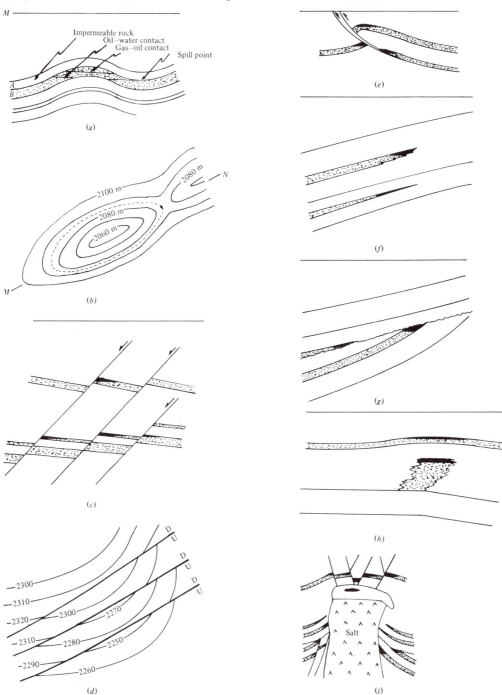

The reef may form a trap for hydrocarbons generated in the reef itself or flowing into it from other beds.

Figure 9.1*i* represents a *saltdome* formed when a mass of salt flows upwards under the pressure resulting from the weight of the overlying sediments. The saltdome bows up sedimentary beds, produces faulting and affects the nature of the beds being deposited. Consequently, traps may be produced over or around the sides of the dome and sometimes within cavities in caprock over the salt, the trapping sometimes resulting from dip reversal, faulting, unconformities, or stratigraphic changes.

The primary objective of a seismic survey for hydrocarbons is usually to locate structures such as those shown in fig. 9.1. However, many structures which provide excellent traps do not contain oil or gas in economic quantities. We also try to derive from the seismic data as much information as possible about the geologic history of the area and about the nature of the rocks in an effort to form an opinion about the probability of encountering petroleum in the structures which we map.

9.2   Interpretation procedures

9.2.1   *Fundamental geophysical assumptions*

Seismic interpretation generally assumes (1) that the coherent events seen on seismic records or on processed seismic sections are reflections from acoustic-impedance contrasts in the Earth and (2) that these contrasts are associated with bedding which represents the geologic structure. Thus, mapping the arrival times of coherent events is related to the geologic structure, and by allowing for velocity and migration effects, we obtain a map showing the geologic structure. We also assume (3) that seismic detail (waveshape, amplitude, etc.) is related to geologic detail, that is, to the stratigraphy and the nature of the interstitial fluids; we examine this assumption further in §9.7.

9.2.2   *Collection and examination of data*

(*a*) *Introduction.* The interpreter gathers together all the data relevant to the interpretation, including geologic, well data, etc. The relevant seismic data usually include seismic sections, a base map, and velocity and other data from the field or generated in processing. Sometimes the interpretation is done concurrently with the field and processing work, so that the interpreter receives additional data while he is carrying out his interpretation, and he may be able to feedback conclusions from his preliminary interpretation so that field or processing procedures can be changed or additional work

can be carried out, in order to prove or disprove points which are not resolved.

Alternative ways of interpreting data are almost always possible. This 'inherent ambiguity' exists with almost any data, although ambiguity in seismic interpretation is less than with most geophysical and geologic data. Ambiguity arises because data are incomplete and/or inaccurate, and the best way to reduce ambiguity is to add more data. The added data might be more seismic data, but it also might be information from surface geology, wells, gravity measurements, etc. The regional geologic setting and concepts about the tectonic stresses to which the region has been subjected should also be used as a check on seismic information.

As an example, one sometimes encounters disruptions in a seismic reflection. If we explain this as caused by faulting, then we must determine what else the fault did. Where did it cut shallower beds, or did it die out? Where did it cut deeper beds, or was the fault displacement absorbed by flowage in mobile salt or shale sediments, or did the fault sole out into the bedding plane? Where is the fault on parallel and intersecting lines, or did it die out laterally? Is the fault a normal fault indicating extension or a reverse fault indicating compression? An interpretation cannot be regarded as complete until such questions have been answered as completely as possible. A fault which dies out both shallower and deeper is difficult to justify (though occasionally this is the correct interpretation). Faults which have not produced effects on nearby lines may also be difficult to justify.

(*b*) *Examining sections.* One of the first tasks of an interpreter is to examine his data for evidences of mislocation (do sections tie properly?) or improper acquisition or processing. Such an examination, while not conclusive, often uncovers gross errors. Unmigrated seismic traces at the intersections of seismic lines ought to copy. When they do not, mislocation or mislabeling of one or both of the lines is a possible explanation, but differences in recording or processing techniques also provide possible explanations. The title blocks of the sections should be examined to see what differences exist. Different size arrays, different mixes of offset distances, or different processing procedures may have resulted in noise contributions being different. Lack of full multiplicity at the ends of lines or where shotpoint spacing is irregular (possibly because of access problems) may have affected data quality. Features which line-up vertically on unmigrated sections are especially suspect because most geologic features are not vertical, whereas the effects of statics errors often are. Occasionally files get mixed up and data get assembled incorrectly. The various data elements should be consistent; if the velocity was assumed to vary, are the assumptions consistent with the structure and the character of the sections? Are certain data which show on sections made as intermediate steps in the processing missing or changed on the final sections? Unexplained differences or departures from what is geologically reasonable should be investigated, so that geologic significance is not attributed to errors in the data.

Figure 9.2 shows the ends of a typical seismic section, including the data block. The data block is often subdivided into parts listing information about line identification, data acquisition, and processing. A generalized line direction and horizontal scale may be given. Where the horizontal scale is not indicated, it can be determined by counting the number of traces per centimeter, the trace spacing being half the geophone group spacing (assuming no horizontal compositing). The locations at which velocity analyses were run are usually indicated, often with the results of the analyses tabulated as time–velocity pairs. Those should be examined for consistency along the line. The locations of changes in line direction or abrupt surface changes (such as elevation differences) should be noted for their possible effects on reflection quality or attitude. Irregularities in coverage are common in land data because of surface or access problems; these often show as irregularities in the first-break patterns and they may also affect reflection quality and the apparent attitude of reflections. The multiplicity involved in each trace is sometimes shown by encoding at the bottom of the section, providing a key to irregularities of coverage.

The vertical scale shown on the section is usually linear in time. Depth equivalents are sometimes given but these are only intended to be approximate; depth determinations can be made more accurately by measuring the arrival times and converting these to depths using the appropriate velocity relations. Where processing or display parameters are changed with arrival time, the locations of changes should be noted so that changes in quality produced by the processing changes are not interpreted as geologic changes. If changes have been made in the midst of the objective section, or if the horizon being mapped varies in depth so as to cross the zone of changed parameters, special care has to be employed to avoid possible misinterpretation.

(c) *Interpretation approaches.* Interpretation involves building a model of the prospect area in the interpreter's mind. Some interpreters hang all their data on their walls,

thus surrounding themselves with the data so that they can look from one section to another to see interrelationships better. They develop much of their interpretation sitting back and pondering about which ideas are feasible, rather than being busy all the time timing events, transferring data to base maps, and drawing contours. Imagination is required in interpretation and it takes time to develop an interpretation which leads to new discoveries.

Two basically different approaches are made to seismic interpretation: the one focuses on objectives, the other gradually builds up a complete interpretation. Often only a few reflectors are considered to be of interest because it is already known that the only prospective reservoirs are in one part of the section. Such areas usually contain wells in which formations of interest are identified and then related to seismic sections, either by synthetic seismograms or, more simply, by using velocity to relate depth in the wells to arrival time on the seismic sections. The associated events on the seismic sections are followed throughout the region to give structural relationships, looking for faults which displace the reflections and watching for reflection character changes which may indicate sand pinchouts, patch reef growths, other stratigraphic variations or hydrocarbon accumulations.

To develop a more complete interpretation of an entire section, an interpreter generally starts with the most obvious features, usually the strongest reflection event or the reflection event which possesses the most distinctive

Fig.9.2. Left and right ends of a marine seismic section. The data were recorded with a 48-channel streamer with numbered sourcepoints every 220 ft, individual traces being spaced 110 ft (33.5 m) apart, which information can be used to give the horizontal scale.
Full 48-fold CDP multiplicity is not achieved for the 47 traces at each end of the line, which may explain some of the data-quality deterioration evident in these regions; the lack of full multiplicity affects multiple attenuation, so that multiples become more likely in these regions. The locations of velocity analyses are indicated by the 'V' at the top of the section and the tabulated data give stacking (labeled RMS), interval and average velocities assuming horizontal velocity layering according to (7.12), (7.14), and (7.13). The solid triangle above SP 27 indicates an intersecting seismic line; TVF stands for time-variant filter, WD for water depth; RMS gain means that the RMS amplitude of each trace has been brought to the same value; LC and HC indicate the frequencies for which the low-cut and high-cut filters introduce 3 dB attenuation. (Courtesy Conoco.)

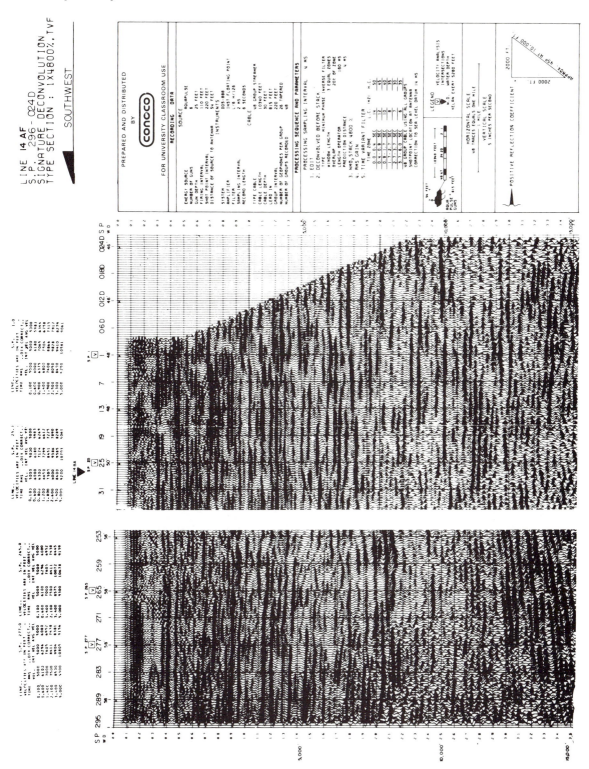

character, and follows this feature as long as it remains reliable. When the feature being followed deteriorates or changes character so that it is not clear what is happening, it is dropped rather than 'pushed' or extrapolated beyond its region of reliability, and the interpreter returns to it subsequently after he has developed other features of reasonable reliability. By following features in the order of their reliability, overall interpretation is developed before attention is concentrated on the objective reflections. This type of approach usually leads to a better interpretation, but it is much more time-consuming. An interpretation of the section shown in fig. 9.3, for example, might start with mapping of the strong reflector AA', distinguished in part by the following zone of poor reflections BB'. Reflector AA' is cut by a normal fault, and following this fault onto intersecting seismic lines might be the next step. Attention might then shift to mapping a shallower horizon, perhaps CC', but here it will be difficult to be sure one is staying with the same reflector. Perhaps one would try to map the base of the BB' zone next; this will also involve uncertainties, and to resolve some of these it may be

necessary to go back and revise the parts interpreted earlier. The petroleum objectives are about the middle of the BB' portion so a map will have to be made here, but this map can be made more reliably with the aid of more certain events already mapped.

9.2.3 *Mapping reflecting horizons*

The horizons which we draw on seismic sections provide us with a two-dimensional picture only. A three-dimensional picture is necessary to determine whether closure exists, the area within the closing contour, the location of the highest point on the structure, and so on. To obtain three-dimensional information, we usually shoot lines in different directions. Most reflection surveys are carried out along a more-or-less rectangular grid of lines, often with common shotpoints at the intersections of lines to facilitate correlating reflections on the intersecting profiles.

Events picked on one section are compared with those on intersecting sections in order to identify the same horizons; identification is made on the basis of character

Fig.9.3. Portion of a line in the US Gulf Coast area. (Courtesy Conoco.)

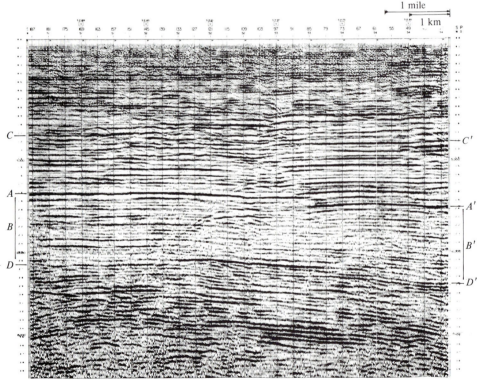

and arrival times. The horizons are now 'carried' along the cross-lines, and ultimately, along all lines in the prospect to the extent that the quality of the data permits.

When a horizon can be carried all the way around a closed loop, we should end up with the same arrival time with which we started. This *closing of loops* provides an important check on reliability. When a loop fails to close within a reasonable error (which depends mainly upon the record quality and the accuracy of the weathering corrections), the cause of the misclosure should be investigated carefully. Migrated sections have to be tied by finding the same reflection on the intersecting sections; such tie points will be displaced from the vertical through the shotpoint by the amount of the migration on each of the lines. Often misclosure is due to an error in correlating along the line or from line to line, possibly because of inaccurate corrections, a change in reflection character, or error in correlating across faults. When the dip is different on the two sides of a fault or the throw varies along the fault, an incorrect correlation across the fault may result in misclosures (but not necessarily). After the sources of misclosure have been carefully examined and the final misclosure reduced to an acceptable level, the remaining misclosure is distributed around the loop.

After horizons have been carried on the sections, maps are prepared. For example, we might map a shallow horizon, an intermediate horizon at roughly the depth at which we expect to encounter oil, if any is present, and a deep horizon. We map on a *base map* which shows the locations of the seismic lines (usually by means of small circles representing shotpoints) plus other features such as oil wells, rivers, shorelines, roads, land and political boundaries, and so on. Values representing the depth of the horizon below the datum plane are placed on the map (*posted*), usually at each shotpoint (although, strictly speaking, they should be posted at the reflecting points on the respective horizons). Other information relevant to the horizon being mapped (depths in wells, locations of gravity anomalies, relevant geologic information, etc.) is also posted. Faults which have been identified on the record sections are drawn on the map and the depth values are then contoured.

After the structural information has been extracted, the next step is to work out as much as possible of the geologic history of the area. Fundamental in this connection is the determination of the ages of the different horizons, preferably according to the geologic time scale, but at least relative to one another. Often seismic lines pass close enough to wells to permit correlating the seismic horizons with geologic horizons in the wells. Refraction velocities (if available) may help identify

certain horizons. Occasionally a particular reflection has a distinctive character which persists over large areas, permitting not only it to be identified but also other events by their relation to it. Notable examples of persistent identifiable reflections are the low-frequency reflections sometimes associated with massive basement and the prominent reflection from the top of the Ellenburger, a limestone encountered in Northern Texas.

9.2.4 *Deducing geologic history*

Seismic sections often subdivide naturally into units. The boundaries between units are often the better reflectors and the units often have angular relations to each other which indicate features of geologic history: periods of tectonism, unconformities, transgressions, etc. The boundaries between units generally indicate a gap in geologic time and often separate sediments deposited in different kinds of environments. Velocity and other seismic measurements, such as of amplitude or instantaneous frequency, and their variations in the direction of the bedding yield additional information. Lithology and/or stratigraphic situations are usually inferred from many evidences which are individually weak but which, taken together, make a coherent pattern.

Isopach maps which show the thickness of sediments between two horizons are useful in studying paleostructure and structural growth. Ideally, only one rock unit should be encompassed by the interval between the horizons, but often the only horizons which can be mapped reliably will be separated by more than one rock unit, so that the resulting isopach map may show more than one period of movement or more than one depositional trend. The interval between the horizons is often measured in terms of two-way traveltime rather than thickness in meters or feet, it being implied that velocity-variation effects are minor compared to thickness variations. Isopach maps are often prepared by overlaying maps of two horizons and subtracting the contour values wherever the contours on one map cross the contours on the other. The differences are recorded on a blank map and then contoured. If the contours show a trend toward increased thickness in a certain direction, it may suggest that the region was tilted downward in this direction during the period of deposition or that the source of the sediments is in this direction. Uniform thickness of a folded competent bed indicates that the folding came after the deposition whereas if the thickness increases away from the crest of an anticline, deposition probably was contemporaneous with the growth of the anticline. Growth during deposition is usually more favorable for petroleum accumulation since it is more likely for reservoir sands to be deposited on the flanks of structures with even slight relief.

Paleosections (*palinspastic sections*) can be made by time-shifting traces to flatten some distinctive horizon which can be assumed to have been deposited horizontally; the objective is to show relationships as they existed at the time of deposition of this horizon. In practice, such flattening is often done in the interpreter's mind rather than by actually manipulating the data because of the cost of reprocessing, but in areas of even moderate complications actual flattening can be worthwhile. Obviously, migration before the flattening is necessary and the flattened horizon should be selected judiciously. Compaction effects and changes in velocity since the deposition of the flattened horizon should be allowed for, but usually the information required to do this is lacking.

The unravelling of the geologic history of the area is important in answering questions such as the following: (*a*) Was the trap formed prior to, during, or subsequent to the generation of the oil and gas? (*b*) Has the trap been tilted sufficiently to allow any trapped oil to escape? (*c*) Did displacement of part of a structure by faulting occur before or after possible emplacement of oil? While the seismic data rarely give unambiguous answers to such questions, often clues can be obtained which, when combined with other information such as surface geology and well data, permit the interpreter to make intelligent guesses which improve the probability of finding oil. Alertness to such clues is the 'art' of seismic interpretation and often the distinction between an 'oil finder' and a routine interpreter.

9.2.5 *Integrating well data into an interpretation*

Wells drilled in the area provide geologic information which must be consistent with the interpretation. Borehole logs (fig. 9.4) are interpreted to determine formation tops, lithology, depositional environment, the location of faults (fig. 9.5) and unconformities with an indication of the amount of section missing, etc. Well logs plotted linearly in time at the seismic section scale aid in correlating (fig. 9.6). While a borehole provides an opportunity for actual measurements, the results available to a seismic interpreter are usually interpretations of measurements. When a disagreement between well and seismic data appears, both well and seismic data should be reexamined to effect a resolution of the problem.

Well information usually has to be projected into seismic data, the wells not having sampled the same subsurface locations as the seismic data. Even where a wellhead is located on a seismic line, dip may result in the

seismic work seeing a different portion of the subsurface than the well. The projection of well information involves an interpretation so that the data are projected up- or downdip by the correct amount (which may vary with depth in both magnitude and direction), making appropriate allowance for faults or other features which intervene. The seismic section is usually plotted in time whereas the well data are usually in depth, so that the choice of an appropriate velocity for converting the one into the other has to be made.

Even when location problems are not present, the well information results only from the rock within a few centimeters of the borehole (which may have been altered by the drilling process) whereas seismic data include contributions from a large Fresnel-zone region. Well and seismic data may be plotted with respect to different

Fig. 9.4. Correlations between well logs within the same field. For each well the left curve indicates the SP (spontaneous potential) response and the right curve resistivity. Several correlation lines have been drawn and numbered. Some intervals are thinner in one well than in another; the interpreter must decide whether intervals are thinner because section is missing (as a result of faulting or an unconformity) or because of stratigraphic variations (or because of miscorrelation). Part of the 3–4 section (60 m) is faulted out of well *C*, 40 m of the 6–7 section is faulted out of well *B*, and horizon 5 marks an unconformity, explaining the thicker 4–5 section in *A*. Obviously, other interpretations are possible so that a seismic interpreter should not regard well information as infallible.

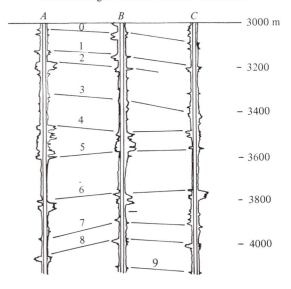

datums and time shifts may have been introduced into the seismic data in recording or processing. Furthermore, most reflections are the interference composites of several reflections, and multiples from shallow reflectors or other noise may also affect the interference. In consequence, relating interfaces to specific seismic events is not easy and is often done incorrectly (see also §9.4.4).

9.2.6   *Drawing conclusions from reflection data*

Structural traps, such as anticlines and fault traps, and structural *leads* (possibilities of traps which require more work to define them completely) are usually evident from examination of maps and sections. Traps resulting from pinchouts and unconformities are more difficult to recognize and non-seismic evidence often must be combined with seismic data to define such features. Nevertheless, careful study of the maps, sections and records plus broad experience and ample imagination will at times disclose variations of dip or other effects which help locate traps.

When the interpretation is finally completed, a report (see Appendix C) is usually prepared, often both for submission in writing and for oral presentation. In some ways this is the most difficult and most important task of the interpreter. He must present his findings in such a way that the appropriate course of action is defined as clearly as possible. The important aspects should not be obscured by presenting a mass of details nor should they be distorted by presenting carefully selected but non-representative maps and sections. Evidences to support significant conclusions should be given. Alternate interpretations should be presented and an estimate given of the reliability of the results and conclusions. Finally, the interpreter should recommend what further action should be undertaken.

9.2.7   *Display techniques: color*

Display techniques strongly affect the ease with which features can be seen. The display which is optimum for one interpretation objective may not be optimum for another. A regional interpretation needs a synoptic view

Fig.9.5. Construction to aid in mapping faults from well picks. Assume that it has been determined that the same normal fault cuts well *B* at 3860 m with 40 m of section missing, well *C* at 3460 m with 60 m missing, and is not seen in well *A*. From regional considerations we expect the fault to dip about 45°. Our mapping horizon is at 3605 m in well *A*, 3570 in *B*, and 3560 in *C*. (*a*) Section through well *B* neglecting dip of the horizon; the mapping horizon will encounter the downthrown side of the fault a distance of (3860 − 3570 = 290 m) from the well and the upthrown side at 330 m from the well. Likewise the fault's upthrown and downthrown sides must lie 100 and 160 m from well *C*. (*b*) Map view

showing locations of wells *A*, *B*, *C* and circles to which fault must be tangent; the fault may strike either NE (trace *FF′*) or SE (trace *GG′*). If the fault strikes SE, well *A* is downthrown and the fault must be at about (3605 + 200 = 3805 m) in *A*; if the fault strikes NE, well *A* is upthrown and we expect the fault at (3605 − 330 = 3275 m). Growth along the fault, dip, and variations of the fault angle introduce uncertainty into the construction. In this instance it is believed that *A* is not faulted at 3805 m but the fault may die out upward before 3275 m so the lack of a fault cut there is not proof that *A* is not upthrown. Hence fault trace *FF′* is preferred to fault trace *GG′*.

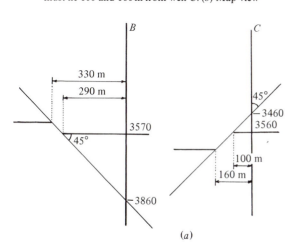

(*a*)

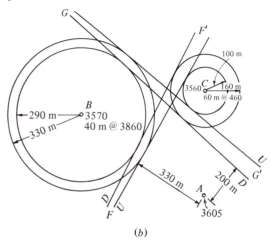

(*b*)

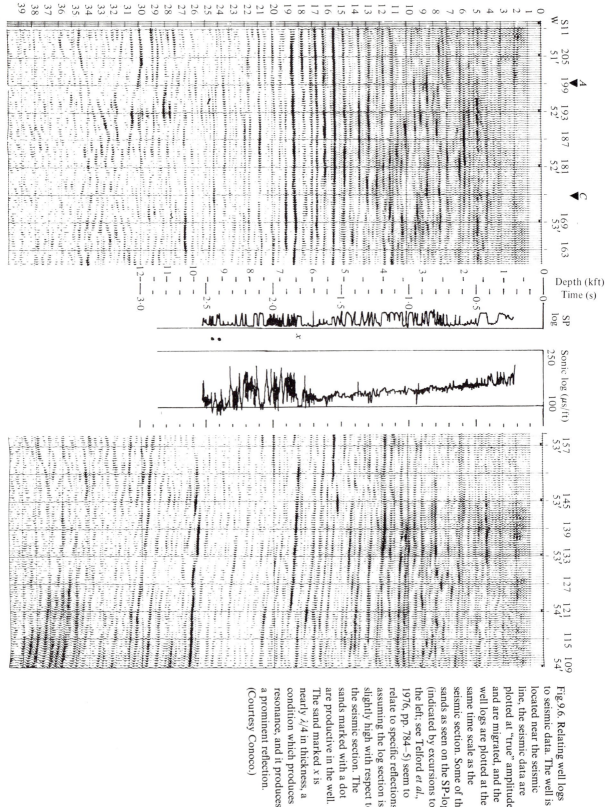

Fig.9.6. Relating well logs to seismic data. The well is located near the seismic line, the seismic data are plotted at "true" amplitude and are migrated, and the well logs are plotted at the same time scale as the seismic section. Some of the sands as seen on the SP-log (indicated by excursions to the left; see Telford *et al.*, 1976, pp. 784–5) seem to relate to specific reflections, assuming the log section is slightly high with respect to the seismic section. The sands marked with a dot are productive in the well. The sand marked ✕ is nearly λ/4 in thickness, a condition which produces resonance, and it produces a prominent reflection. (Courtesy Conoco.)

and reduced sections so that the features over a large area can be seen relative to each other. The mapping of a prospect requires larger sections in order to see detail to resolve faults and structures. Locating stratigraphic traps associated with an unconformity requires a full-waveform display of the unconformity reflection on a fairly large scale. Evidences of hydrocarbon accumulations may require displays at very low amplitude (so that 'bright spots') become evident), displays with reversed polarity, and displays of attributes such as frequency and velocity. A display of velocity and other attributes may also assist in lithologic identification. Stratigraphic variations may be more evident if appreciable vertical exaggeration is employed, whereas structural interpretation is usually easier if horizontal and vertical scales are nearly the same. The varying requirements in interpretation call for a variety of displays of seismic data. Sheriff and Farrell (1976) show a section displayed with various plotting parameters. Feagin (1981) discusses plotting parameters.

Most often interpretations are based on migrated sections; features on these should be checked against unmigrated sections to guard against possible migration errors and to effect ties to intersecting seismic lines. Where an appreciable range of depths is of interest, sections plotted so that the vertical scale is linear with depth rather than with time are useful, especially in working out structural problems. The velocities used in the time-to-depth conversion should be checked, especially if the velocities vary horizontally. Where several outputs from data-processing are available, for example where there are outputs employing different filtering or special processing, these should be examined to see what differences they produce. Different displays may prove better for mapping different horizons, but waveshapes and time delays may change with processing parameters and these changes have to be taken into account in the mapping.

The various products used to control processing should be examined so that the interpreter understands more clearly exactly what was done in the processing and how the decisions made there affect the final product. Velocity analyses should be especially studied for consistency in picking and to yield clues regarding lithology, high-pressure zones, etc. Where velocity data exist only in tabular forms, one might wish to redisplay them graphically to facilitate understanding the significance of variations, especially where velocity varies along the seismic line.

One of the problems with interpretation is that there is often too much data to be examined and comprehended. Superimposed displays may help in seeing the interrelationship of various data aspects. Color

overlays provide one way of adding an additional variable to a display, and color overlays are employed to show amplitude, velocity, frequency and other aspects (Balch, 1971, Taner and Sheriff, 1977, and Taner *et al.*, 1979).

9.3   Evidences of geologic features
9.3.1   *Concepts from structural geology*

(a) *Structural style and plate-tectonic setting.* Many areas have been subjected to fairly simple stress fields which have determined the types and orientations of structural features present, that is, the *structural style.* The stress field may have changed with time, so that the structural style may be different for different portions of the section, or later patterns may be superimposed on older, different patterns. Because the interpretation of features on seismic sections often involves some ambiguity, knowledge of the structural style appropriate to the region can aid in selecting the most probable interpretation and in making the interpretation consistent with everything known about the area, not merely with the seismic data alone. It is also important in programming data acquisition in the most economical manner.

Structural style depends on the tectonic setting, especially the location with respect to plate boundaries and the type of boundary. The types of plate boundaries (fig. 9.7) relate to the relative movement of the plates involved: (1) a pull-apart zone where plates are separating, (2) a collision or subduction zone where they are coming together, (3) a strike-slip zone or transform boundary where they are sliding past each other. The latter may involve transform-faults, a kind of tear fault that accommodates the changes where different portions of

Fig.9.7. Plate-tectonic model. Material upwells in rift zones where plates move apart; in convergence (subduction) zones one plate plunges under another and eventually gets so deep that it melts. Transform faults (*T*) link offsets in rift or convergence zones; transform faults may involve plates sliding by each other only over part of their length, for example, only between the active rifts, other portions having been formerly active before the plates grew at the rift zones. (After Isaacs *et al.*, 1968.)

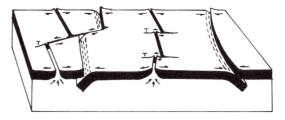

Table 9.1. *Structural styles and plate-tectonic habitats* (After Harding and Lowell, 1979.)

	Structural style	Characteristics	Dominant deformational stress	Plate-tectonic habitat	Typical profile
BASEMENT-INVOLVED STYLES	Pull-apart zones	Fairly high-angle normal faults dipping 60–70° in either direction Rotated fault blocks	Extension	Divergent boundaries (1) at spreading centers (2) aborted rifts Intraplate rifts Transform boundaries with component of divergence Secondary at convergent boundaries: (1) Trench outer slope (2) Arc massif (3) Stable flank of foreland and fore-arc basin (4) Back-arc marginal seas	
	Compressive faults and basement thrusts	High-angle reverse faults, upward imbricating of faults	Compression	Convergent boundaries (1) Foreland basins (mostly) (2) Orogenic belt cores (3) Trench inner slopes and outer highs Transform boundaries with component of convergence	
	Wrench-fault assemblages	Strike-slip faulting is primary, secondary features at about 30° angle to main trend Fairly narrow trend Faults generally steepen with depth	Couple	Transform boundaries Convergent boundaries at an angle: (1) Foreland basins (2) Orogenic belts (3) Arc massifs Divergent boundaries with offset spreading centers	
	Basement warps	Gentle structure: domes, arches, sags	Isostatic adjustment Heat-flow	Plate interiors Passive boundaries Other areas	
BASEMENT-DETACHED STYLES	Thrust assemblages	Faults sole out at décollement in incompetent rocks	Compression	Convergent boundaries (1) Inner slopes of trenches and outer highs (2) Mobile flank of forelands (orogenic belts) Transform boundaries with component of convergence	
	Growth faults and other normal fault assemblages	Downthrown toward basin or toward center of uplift Dip often lessens with depth (for growth faults) Often contemporaneous with deposition	Extension	Passive boundaries Secondary to uplifts (folds, saltdomes)	
	Salt structures	Pillows, domes, salt walls	Plastic flow Solution	Divergent boundaries (rifts provide venue for salt deposition)	
	Shale structures		Plastic flow (often involving overpressuring produced by rapid burial)	Passive boundaries	
	Drape features		Differential compaction	Subsiding basins Over reefs	
	Volcanic plugs		Igneous intrusions		

pull-apart or subduction zones are offset, or the changes between boundaries of different types. Subsidence and isostatic adjustment are important factors influencing structural style away from plate boundaries, for example, at the 'passive margins' of continents.

First-order or primary structural features relieve the main stresses produced by plate movements; they also generate secondary stresses which produce secondary structural features, and these generate tertiary features, and so on.

(b) Types of structural style. The major structural styles are listed in table 9.1, based on a classification by Harding and Lowell (1979). The first-order distinction is between styles which involve the basement and those which are detached from it.

The beginning of seafloor spreading seems to be a local uplift area which produces three grabens radiating from a triple-junction (fig. 9.8), two of these generally take over and form the rift zone which subsequently forms a new ocean. (Local uplifts, such as over saltdomes, develop similar faulting patterns.)

Seafloor spreading (fig. 9.9) produces many more-or-less parallel normal faults trending perpendicular to the direction of extension, the faults becoming older as distance from the spreading centers increases. Except near

the spreading centers, the faults are generally inactive and their upper ends often terminate abruptly at an unconformity. Fault blocks are often rotated, occasionally producing a high edge which may become the venue for reef development. In intraplate areas, spreading may have occurred for only a short time. Extensional features are also found as secondary features at convergent boundaries.

Compression produces high-angle reverse faults and basement thrusts at convergent boundaries. Secondary compression is sometimes produced in other settings.

A complex array of features can be produced at convergent boundaries. In the subduction of an oceanic plate under a continental plate, as in fig. 9.10, portions of the oceanic plate may be scraped off in a bulldozer-like action to produce a melange wedge with thrust faulting. The composition of the melange will include sediments derived from the continental plate, as well as oceanic crust and also deep-water sediments originally remote from the subduction zone. Thrusting and a variety of other effects will also be seen in the forward portions of the continental plate and the affected zone may be very wide. The oceanic crust may change to eclogite as it plunges deeper and gets hotter and under more pressure, the eclogite perhaps

Fig.9.8. Rupturing produced by an uplift. (*a*) The tension resulting from a localized uplift tends to produce three sets of normal faults which produce three grabens. (*b*) As stresses continue, two of these (*A, B*) tend to become pull-apart zones and the third (*C*) is called the *failed arm* or *alaucogen*.

Fig.9.9. Structure associated with rifting and sea-floor spreading. (*a*) Early phase of rifting: normal faulting occurs, down toward a central graben; infilling sediments are mainly continental. (*b*) Transition from rifting to drifting; with continued spreading, crustal thinning and new oceanic crust result in isostatic subsidence, restricted circulation and evaporite deposition. (*c*) As more oceanic crust and subsidence result from the plates drifting apart, an ocean is formed.

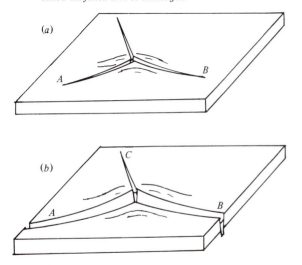

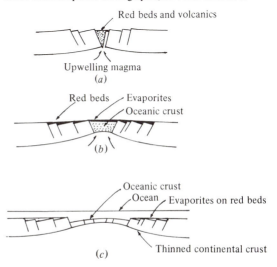

becoming more dense than the mantle of the continental plate, thereby facilitating the subduction process. The materials carried down by the subducting plate may melt and produce a volcanic arc parallel to the subducting edge but an appreciable distance from the trench. Back-arc basins and other features may also develop. Occasionally pieces of the oceanic arc or continental blocks riding on the subducting plate may adhere to the continental plate and the subducting edge may jump to a new location, so that a variety of complications is possible.

Wrench-fault assemblages are most commonly associated with transform boundaries (fig. 9.7). While the predominant motion is strike-slip, a small vertical component of throw is often the most evident aspect. Features are usually confined to a relatively narrow linear zone along the principal strike-slip direction. Some associated secondary features are illustrated in fig. 9.11. These secondary features are often fairly straight and arranged

en echelon. Fault traces are generally straight and the faults tend to steepen with depth. The main stresses may have components of extension or compression perpendicular to the main strike-slip motion, and irregularities or kinks along strike-slip faults also produce extensional or compressional features. Where compression is associated with strike-slip motion, *flower structures* (upward-spreading fault zones) may occur (fig. 9.12).

Basement warps are often solitary, very gentle features, sometimes with associated normal faulting. They may be of basin size (Williston Basin), or occur as regional arches or local domes. They tend to persist over long periods of time and hence localize truncation, unconformities and various types of stratigraphic traps.

Thrusts tend to exist as a set of sub-parallel salients on the overriding plate at subduction zones. Thrust zones may be very broad and of somewhat different forms (fig. 9.13). Thrusts generally parallel the bedding in incompet-

Fig.9.10. Tectonic features associated with subduction of an oceanic plate under a continental plate.

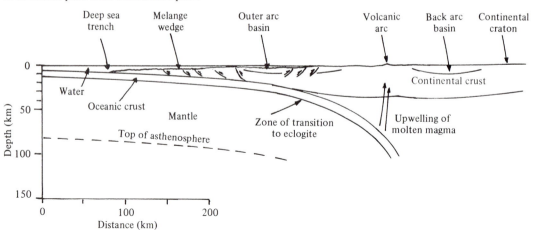

Fig.9.11. Features secondary to a wrench fault. The angles of secondary features are ordinarily about 30° or 60° to the primary fault. The secondary features can in turn produce tertiary features, etc. (*a*) Map view (for symbols, see appendix D); (*b*) isometric diagram showing some of the features indicated on the map.

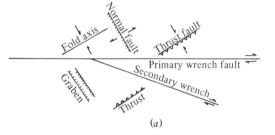

(*a*)

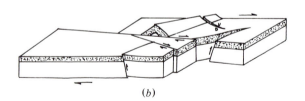

(*b*)

ent rocks and cut across the bedding in competent rocks; anticlines often overlie where thrusts cut across competent members (fig. 9.14). Abnormal pressure zones (§7.2.4) probably provide the décollement (detachment) zones along which the gliding takes place, the overthrust in effect 'floating' into place (Gretener, 1979).

Where massive carbonates form a significant part of the section, thrust sheets often involve repeated slabs (fig. 9.13*a* through 9.13*c*); where ductile formations predominate, hanging-wall folds are common (fig. 9.13*d*). Dips often decrease with depth. The shortening associated with thrusting and folding may be distributed among features in the strike direction by perpendicular tear faults (fig. 9.15*a*) or by transfer zones in which parallel features grow or decrease in magnitude (fig. 9.15*b*).

In folding, the length and volume of beds tend to remain constant; however, often both cannot be conserved at the same time, especially with intense folding, and flow and/or faulting occurs in some beds (fig. 9.16). Folding cannot persist to great depths but must give way to flowage or faulting mechanisms.

Normal faults detached from basement occur adjacent to subsiding basins, especially at the passive margins of continents. They are generally downthrown toward the basins and usually grow contemporaneously with deposition as basins subside, as evidenced by thickening into faults on the downthrown side (fig. 9.17). The fault planes generally decrease in dip with depth. Fault planes not associated with basement faulting are generally concave upward. Rotation of the downthrown block in growth faulting produces *rollover* into the fault, the consequent dip reversal possibly making hydrocarbon traps. Growth faults are not only curved in cross-section but also in plan view (fig. 9.17), generally being concave toward the basin. Growth faults frequently die out upward. The throw on growth faults often increases with depth. They sometimes sole-out in the bedding, the movement parallel to the bedding sometimes producing *toe structures* (fig. 9.18).

Normal faults are often secondary elements of other structural styles; for example, normal faults occur on the crests of folds and above diapirs. They are often associated with shale or salt flowage, especially overpressured shale. Normal faults are probably the most common structural feature because rocks are especially weak under tension.

Salt and shale structures can be of a variety of types involving flow and withdrawal structures, and sometimes collapse features resulting from salt removal by solution. In fig. 9.19 the system of growth faults developed just beyond the underlying shelf edge, where presumably shale under abnormal pressure acted as a fluid and flowed to the left and up to form shale diapirs. Salt and shale often provide the detachment surfaces associated with detached faulting (as with thrusting) and folding. Buoyant salt structures are especially common along passive continental margins, the basinward progression often being salt-withdrawal structures, pillows, non-piercement and piercement domes, and finally, a 'wall' of salt (figs. 9.20 and 9.21), the salt often having moved considerable distances basinward as deposition continues and basin subsidence moves seaward.

Depositional and structural features are often interrelated. Reefs may grow over shelf edges or on the upturned edges of rotated fault blocks. Differential compaction may produce drape over reefs and shelf edges. Progradation beyond a shelf edge may produce growth faults. The weight of sediments deposited by a river delta system may produce subsidence because of isostatic adjustment, affecting faulting patterns and causing salt and shale movement.

9.3.2   *Faulting*

(*a*) *Introduction.* Ideally, reflection events terminate sharply as the point of reflection reaches the fault plane and then they resume again in displaced positions on the other side of the fault. In addition, ideally the reflection has a sufficiently distinctive character that the two portions on opposite sides of the fault can be recognized and the fault throw determined. In practice, diffractions usually prolong events so that the locations of fault planes are not clearly evident, although occasionally it is possible to observe sharp terminations. Moreover, although sometimes the same reflection can be identified unequivocally on the two sides of a fault, in many cases we can make only tentative correlations across faults. Campbell (1965) discusses criteria for detecting faults on seismic sections.

Fig.9.12. Production of upthrust flower structure as a result of strike-slip motion which involves a component of compression. (After Lowell, 1972.)

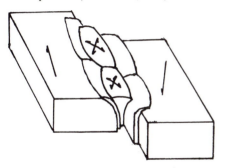

Fig.9.13. Overthrusting. (*a*) Seismic profile 32 km long in
Valley-Ridge province of East Tennessee (from Harris
and Milici, 1977); (*b*) interpretation of (*a*) (from Harris
and Milici, 1977); (*c*) thrusting in Canadian Rockies
producing the Turner Valley structure (from Gallup,
1951); (*d*) décollement of the Jura Mountains (from
Buxtorf, 1916).

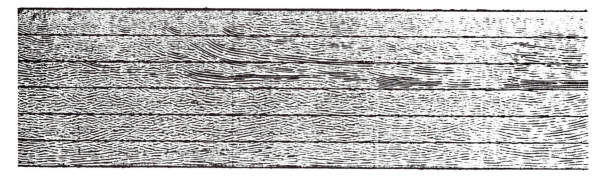

(*a*)

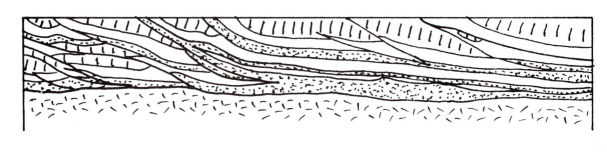

[/ [[ ] Chickamagua Group	[ ] Rome formation
[ ] Conasuaga Group	[ ] Precambrian basement

(*b*)

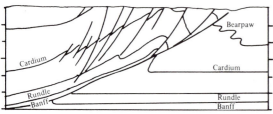

(*c*)

(*d*)

Fig.9.14. Diagram of the southern Appalachian
décollement. (After Harris and Milici, 1977.) (*a*) Initial
low-angle thrust; (*b*) after further movement, anticlines
develop over the ramp portions of the décollement
(detachment zone).

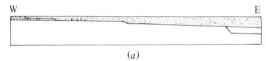

(*a*)

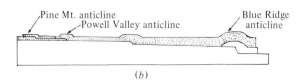

(*b*)

Fig.9.15. Transfer of throw from one thrust fault to
another (After Dahlstrom, 1970.) (*a*) Tear faults separate
a thrust into separate thrusts or folds; (*b*) thrusts grow
or die laterally.

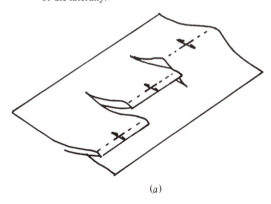

(*a*)

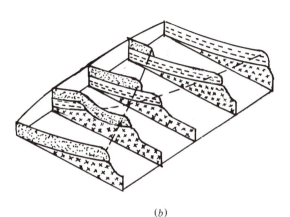

(*b*)

Fig.9.16. Mechanisms to maintain bed length and
volume in interpretation of folding. (*a*) 'Concentric'
folding with flow or severe distortion as the mechanism
(after Goguel, 1962); (*b*) 'similar' folding with bedding-
plane shear and thickness variations; (*c*) combination
of folding with faulting as the mechanism for competent
members; other members must undergo flow or other
distortions (after Hobbs, *et al.*, 1976).

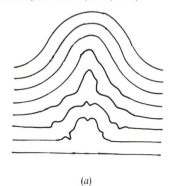

(*a*)

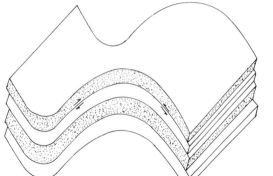

(*b*)

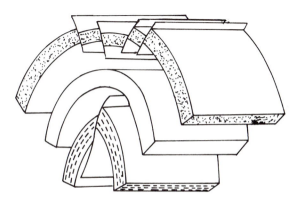

(*c*)

Fig.9.17. Growth fault, with part of down-dropped block cut away. Poorly-consolidated sediments slide toward the basin along a concave-upward (listric) fault plane. Rotation of the down-dropped block contemporaneous with deposition results in thickening into the fault and rollover to produce an anticlinal axis parallel with the fault plane. The fault is arcuate in plan view with the throw gradually diminishing away from the center of the fault. Such a fault is shown in fig. 9.24.

Fig.9.19. (*opposite*) Seismic section showing shale flowage. (Courtesy Exxon.) (*a*) CDP seismic data; the vertical exaggeration is 5x to 8x, decreasing with depth; (*b*) interpretation; the lines show the attitude of reflections without any attempt to correlate the same event across faults, except for the events marked '*U*', an angular unconformity. At their lower ends the faults encounter an abnormally-pressured shale which flowed to the left and up into shale diapirs '*D*'.

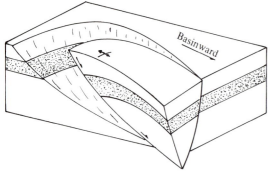

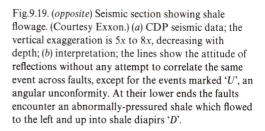

Fig.9.18. Seismic section showing toe structures. (Courtesy Exxon.) (*a*) CDP seismic data; the marks at the top are 2 km apart, the vertical exaggeration is of

the order of 11x, and the seafloor slope is less than 1°; (*b*) interpretation of fault glide planes; the seafloor multiple is dashed.

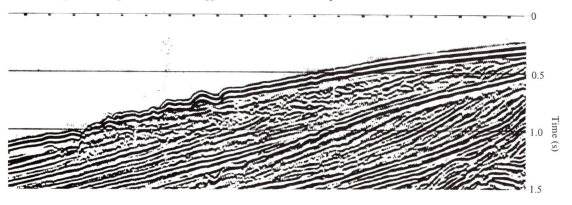

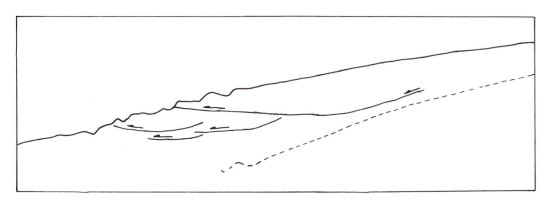

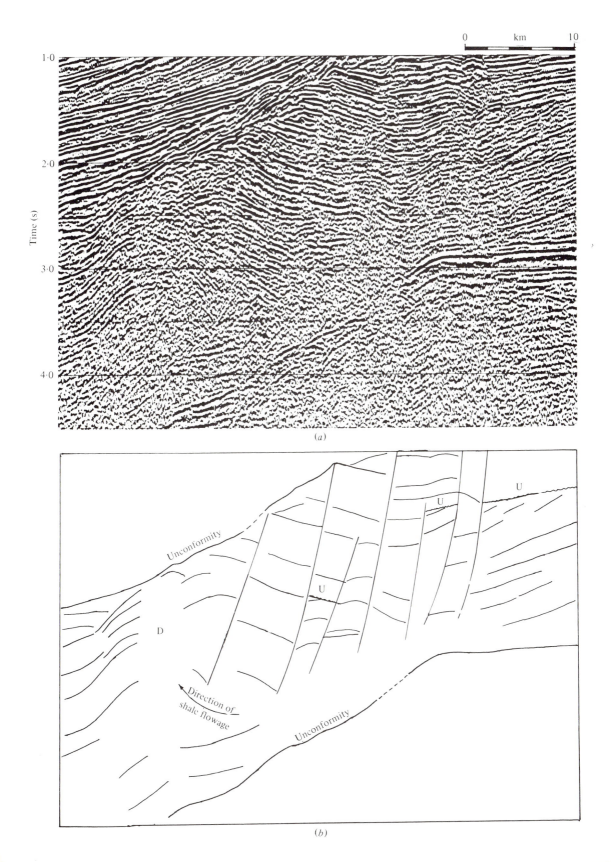

0 km 10

(a)

Time (s)

1·0
2·0
3·0
4·0

Unconformity

D

U

U

U

Direction of
shale flowage

Unconformity

(b)

(*b*) *Faulting example.* The two record sections in fig. 9.22 join at their north and west ends at right angles. On the N–S section (fig. 9.22*a*) the reflection band consisting of four strong legs marked Σ can be correlated readily across the normal fault. This event is downthrown to the south by slightly less than 2 cycles (about 65 ms) at its 1.6 s arrival time; at a velocity of 2300 m/s this represents a vertical throw of about 75 m. The exceptionally strong event near 2.3 s (marked χ) indicates a throw of about 3 cycles (about 120 ms, the dominant frequency having become slightly lower); at a velocity of 3000 m/s this represents 180 m of throw so that the fault appears to be growing rapidly with depth.

Although the evidence suggests that the fault is a simple break in the shallow section, at greater depths there may be a fault zone with subsidiary faults (shown dashed in the figure). If the deeper correlations across the fault(s) are correct, the downthrown event Ω at 3.5 s is found at around 2.9 s on the upthrown side; assuming a velocity of 3500 m/s at this depth, we get a vertical throw of 1050 m.

The correlation across the fault for the shallow event in fig. 9.22*a* is based on reflection character; for the deeper event it is based upon intervals between strong reflections, systematic growth of throw with depth, and time-ties around loops. Sometimes the displacement of an unconformity or other recognizable feature will indicate the amount of throw. Often, however, the throw cannot be determined clearly from the seismic data.

If the data in fig. 9.22 are transformed into a depth section, we get fig. 9.23. The components of fault-plane dip in fig. 9.23 are around 55° and 48°. Note that a fault which is nearly straight on a depth section is concave upwards on a time section because of the increase in velocity with depth. If the fault surface is actually concave upwards, the curvature will be accentuated on a seismic section. Where the fault was most active (indicated by the most rapid growth in fault throw), the fault surface is most curved.

The fault has not completely died out by the north end of the line and hence the fault trace should appear on the intersecting line (fig. 9.22*b*). As picked on the E–W section, the fault offsets the event at 1.6 s by only about 30 m, indicating that the fault is dying out rapidly toward the east. The fault plane has nearly as much dip in the E–W section so that the strike of the fault plane near the intersection of the two lines is NE–SW and the fault plane dips to the southeast. The true dip of the fault plane is about 62° (the apparent dip on sections is always less than the true dip unless the line is perpendicular to the strike of the fault). Fault indications are not evident below about 2 s on the E–W section so that the fault appears to have died out at depth towards the east. In poorly consolidated sediments such rapid dying out of faults is common. In this instance we are dealing with a radial fault from a deep salt-cored diapir located just south and slightly west of these lines; such radial faults often die out rapidly with distance from the uplift.

(*c*) *Evidences for faulting.* A number of the more common faulting evidences can be seen in the foregoing example. Several diffractions can be identified along the fault trace in fig. 9.22*a* between 1.9 and 2.5 s. If we had been dealing with migrated sections these diffractions would have been nearly collapsed (but not completely because the fault is not perpendicular to the lines). Terminations of events

Fig.9.20. Salt structures grow basinward as mother salt increases in thickness. (From Trusheim, 1960.)

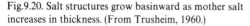
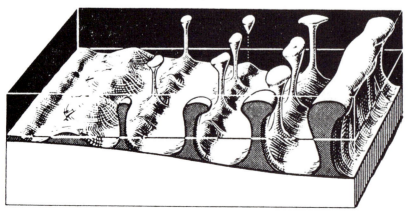

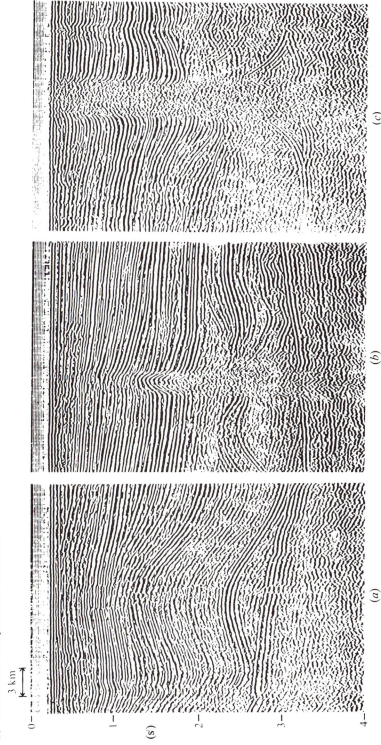

Fig.9.21. Portions of a section in the North Sea. (Courtesy Seiscom Delta.) (*a*) Salt swell or pillow; (*b*) saltdome which has pierced some of the sediments; (*c*) saltdome which has pierced to the seafloor.

Fig.9.22. Intersecting unmigrated seismic sections showing faulting. (Courtesy GSI.) (*a*) N–S section; (*b*) E–W section.

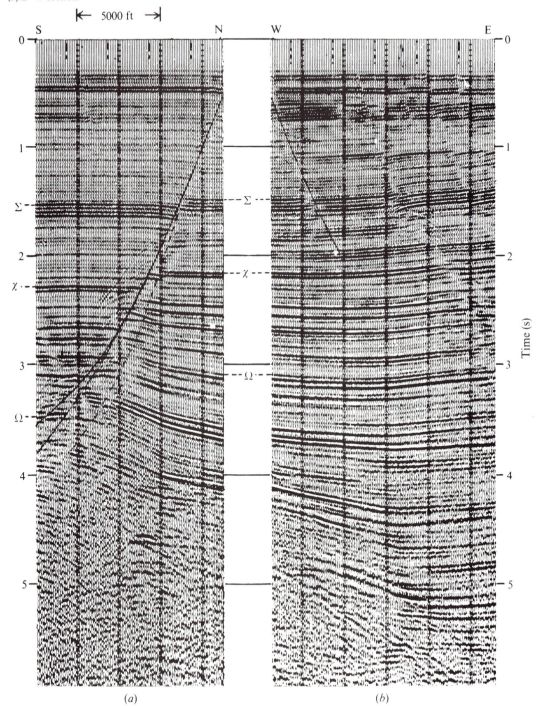

(*a*)   (*b*)

and offset of reflections (and non-reflection zones) across the fault are other important faulting evidences.

Different reflection dips are often seen on the two sides of the fault. Some of these dip changes are real, involving slight rotation of the section as the fault moved along a slightly-curved fault plane, drag and other real phenomena. On the other hand, some (especially those in the upthrown block) are distortions resulting from ray-path bending (refraction) in passing through the fault plane, because there are local velocity changes at the fault. While the upthrown sediments are most apt to have the higher velocity at any given level, the polarity and magnitude of the contrast vary down the fault plane as different units are juxtaposed against other units, so that the nature of the distortion varies from one place to another. In fact, the distortions may be so great and may change so rapidly as to cause marked deterioration of data quality below the fault, sometimes so great that reflections are almost entirely absent (a 'shadow zone') below the fault.

Occasionally the fault plane itself generates a reflection but generally the fault plane is a highly variable reflector because of the rapid changes in velocity contrast along the fault plane. Also, faulting is often distributed over a zone and involves many fracture surfaces. Furthermore, most reflection recording and processing discriminate against fault-plane reflections because of the use of arrays and of stacking velocities which do not optimize such events. In addition, fault-plane reflections are usually displaced an appreciable distance from the fault on unmigrated data and often the traveltime to them is so great (because of the long slant path) that they are not recorded and processed. Many of the above evidences for faulting can be seen on fig. 9.24, a growth fault of the type illustrated in fig. 9.17. Other faults are also present on fig. 9.24.

(d) *Characteristics of faulting.* Faults are produced by unbalanced stresses which exceed the strength of rocks, the type of fault depending largely on whether the vertical or horizontal stresses are the larger (fig. 9.25). Normal faults result when the maximum compressive stress is vertical and the minimum horizontal (fig. 9.25a), often with dip of the order of 50°–60°. When the maximum compressive stress is horizontal, thrusts result, often with a fault plane dip of 30°–40°. Where the maximum and minimum stresses are both horizontal, wrench faults result, the faulting often being at about 30° to the maximum stress direction.

Because velocity ordinarily increases with depth, the same vertical distance is represented by less time as depth increases; in consequence a post-depositional fault with constant attitude and throw usually has concave-upward curvature on a time section. Furthermore, constant throw is represented by fewer wavelengths as depth and velocity increase and so constant-throw faults appear to die out with depth on a time section. Post-faulting compaction with increased depth of burial also produces concave-upward curvature. Thus, fault traces are rarely straight on seismic sections.

The locations of faults often are determined by underlying features. An underlying uplift places the overlying sediments under tension as distances are stretched to accommodate the drape over the structure. Graben faulting forms to relieve the stress, but if the uplift is three-dimensional rather than two-dimensional, radial faulting is also required (fig. 9.8). Normal faulting commonly accommodates stretching above a hinge line or at a shelf edge where the basin side of the shelf edge subsides more rapidly (fig. 9.19). The location of faulting can be the key to underlying features, and conversely the underlying features can aid in connecting sometimes-confusing faulting evidences into a probable pattern.

9.3.3   *Folded and flow structures*

When subjected to stress, rocks may fault, fold or flow, depending on the magnitude and duration of the stresses, the strength of the rocks, the nature of adjacent rocks, etc. The folding of rocks into anticlines and domes provides many of the traps in which oil and gas are found.

Figure 9.26 shows a migrated seismic section across an anticline. Some portions such as *A* which are composed of the more competent rocks (for example limestones and consolidated sandstones) tend to maintain their thickness as they fold. Other portions such as *B*

Fig. 9.23. Depth sections of faults and mapped horizons of fig. 9.22. (a) N–S section; (b) E–W section.

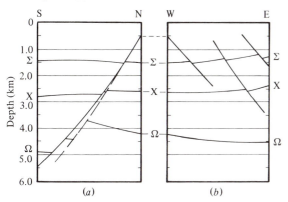

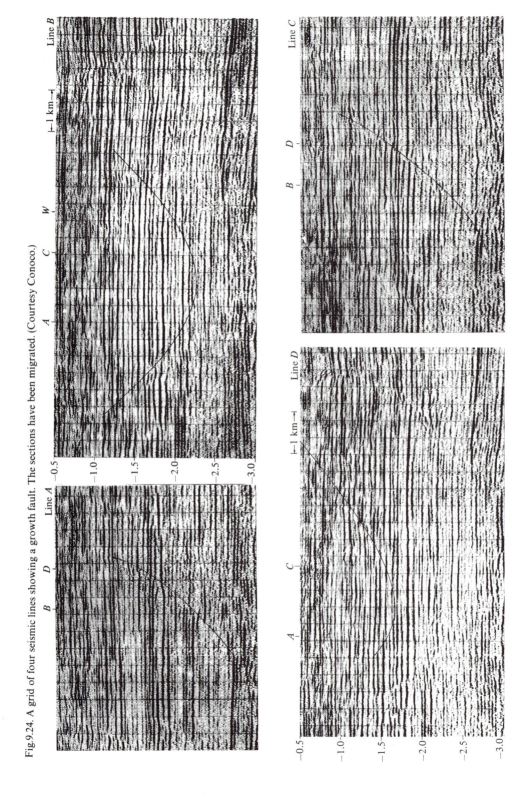

Fig.9.24. A grid of four seismic lines showing a growth fault. The sections have been migrated. (Courtesy Conoco.)

Fig.9.25. Types of faulting depend on the directions of
the maximum and minimum stresses. (*a*) Normal
faulting; (*b*) thrust faulting; (*c*) strike-slip faulting.

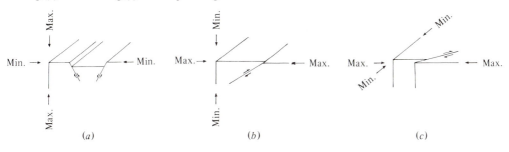

(*a*)   (*b*)   (*c*)

Fig.9.26. Migrated seismic section showing anticlinal
structure in the Central Valley of California. (Courtesy
Getty Oil and Geocom.)

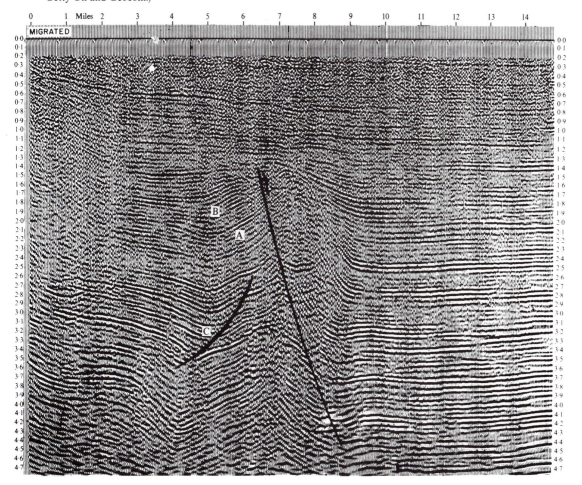

which contain less competent rocks (often shales and evaporites) tend to flow and slip along the bedding, resulting in marked variations in thickness within short distances. Geometry places limits on the amount of folding which is possible and folded structures almost always involve faulting (fig. 9.16c). Note at *C* on fig. 9.26 how a fault is involved with the folding.

Arching causes extension so that often the sediments break along normal faults and produce graben-type features on the top. Folding must disappear by faulting or flowage at some depth. Anticlinal curvature tends to make seismic reflections weaker as well as increase the likelihood of faulting and flowage, so that data quality commonly deteriorates over anticlines. Such deterioration is evident in fig. 9.27.

Salt flow often produces anticlines and domes. In many parts of the world thick salt deposits have been buried fairly rapidly beneath relatively unconsolidated sediments. The sediments compact with depth and so increase their density whereas the salt density remains nearly constant. Thus, below some critical depth the salt is less dense than the overlying sediments. Salt behaves like a very viscous fluid under sufficient pressure, and buoyancy may result in the salt flowing upward to form a

saltdome, arching the overlying sediments and sometimes piercing through them (fig. 9.21). Piercement does not necessarily imply uplift, however, because subsidence of the sediments surrounding a salt plug accomplishes nearly the same structural result. Often the velocity in 'uplifted' rocks is nearly the same as that in laterally-adjacent non-uplifted rocks, implying that neither was ever buried deeper; if they had been, they would have irreversibly lost porosity and attained a higher velocity.

Grabens and radial normal faults (whose throw decreases away from the dome) often result from arching of the overlying sediments (fig. 9.8), to relieve the stretching which accompanies the arching. Saltdomes tend to form along zones of weakness in the sediments, such as a large regional fault. The side of a saltdome may itself be thought of as a fault.

Figure 9.28 shows a seismic section across a saltdome. Shallow saltdomes are apt to be so evident that they can scarcely be misidentified. Because of the large impedance contrast, the top of the saltdome (or caprock on top of the dome) may be a strong reflector. Steep dips may be seen in the sediments adjacent to the saltdome as a result of these having been dragged up with the salt as it flowed upward. The sediments often show rapid thinning

Fig.9.27. Section in Ardmore Basin of Oklahoma (single-fold coverage). (Courtesy GTS.)

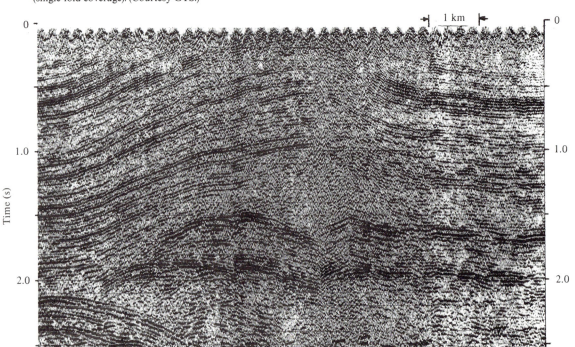

Fig.9.28. Unmigrated section across a saltdome of intermediate depth. Note the down-toward-the-dome normal (graben) faulting over the dome. The thinning of section as the dome is approached and the location of the rim syncline with depth can be used to work out the history of the dome's growth. The uplift area and conflicting dips toward the right-hand side of this section probably indicate another dome off to the side of the seismic line (dome sideswipe). (Courtesy Western Geophysical.)

1 km

toward the dome. The salt itself is devoid of primary reflections, although multiples often obscure this feature, especially if AGC is used.

Defining the flank of a saltdome precisely is often important economically and at the same time difficult seismically. Frequently, the oil is in a narrow belt adjacent to the flank of the dome but since the flank is usually nearly vertical it rarely gives rise to a recognizable reflection. Fortunately, the velocity distribution is often only slightly affected by the growth of the dome (except for the velocity in the salt and the caprock) so that the steep dips of the sediments adjacent to the flanks can be migrated fairly accurately and the flank outlined by the terminations of these reflections. Nevertheless, there remains much art and experience in defining saltdome flanks.

The salt in a saltdome generally has come from the immediately surrounding region. The removal of the salt from under the sediments around the dome has allowed them to subside, producing a rim syncline. The seismic data over such synclines are often very good and aid in mapping the adjacent dome by indicating the volume of salt involved, when movement took place (by sediment thickening), etc. Such synclines may also help provide closure on neighboring areas where the sediments continue to be supported by residual salt.

Figure 9.21 is a portion of a section in the North Sea (the horizontal scale has been compressed so as to display a long line on a short section, producing considerable vertical exaggeration). This line shows deep salt swells which have not pierced through the overlying sediments (fig. 9.21a), salt which has pierced through some of the sedimentary section (fig. 9.21b), and also salt which has pierced all the way to the seafloor (fig. 9.21c). The reflection from the base of the salt is generally continuous and unbroken but distortions produced by the variable salt thickness above it at times interrupt this reflection. Because the salt velocity is greater than that of the adjacent sediments, the base of the salt event appears to be pulled up where the salt is thicker. In other areas where the salt velocity is lower than that of the surrounding sediments, flat reflectors beneath the salt may appear to be depressed where the overlying salt is thicker.

Figure 9.29 shows a salt uplift at a shelf edge. Most of the salt movement occurred prior to the unconformity U_2, although the right side continued to subside somewhat even after U_1, producing the monocline in the shallower sediments. Note the graben faulting seen most prominently around 2.0 s.

Occasionally substances other than salt form flow structures. Poorly consolidated shale may flow, forming structures which strongly resemble saltdomes on reflection sections; also, at times shale flows along with salt, producing a saltdome with a sheath of shale. Magma also sometimes flows into a sedimentary section to produce structural uplifts, including piercement domes.

9.3.4 *Reefs*

The term 'reef' as used by petroleum geologists comprises a wide variety of types, including both extensive barrier reefs which cover large areas and small isolated pinnacle reefs. It includes carbonate structures built directly by organisms, aggregates comprising limestone and other related carbonate rocks, as well as banks of interstratified carbonate (and sometime also noncarbonate) sediments. Reef dimensions range from a few tens of meters to several kilometers, large reefs being tens of kilometers in length, a few kilometers wide and 200–400 m or more in vertical extent. Some reefs grow at the boundary between different environments, such as the shelf-margin and barrier types shown in fig. 9.30, whereas others, such as patch and pinnacle reefs, are surrounded by the same environment.

We shall describe a model reef so that we may develop general criteria by which reefs can be recognized in seismic data, keeping in mind that deviations from the model may result in large variations from these criteria. Our model reef forms in a tectonically-quiet area characterized by flat-lying bedding which is more-or-less uniform over a large area. The uniformity of the section makes it possible to attribute significance to subtle changes produced by the reef which might go unnoticed in more tectonically-active areas. The reef is the result of the build-up of marine organisms living in the zone of wave action where the water temperature is suitable for sustaining active growth. The site of the reef is usually a topographic high which provides the proper depth. Although the topographic high may be due to a structure in the underlying beds or basement, such as a tilted fault block, more often it is provided by a previous reef; as a result, reefs tend to grow vertically, sometimes achieving thicknesses of 400 m or more and thereby accentuating the effect on the seismic data. In order for the reef to grow upward, the base must subside as the reef builds upward, maintaining its top in the wave zone as the sea transgresses. The reef may provide a barrier between a lagoonal area (the *backreef*) and the ocean basin (the *forereef*), so that sedimentation (and consequently the reflection pattern) may be different on opposite sides of the reef. The surrounding basin may be *starved* (that is, not have sufficient sediments available to keep it filled at the rate at which it is subsiding); at times only one side, more often

Fig.9.29. Migrated section across a fairly deep saltdome.
Note evidences for unconformities (U_1, U_2). Stacking
velocity data for this line were plotted in fig. 8.14.
(Courtesy Seiscom Delta.)

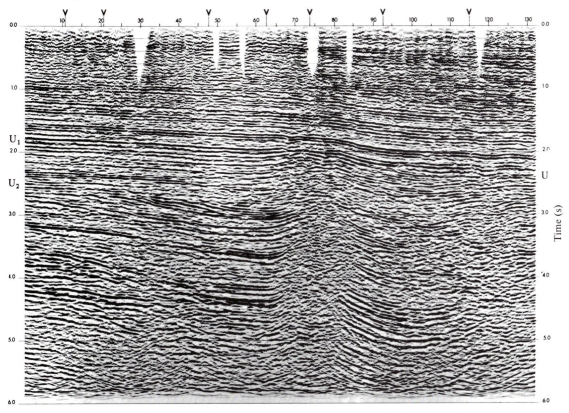

Fig.9.30. Types of reefs. (From Bubb and Hatlelid,
1977.)

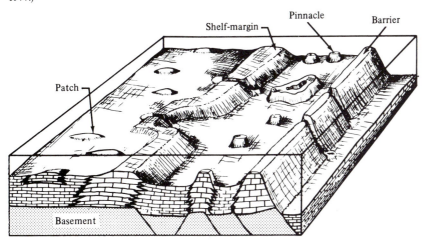

the ocean side of the reef, may be starved. Alternatively, the reef may not be a barrier to movement of the sediments and in this case it will be surrounded by the same sediments. Erosion of the reef often provides detritus for deposition adjacent to the reef, resulting in *foreset beds* with dips up to 20°, but usually with smaller dips, often of only 1–2°. The primary reef may possess considerable porosity which makes it a good potential reservoir rock, but other organisms such as sponges will penetrate into and replace much of the reef rock, altering the porosity in the process. Usually the actual *biohermal* portion of the reef (the portion produced by reef-building organisms) cannot be distinguished from other portions except by the examination of samples, and the entire complex associated with the reef is called the 'reef', only perhaps 15% of which is biohermal. Eventually the environment for the reef organisms will change so that they can no longer continue to live and build the reef; this might come about because of changes in the water temperature, an increase in the rate of subsidence so that the organic build-up cannot keep pace (called *drowning* of the reef), or various combinations of circumstances. Subsequently the reef may become buried by deep-water shales which may provide both an impermeable cap to the porous reef and sufficient hydrocarbons that the reef becomes a petroleum reservoir. Additional sediments may continue to be deposited, their weight compacting the sediments which surround the reef more than they compact the relatively rigid reef; thus the overlying sediments which were deposited flat may develop a *drape* over the reef. The interior of the reef may be more porous and less rigid than the edges so that some differential compaction may occur over the reef itself.

Based on the foregoing model, we develop the reef criteria illustrated in fig. 9.31. The reef may be outlined by reflections (fig. 9.31a) but its interior is apt to be a reflection void (fig. 9.31b). We may see diffractions from the top and/or flanks of the reef (fig. 9.31c). Abrupt termination of reflections from the surrounding sediments may indicate the location of the reef (fig. 9.31d). If the reef provided a barrier to sedimentation, the entire reflection pattern may differ on the two sides of the reef reflecting the different sedimentary environments (fig. 9.31e). Overlying reflections may show small relief (usually only a few milliseconds in magnitude) because of differential compaction, the effect decreasing with distance above the reef (fig. 9.31f). A velocity difference between the reef materials and the surrounding sediments may cause the traveltime to flat-lying reflections below the reef to vary (Davis, 1972), and this velocity difference may produce pseudo-structure on reflecting horizons below

the reef. Usually the velocity in the reef limestone is greater than that in the surrounding shales so that the reef may be indicated by time thinning between reflections above and below the reef and by a pseudo-high under the reef (fig. 9.31g); the magnitude of such an anomaly is small, usually less than 20 ms. Sometimes, however, the reef may be surrounded by evaporites or other rocks with higher velocity than that of the porous reef limestone so that the time anomaly is reversed (fig. 9.31h). The hinge-line or high which localized the reef development may also be detectable (fig. 9.31i,j).

Reef evidences are often so subtle that seismic mapping of reefs is feasible only in good record areas. Of importance is geologic information about the nature of the sediments and the environment of deposition, so that one knows beforehand in what portion of the section reefs

Fig. 9.31. Criteria for reef identification. (After Bubb and Hatlelid, 1977.) (a) Reef outlined by reflections, (b) by reflection void, (c) by diffractions from reef edges, (d) by abrupt termination of reflections, (e) by change in reflection pattern on opposite sides of the reef, (f) by differential compaction over the reef, (g) by velocity anomaly underneath the reef where $V_{reef} > V_{surrounding}$ or (h) where $V_{reef} < V_{surrounding}$; (i) reef located on hingeline and (j) on structural uplift.

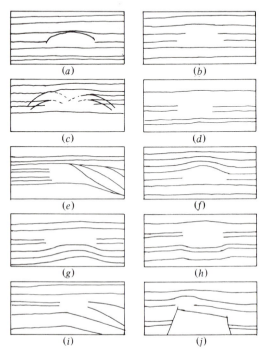

are more likely to occur. Subtle features in the part of a seismic section where reefs are expected may be interpreted as reefs whereas similar features elsewhere are ignored.

Similarities between reefs and salt features cause problems at times. The lagoonal areas behind reefs often provide proper conditions for evaporite deposition so that salt is frequently present in the same portion of the sedimentary column. The amount of salt may not be thick enough to produce diapiric (flow) features but differential solution of salt beds followed by the collapse of the overlying sediments into the void thus created may produce seismic features which are similar in many ways to those which indicate reefs.

A seismic line across a barrier reef is shown in fig. 9.32; note the change in reflection pattern across the reef, the differential compaction and velocity uplift evidences, and the change in regional attitude of reflections beneath the reef which indicates a weak hinge-line. A line across patch reefs is shown in fig. 9.33. Patch reefs are usually much smaller in vertical extent than these and, consequently, often difficult to locate.

9.3.5 *Unconformities and channels*

Unconformities represent a missing sequence of rock, a time period during which rocks were being eroded away, or at least not deposited. Conditions probably changed during the hiatus, so that the nature of the sediments above the unconformity are often different from those below and an acoustic-impedance contrast exists at the unconformity. Hence unconformities are usually good reflectors. They frequently involve some angularity between the bedding below and above and this also tends to make them stand out as reflectors. The result is that unconformities are often among the easiest and most distinctive reflectors to map. On the other hand, the rocks which an unconformity separates often vary from one location to another so that the contrast at the interface changes and hence the unconformity reflection varies in amplitude (see fig. 9.34) and sometimes even in polarity (see fig. 9.35). There may be large regions over which the beds above and below the unconformity parallel the unconformity so that there is no angularity to distinguish the unconformity reflection from other reflections. In such regions, the unconformity has to be

Fig.9.32. Section across Horseshoe Atoll in West Texas. *R* denotes the portion of the section which contains the reef (just left of center). The backreef area of flat-lying, strong, continuous reflections (*A′*) is to the right; the forereef showing an entirely different reflection pattern (*A*) is to the left. (Courtesy Conoco.)

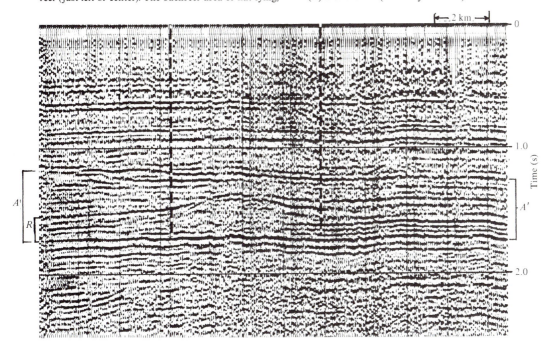

mapped by correlating it along the bedding with places where the unconformity can be identified by angularities on the seismic section or where it can be identified from well or other types of data.

Rather prominent unconformities can be seen on figs. 8.26 and 9.29, mainly evidenced by angularities and fairly strong reflections from the unconformities themselves.

Various types of hydrocarbon traps are associated with unconformities – both (1) pinchouts and truncations of reservoir beds below the unconformity where the unconformity constitutes a seal, and also (2) stratigraphic variations in the sediments laid down on the unconformity (see also fig. 9.38). Unconformities are involved in most stratigraphic traps. Streams flowing across the unconformity surface may have eroded valleys into the surface and the stream deposits may constitute the reservoir or sometimes the seal.

Ancient stream channels are involved with a number of oil and gas accumulations in the Central United States and elsewhere. The relief associated with large river valleys may be sufficient to give structural evidences but most often the seismic effects will be slight.

Figure 9.36 represents a shallow seismic time slice (see figs. 5.35 and 5.36) which shows the pattern of a meandering stream, suggesting that evidence of very minor features is contained in seismic data if we can but find an economical, feasible way of extracting it. An example of a search for accumulations associated with stream channels is discussed in §9.7.4.

Profiler surveys in deep water often reveal channel cuts and fill (fig. 9.37) which indicate that channels are important in the deep marine environment as well as on land. Turbidity currents at places have clearly eroded deep channels and have sometimes built up extensive levee systems under deep marine conditions. (A turbidity current is a density current in water caused by different amounts of solids in suspension; they are important in submarine erosion and deposition.) Some channels undoubtedly result from sealevel lowering and some from marine erosional processes. Brown and Fisher (1980) develop a 'destructive shelf' concept relating the erosion of channels into shelves to the availability of new material for deposition: this concept has helped discover accumulations in fans on the slopes offshore Brazil.

Fig.9.33. Two patch reefs in the Etosha Basin of Southwest Africa. *C* denotes the carbonate portion of the section, *B* the base of the reefs. The region of reefs is indicated approximately by the arrows below the section. The reef to the left has about 85 ms thickness (210 m), the one to the right 120 ms (300 m). (Courtesy Etosha Petroleum.)

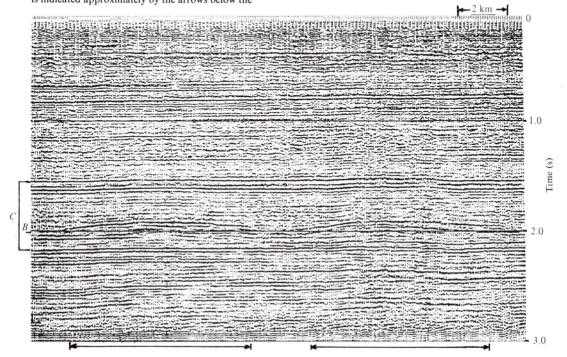

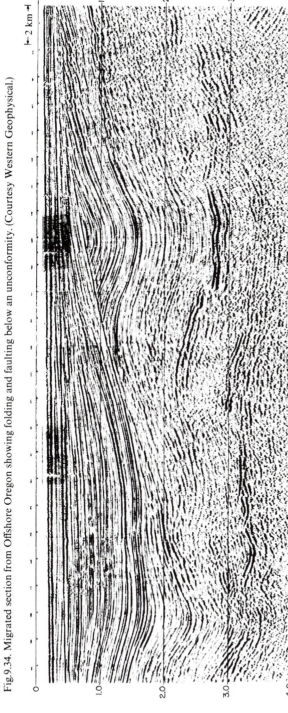

Fig.9.34. Migrated section from Offshore Oregon showing folding and faulting below an unconformity. (Courtesy Western Geophysical.)

9.3.6 *Stratigraphic traps*

Rittenhouse (1972) gives a classification scheme for stratigraphic traps, summarized in table 9.2. His first-order division depends on whether the traps are or are not adjacent to unconformities. Illustrations of some of these are given in fig. 9.38.

Fig.9.35. Portion of migrated section showing change in polarity of unconformity reflection UU' because of change in velocity of subcropping beds. (Courtesy Seiscom Delta.)

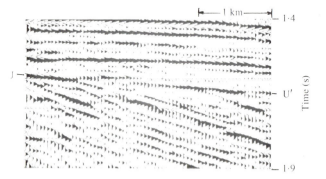

Fig.9.36. Horizon-slice map showing reflectivity variations along the same reflector. The three-dimensional data set is sliced along one reflection, much as a time-slice map (figs. 5.35 and 5.36) is sliced along constant arrival time. The pattern shows a meandering stream channel. (From Brown *et al.*, 1981.)

A large portion of the hydrocarbons remaining to be found probably involve stratigraphic traps at least in part. Marr (1971), Lyons and Dobrin (1972) and Dobrin (1977) discuss seismic evidences of stratigraphic traps drawn from published case histories. The picture they painted was discouraging; most stratigraphic accumulations documented in the literature were found by searching for something else, that is, by serendipity. While this is the conclusion to be drawn from the literature, we know of deliberate searches for stratigraphic traps which have been successful but their stories are not in the public domain. We believe that present techniques combined with well data and geologic insight are adequate for the search under many circumstances and we believe that improvements in techniques will be forthcoming which will expand considerably the circumstances under which this can be done successfully.

Sheriff (1980) includes the following quote:

Nevertheless, stratigraphic case histories had one important moral. While the discovery of stratigraphic accumulations was not generally attributed to a sound exploration program, the genius lay in being alert when a surprise occurred. Often

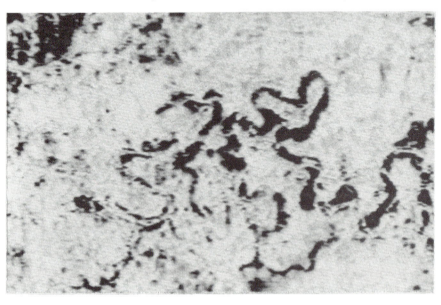

0   5

km

the surprise occurs in the record from a borehole; some portion differs from what we expected in such a way as to suggest the possibility of a stratigraphic trap nearby. But where? This is where reflection-character analysis comes into its power; it can help us locate the nearby accumulation that the unexpected in a well suggests. It can help us search for stratigraphic traps directly rather than relying on luck and statistics.

9.3.7   *Integration with other geophysical data*

An interpreter should utilize all data available to him in his interpretation. Other types of geophysical data, especially gravity and sometimes magnetic data, may be available and should be examined to see if they suggest anything not otherwise evident and whether the gravity and magnetic fields are consistent with the mapped features. The nature of a diapir is not always evident from examination of seismic data alone and other data may reduce the ambiguities. A gravity model can be constructed from a seismic structural interpretation by assigning density values to various portions of the section; the gravity field calculated from this model then can be compared with the measured gravity field. Such a comparison may reveal areas of disagreement which call for a re-examination of the seismic interpretation. The depth of basement, which may not be evident from seismic data, may be indicated by magnetic data. Refraction velocities may help in the identification of the nature of certain reflectors. Especially where seismic record quality is poor, such as in areas of karst or volcanics on the surface, magnetotelluric soundings may be useful in reducing interpretational ambiguities.

9.4   Modeling
9.4.1   *Introduction*

Interpretation of seismic data invariably involves a conceptual 'model' of the portion of the Earth involved in seismic measurements. The model is a simplification of the actual Earth in which the only elements included are those expected to be most important in affecting the measurements. For example, the identification of stacking velocity with rms velocity is based on a model in which velocity does not vary in the horizontal direction, and statics corrections are based on a model in which travel through the weathering is vertical regardless of raypath direction below the weathering. A model may be an actual physical model, mathematical expressions, or merely a rather vague mental picture.

Modeling is often subdivided into two types, forward and inverse. *Forward or direct modeling* involves computing the effects of a model and *inverse modeling* involves calculating a possible model from observation of effects. Inverse modeling in a sense includes the entire interpretation process and invariably involves uncertainty and ambiguity. Often 'modeling' without a preceding adjective implies forward modeling.

In forward modeling expected values are calculated from the model and compared with actual measurements, differences ('errors') being attributed to either inaccuracies in the model or factors not accounted for. Modeling is usually iterative, the model being altered in an effort to account for the error, a new error calculated, etc., until the error has been reduced to what is considered acceptable. Adequate agreement, however, does not 'prove' that the model corresponds to the actual Earth; a different model might also provide adequate agreement.

Fig.9.37. Tracing of a profiler section showing channel cut and fill. A present-day channel can be seen in the seafloor reflection with indications of a natural levee to the left of it, both being created in about 750 m of water. Other earlier channels and levees are also indicated.

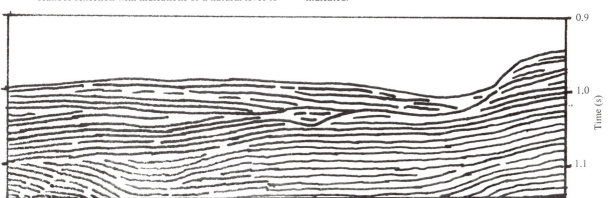

Time (s)

0.9

1.0

1.1

9.4.2 *Physical modeling*

Many geologic phenomena are too complicated to be amenable to theoretical treatment, hence modeling sometimes involves experiments with miniature physical models (fig. 9.39). However, models must be geometrically, kinematically and dynamically similar to the systems being modeled (Hubbert, 1937) if the results are to be useful. Geometric similarity is achieved by making angles in the model equal to those in the system and corresponding lengths proportional. If λ is the ratio of lengths, the

Fig.9.38. Some types of stratigraphic traps. Reservoir rocks are dotted, impermeable rocks are clear. (After Rittenhouse, 1972.) (*a*) Accumulation of sands on flanks of growing structure resulting from winnowing and lateral transport of sand; (*b*) sand body formed at edge of shelf resulting from lowering of sea level; (*c*) accumulation of sand over growing structure resulting from winnowing; (*d*) reservoir beds subcropping at an unconformity; (*e*) trapping against impermeable sediments in valley fill; (*f*) reservoir sediments in valley (or canyon) fill; (*g*) trapping against hill or other topography on unconformity; (*h*) accumulation against lake or coastal cliff; (*i*) reservoir sands onlapping an unconformity; (*j*) accumulations subcropping at sides of valley fill.

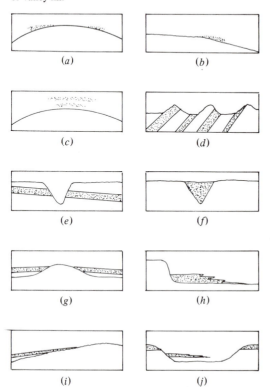

(*a*)   (*b*)

(*c*)   (*d*)

(*e*)   (*f*)

(*g*)   (*h*)

(*i*)   (*j*)

ratios of areas and volumes are proportional to λ^2 and λ^3. Kinematic similarity concerns the ratio of times, τ, required to effect similar changes in position or shape. The ratios of velocities and accelerations will be λ/τ and λ/τ^2, respectively, and angular velocity and angular acceler-

Table 9.2. *Classification of stratigraphic traps. (After Rittenhouse, 1972.)*

(I) Not adjacent to unconformities
    (*A*) Facies-change traps involving current-transported reservoir rock
        (1) Eolian (dunes or sheets)
        (2) Alluvial fan
        (3) Alluvial valley (braided stream, channel fill, point bar)
        (4) Deltaic (distributary mouth or finger bars, sheet, channel fill)
        (5) Non-deltaic coastal (beach, barrier bar, spit, tidal delta or flat)
        (6) Shallow marine (tidal bar, sand belt, washover, shelf edge, shallow turbidite or winnowing)
        (7) Deep marine (marine fan, deep turbidite or winnowing)
    (*B*) Non-current-transported reservoir rock
        (1) Gravity (slump)
        (2) Biogenic carbonate (shelf-margin reef, patch reef, algal build-up or blanket)
    (*C*) Diagenetic traps
        (1) Change from non-reservoir to reservoir
            (*a*) Replacement and leached (dolomitized)
            (*b*) Leached
            (*c*) Brecciated
            (*d*) Fractured
        (2) Change from reservoir to non-reservoir
            (*a*) Compaction (physical or chemical)
            (*b*) Cementation
(II) Adjacent to unconformities
    (*A*) Traps below unconformities
        (1) Seals above unconformity
            (*a*) Subcrop at unconformity
            (*b*) Topography (valley flank or shoulder, dip-slope, escarpment, valley, beveled)
        (2) Seal below unconformity
            (*a*) Mineral cement
            (*b*) Tar seal
            (*c*) Weathering product
    (*B*) Traps above unconformities
        (1) Reservoir location controlled by unconformity topography
            (*a*) On two sides (valley, canyon, fill)
            (*b*) On one side (lake or coastal cliff, valley side, flank of hill or structure)
        (2) Transgressive

ation ratios will be $1/\tau$ and $1/\tau^2$. Dynamic similarity concerns the ratio of mass distributions, μ; this fixes density ratios, μ/λ^3. Forces acting on corresponding mass elements must be such that the motions and changes in shape produced are geometrically and kinematically similar; force ratio is $\mu\lambda/\tau^2$. Dimensionless quantities (like Poisson's ratio) must have the same numerical value. Thus there are only three independent values, λ, τ and μ.

We might, for example, wish to represent 1 km in the Earth by 10 cm in a model, hence the ratio of model to actual distance is $\lambda = 10^{-4}$. In practice seismic velocity is restricted by available materials and the ratio of model to actual velocity can range only by a very small amount, that is, $\lambda/\tau \approx 1$. Since λ has already been selected, this restricts τ. If the model material has the same velocity as the Earth, $\tau = 10^{-4}$ and we must use frequencies 10^4 times what is used in the Earth (since frequency ratios depend

on $1/\tau$). The density of available model materials is probably about the same as that of Earth materials; since the density ratio $\mu/\lambda^3 \approx 1$, this determines the mass ratio $\mu \approx 10^{12}$. If we wish to model several types of things simultaneously, such as various modes of wave propagation, attenuation, etc., we must be sure that the relevant physical properties are consistent with our model ratios λ, τ and μ. Examples of physical modeling are shown in figs. 4.8 and 4.28.

9.4.3 *Computer modeling*

More commonly, modeling is done by computer and several examples were given in volume 1. Many types of algorithms are used in computer modeling, ranging from simply convolving a wavelet with a sequence of reflection coefficients, to tracing rays through models where the raypaths bend in accordance with Snell's Law,

Fig.9.39. Tank for seismic modeling at the University of Houston. Seismic models with horizontal dimensions of 30–60 cm and vertical dimensions of 5–10 cm made of layers of resins or other materials simulate three-dimensional layered structures; they are immersed in the water-filled tank. Sources and receivers are moved over them to obtain the seismic data, motions being controlled by a computer to simulate various field recording arrangements. (Courtesy Seismic Acoustics Laboratory of Univ. Houston.)

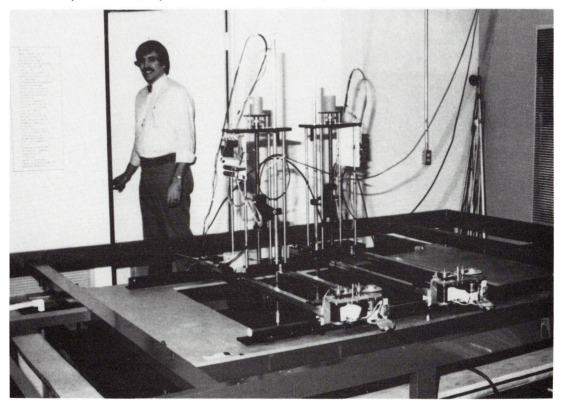

to full-waveform methods based on relations such as Kirchhoff's equation (2.34) or wave-equation methods such as used in migration (§8.3.3 or 8.3.4) which include diffractions (Gazdag, 1981). Synthetic seismograms (§9.4.4) may help in determining how stratigraphic changes might affect seismic records, and raypath modeling (§9.4.6) in determining the distortions which complicated velocity distributions produce. Where the algorithms and models are good, the resemblance to actual seismograms is good. Modeling is an invaluable pedagogical tool (see Hilterman, 1970), but it invariably involves assumptions and approximations which should not be forgotten when drawing conclusions.

9.4.4. *Synthetic seismograms*

(a) *One-dimensional synthetic seismograms.* A seismic trace, as expressed by the convolution model, eq. (8.29), is simply a wavelet convolved with the Earth's reflectivity, and we can determine the Earth's reflectivity if we know the Earth layering and the distribution of velocity and density. Thus we can use (8.29) to create a *synthetic seismogram* based on velocity and density values measured from borehole logs in a well and a knowledge of the wavelet shape. Wavelet-processing methods can give an estimate of the average wavelet. We assume that the wavelet impinges on the first interface where its energy is partitioned among transmitted and reflected waves. Each of the waves is then followed as it travels to other interfaces where additional waves are generated, and so on. The resulting seismic record is simply the superposition of those waves which are ultimately reflected back to the geophone station. Since Snell's law determines raypaths and Zoeppritz' equations determine energy relationships, the problem is completely determined and the solution is straightforward. However, actually solving the problem is a formidable task because of the tremendous number of waves generated for a realistic sequence of layers; for this reason various simplifications are made.

The most common simplification involves making a *one-dimensional synthetic seismogram* where only vertical travel is allowed and dip is neglected. Reflection and transmission coefficients are thus those for normal incidence. Diffractions and other modes are usually ignored, although multiples may be included. The data about the sequence of reflecting interfaces are usually derived from borehole logs. Often only sonic logs are available and density variations are either ignored or assumed to bear some functional relation to velocity (such as (7.2)). Small velocity variations are often lumped together into larger steps to reduce the number of interfaces to be considered, and sampling is usually on a

regular traveltime interval (rather than on a regular depth interval, as with logs). Amplitude-changing factors other than those involving reflection coefficients are often ignored.

The effects of multiples are sometimes incorporated as successive modifications of the propagating waveform (Vetter, 1981), although often they are ignored also. Consider two adjacent interfaces such that the one-way traveltime between them is the sampling interval, Δ, the reflection coefficients for a downgoing wave being R_i and R_{i+1}. If $w(t)_i$ is the downgoing waveform approaching the interface R_i, the upgoing waveform reflected by R_i will be $u(t)_i = R_i w(t)_i$. Neglecting the very slight loss on transmission through R_i, the downgoing wave at R_{i+1} will be $w(t + \Delta)_i$ plus the peg-leg multiples generated between R_i and R_{i+1}, that is,

$$w(t)_{i+1} = w(t + \Delta)_i - R_i R_{i+1} w(t + 3\Delta)_i$$
$$+ (R_i R_{i+1})^2 w(t + 5\Delta)_i - \ldots,$$

the second term on the right being the one-bounce peg-leg multiple, the next the two-bounce, etc. Thus we can include the effects of multiples by modifying the downgoing waveform at each step. We need to likewise modify the upgoing (reflection) waveform for the peg-leg multiples which it generates, and for its contribution to the downgoing wavelet, $w(t)$, as a result of reflections at the underside of interfaces.

The geologic model is usually derived from sonic logs. However, sonic logs do not always yield correct formation velocities because of inadequate penetration into the formation, the effects of hole caving, the picking of later cycles (cycle skips), incorrect depth measurement because of cable stretch, etc. Editing (§7.3.2) is usually essential in order to achieve a good match between synthetics and actual seismic data.

One-dimensional synthetic seismograms made from a single borehole log are useful in identifying reflections with particular interfaces and in distinguishing primary from multiple reflections. Seismic sections often involve time or phase shifts (including polarity reversal) of unknown magnitude (which are sometimes time-dependent) so that the ability to match synthetics to actual data adds considerable confidence to an interpretation. Figure 9.40 shows the synthetic seismogram procedure and fig. 9.41 shows a match to actual seismic data.

Where the match between synthetics and actual data is good, the reflecting sequence can be modified according to stratigraphic changes which might occur so that the effects of such changes on the seismic data can be ascertained. For example, we might assume that a sand unit shales-out in a facies change, that the fill in a stream channel differs from that in the adjacent uneroded formation, that the formation subcropping under an unconformity has changed, or that a small reef grew. Synthetics then give the interpreter a better idea of what to look for in order to locate the hypothesized facies change, channel, subcrop or reef. This usage of synthetics provides one of the main methods for the stratigraphic interpretation of seismic data (§9.7).

The model for manufacturing one-dimensional synthetic seismograms may actually be two- or three-dimensional. A model might simulate the changes expected along a seismic line connecting two or more wells where the changes between the wells are explained on the basis of facies changes, unconformities or faults. The synthetic seismogram from such a model is still considered one-dimensional, however, unless non-vertical travel paths are considered.

(*b*) *Two- and three-dimensional synthetic seismograms.* *Two-dimensional synthetic seismograms* not limited to vertical travel are required to model dipping reflectors,

diffractions, and the dependence of arrival time on offset. Sometimes only arrival times are modeled, as with the ray-tracing methods described in §9.4.6. Sometimes one attempts to model full-waveforms with correct amplitude relationships. The latter usually employ simplifications (such as the scalar form of the wave equation allowing for P-waves only). Trorey (1970, 1977) approximated reflectors by a series of semi-infinite plane strips and based his method on the Kirchhoff equation (2.34). Other methods utilize wave-equation methods of the type used in migration processing (§8.3.3 and §8.3.4). A common method is the *exploding reflector method*: each reflecting interface is assumed to be a distributed source detonated at time $t = 0$, the source density being proportional to the reflectivity at the interface; seismic waves are radiated upward at half the actual velocity (to give the traveltime for two-way travel). The record received at the surface simulates a common-depth-point section in many (but not all) regards. The waves may be tracked by any of the wave-equation methods. More elaborate methods allow for mode conversion, the variation of reflectivity with incident angle, surface waves, head waves, etc., but generally they are time consuming and expensive.

Fig.9.40. Procedure for manufacturing synthetic seismogram. A reflectivity model is constructed from geologic data, usually well data, and convolved with a wavelet which is often extracted from actual seismic data; the modeled trace is then compared with a trace from the actual data and the difference (error) between them determined. The error trace is often used to modify the model or sometimes the wavelet until the degree of match is judged adequate. (From Stommel and Graul, 1978.)

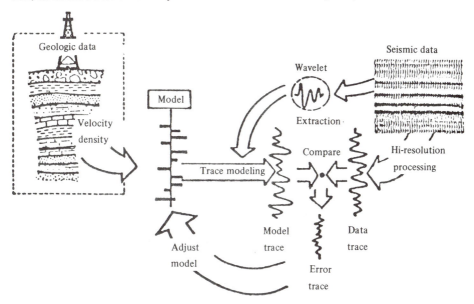

Fig.9.41. Synthetic seismogram (right half) compared to
actual seismic data (left half). A seismic wavelet shape
was extracted from each seismic trace and then
convolved with the reflectivity determined from a sonic
log and the assumption that the density obeyed
Gardner's rule (7.2). (Courtesy Seiscom Delta.)

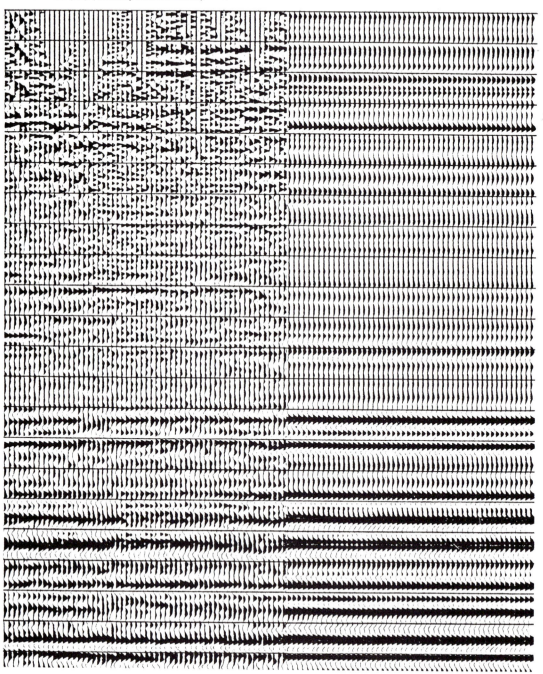

9.4.5   *Inversion to seismic logs*

We write the equation for the reflection coefficient at normal incidence (see (2.129)) as

$$R_i = \frac{Z_{i+1} - Z_i}{Z_{i+1} + Z_i} = \tfrac{1}{2}\Delta\{\ln(Z)\} = \tfrac{1}{2}k\,dA_i, \qquad (9.1)$$

where Z_i, Z_{i+1} are acoustic impedances on opposite sides of the interface and $\Delta\{\ln(Z)\}$ is the change in the logarithm of the acoustic impedance. The last portion of (9.1) assumes that the reflection amplitude change dA_i is proportional to the reflection coefficient R_i, the constant of proportionality k being called the *scaler*. This implies that the seismic trace amplitude is not influenced by noise or by the equivalent wavelet shape. Therefore inversion is usually preceded by processing which removes as much of the noise as is possible and by wavelet processing which attempts to remove wavelet-shape effects. Obviously these can not be accomplished perfectly and remaining noise and/or wavelet effects usually limit the success of inversion.

Equation (9.1) can be solved for the acoustic impedance below an interface in terms of that above the interface,

$$Z_{i+1} = \frac{1 + \tfrac{1}{2}kA_i}{1 - \tfrac{1}{2}kA_i} Z_i. \qquad (9.2)$$

If the density and velocity, ρ_1 and V_1, for the shallowest layer are known, then $Z_1 = \rho_1 V_1$ is known and Z_2 can be found from the amplitude of the first reflection A_1, Z_3 can be found from Z_2 and A_2, etc. The resulting recursively-derived value of acoustic impedance as a function of arrival time or depth is called a *seismic log* or *synthetic acoustic-impedance log*.

A more common way of solving for Z is as a time function $Z(t)$:

$$\int_0^t \Delta\{\ln Z(t)\} = \int_0^t d[\ln\{Z(t)\}]$$

$$= \ln\{Z(t)\} - \ln\{Z(0)\} = \int_0^t k\,dA(t);$$

$$Z(t) = Z(0)\exp\left\{k\int_0^t dA(t)\right\}. \qquad (9.3)$$

The 'constant' of integration, $Z(0)$, is often called the 'low-frequency component'. A seismic trace $A(t)$ is missing low-frequency components which are necessary to determine $Z(t)$. To partially compensate for the missing components, $Z(0)$ may be made time-dependent rather than a constant. A seismic trace is also missing high-frequency components and this limits the amount of detail and resolution achievable.

A seismic log is often expressed as a velocity – time series, that is, as a *synthetic sonic log*. This requires either knowledge of the density distribution or an assumption about the relation between the velocity and the density. Most commonly we assume either that the density is constant, in which case it drops out, or that Gardner's rule (7.2) holds. In the latter case we have

$$Z = \rho(V)V = aV^{\frac{5}{4}}$$

$$V(t) = V(0)\exp\{(4k/5)\int dA(t)\}, \qquad (9.4)$$

which is the same form as (9.3) except that a new constant, $4k/5$, replaces k.

Neither $V(0)$ nor the scaler k can be determined from the seismic trace. If an actual sonic log should be available, it may be used to determine these 'constants'. Stacking velocity may be used to determine $V(0)$; the procedure might be to (a) calculate $V(t)$ from the amplitude information with (9.4), (b) then the rms velocity $\bar{V}$ from $V(t)$ using (7.12), (c) calculate the stacking velocity V_s from normal moveout measurements by the procedures of §8.2.3 (in effect variations of (7.11)), (d) compare the results in (b) and (c), and (e) then iterate the procedure with different $V(0)$ and k values until the $\bar{V}$ and V_s values are close enough. The value of the scaler is usually guessed, the constraint being merely that $V(t)$ should have reasonable values.

The principal limitation with seismic logs is the assumption of a linear relation between reflectivity and amplitude, which in effect assumes a noise-free seismic record (showing the effect of primary reflections only) plus recording and processing that have preserved amplitudes faithfully. The successful manufacture of seismic logs requires excellent reflection records. The additional limitations involving the determining of $V(0)$ and k are minimized where the application is to interpolate velocity information between wells or to extrapolate velocity information in the immediate vicinity of well control. Seismic logs constitute a powerful tool for locating stratigraphic changes, porosity changes and hydrocarbon accumulation under these circumstances (Lindseth, 1979). Figure 9.42 shows synthetic sonic-log traces and an actual sonic log for comparison.

9.4.6.   *Ray–trace modeling*

Where velocity varies in other than a very simple way, the tracing of raypaths through a model obeying Snell's law at each velocity change is one way of developing an understanding of how a seismic section relates to a portion of the Earth where velocity complications exist. Horizontal changes in velocity especially can distort

Fig.9.42. Synthetic sonic logs, each derived from a different seismic trace. The vertical scale is linear with depth. The horizontal scale is linear with transit time (the inverse of velocity), the reference values for each trace moving according to the trace spacing; the scale for the heavy trace is indicated at the top. At the right one synthetic log is shown heavier along with an actual sonic log for comparison. (Courtesy Tecknica.)

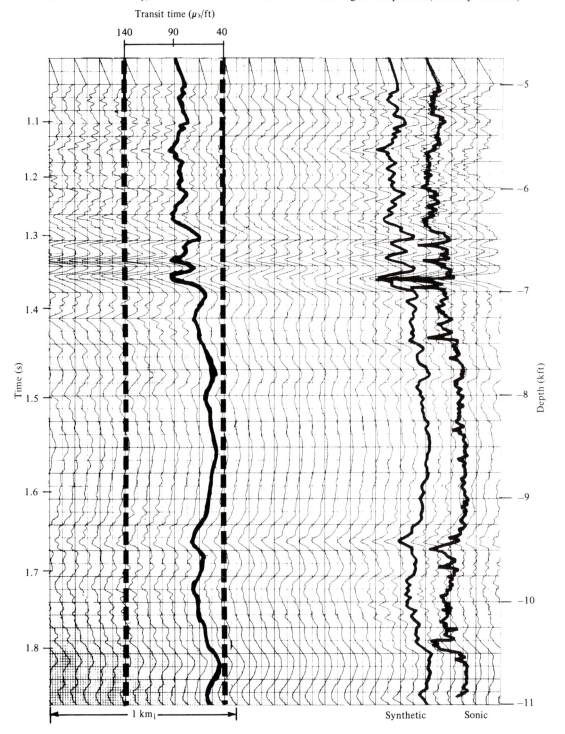

Transit time (μs/ft)

140   90   40

Time (s)

Depth (kft)

Synthetic   Sonic

structural pictures (§9.5) and make it difficult to appreciate the significance of structural evidences. Ray-path modeling is especially useful in seeing how stacking velocity measurements get distorted by velocity complications (May and Covey, 1981).

Taner *et al.* (1970) carried out ray-tracing for several models, one of which is shown in fig. 9.43. Downward ray-tracing is not feasible except on a trial-and-error basis because initially we do not know the starting direction. Where source and detector are coincident, as is assumed in common-depth-point sections, the raypath to and from a reflector must be coincident and so must strike the reflector at right angles. This makes it easy to trace rays upwards. If increments along the reflectors are constant, then the density of emergent raypaths at the surface will also indicate amplitude variations (neglecting diffraction effects). Note the buried-focus effects in regions *B*. Ray-trace modeling is also used in depth migration (see fig. 8.29*c*).

Taner *et al.* traced rays for non-normal incidence at reflectors in order to obtain simulated gathers on which to make velocity analyses (fig. 9.44). Ray-tracing for this situation is generally done iteratively by trial-and-error because with a common midpoint the reflecting point shifts updip with offset in a manner often not easily predicted. The stacking velocities determined for fig. 9.44 do not bear any simple relation to rms velocities, as was pointed out in §7.3.3, and in fact the curve of arrival time versus offset is not even a hyperbola so that values of the stacking velocity obtained from best-fit hyperbolae depend on the mix of offset data used in the calculation.

9.5   Lateral variations in velocity
9.5.1   *Gradual lateral changes*
Often velocity variations in the horizontal direction are sufficiently gradual that their effects can be treated as a

second-order correction. This situation is especially common in Tertiary basins filled mainly with clastic sediments which have not been subject to uplift. The horizontal variations often result from gradual changes in the lithology, for example as distance from the source of the sediments increases. Sometimes the vertical velocity function is changed slightly from location to location and the horizontal gradient is otherwise ignored in plotting data. A common variation of this technique is to map reflectors using a single function for the area and then adding location-dependent depth corrections to the mapped values.

Lateral effects also result from changes in the thickness of a water layer on top of the sediments. The change in velocity with depth effectively begins at the seafloor, variations within the water layer usually being insignificant. The velocity of the sediments is not affected greatly by the amount of water overburden; the difference between overburden and interstitial pressures is usually the factor determining velocity (see §7.1.5) and, since the water layer increases both pressures by the same amount, it does not change the differential pressure or velocity. Of course, the average velocity down to a reflector is affected by inclusion of more travel path at water velocity. Figure 9.45 shows a seismic line which goes from shallow to deep water; much of the apparent dip is a velocity effect rather than real dip (compare fig. 9.44*a*, *b*). The apparent dip can be corrected by changing the velocity function with water depth when making depth calculations.

Lateral velocity changes also affect the horizontal positions of features (see fig. 8.29). This is illustrated for a diffracting point and a simple two-layer model in fig. 9.46 (see also problem 9.10). The crest of a diffraction usually locates the diffracting point but lateral changes of velocity shift the crest of the diffraction. If we consider more complicated models, for example two dipping layers with

Fig.9.43. Ray-trace model for reflections from the top of the salt layer in the North Sea. (From Taner *et al.*, 1970.)

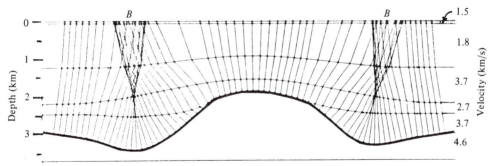

different strikes, serious distortions exist which would be very difficult to unravel from seismic data. Such a situation could easily result where the section and the seafloor dip in different directions.

Correction for gradual velocity changes usually hinges on being able to determine the velocity changes

Fig.9.44. Use of ray-trace modeling for velocity analysis. (From Taner *et al.*, 1970.) (*a*) Depth model with raypaths from deeper reflector; (*b*) time model for (*a*); for the lower reflector, stacking velocity calculations are 2.46 km/s at *A*, 1.69 at *B* and 2.76 at *C*. (*c*) Model showing rms velocities $\bar{V}$ and stacking velocities V_s.

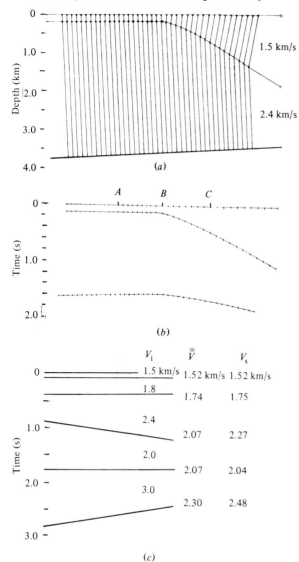

(a)

(b)

(c)

with sufficient reliability. Often velocity has to be determined from the seismic data themselves (see §8.2.3), but velocity analyses (although they may be adequate for use in stacking) often have appreciable uncertainty which may make them unsuitable for such corrections without smoothing. Displays of the type of fig. 8.14 are often especially helpful. Velocity variations are usually reasonably systematic with structure, although in attenuated fashion, that is, velocity relief is usually not as great as the structural relief. Usually data should be smoothed with the geology in mind.

9.5.2 *Sudden lateral changes*

Where lateral changes in velocity are more sudden, correction may not be simple. Consider the effects of the seafloor relief in fig. 9.47. The velocities of the sediments immediately below the canyon probably are markedly different from those of their lateral equivalents because of the differences in overburden, but at large depths the effects of the canyon probably vanish. Furthermore the sediments below the bottom of the canyon may be in fluid-pressure equilibrium with their lateral equivalents, thus having fluid pressure appropriate to the uneroded thickness while their overburden pressure is less because of the erosion, so that they are overpressured. A 'correct' method of removing the velocity effect is not evident and usually the method adopted is the empirical one which produces the most sensible results. In areas of purely erosional relief, such as that in fig. 9.47, this may not be too difficult, but where structural complications accompany, and perhaps cause, the seafloor relief, objective criteria may be lacking.

The flow of salt into lenses and domes may produce velocity anomalies in the section below them. Salt velocity, about 4.5 km/s, may be either higher or lower than that of the laterally adjacent sediments and so the velocity anomalies resulting from a salt lens may be either a *pull-up* or a *push-down*. A pull-up is most common since lower-velocity clastic rocks are the most common lateral equivalent, but lime-rich sediments, anhydrite or other high-velocity rocks may produce push-down, and in some areas (such as Mississippi) both can occur. Figure 9.48 shows a salt pillow with a consequent pull-up. Similar velocity effects can result from other situations, such as reefing (see fig. 9.31*g*, *h*), where either pull-up or push-down can occur, depending on how the velocity in the reef (which depends in part on the reef's porosity) compares with the velocity of the lateral equivalents.

Velocity complications may be very drastic in regions of compressional or thrust tectonics. Figure 9.49 shows velocity effects resulting from the overthrusting of

Fig.9.45. Marine seismic line perpendicular to the
continental slope. Variation of the water depth creates
false dip. (From Sheriff, 1978.)

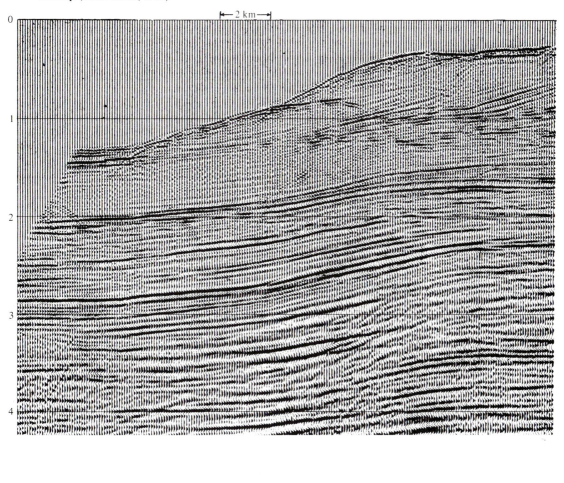

Fig.9.46. Distortion of diffraction arrival-time curve
when $V_2 > V_1$. The earliest arrival of the diffraction is at
A rather than over the diffracting point P. (From
Larner *et al.*, 1981.) (*a*) Depth model; (*b*) time section.

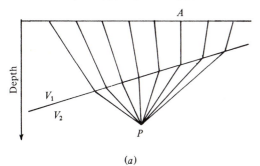

(*a*)

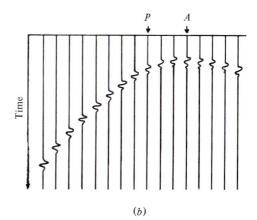

(*b*)

Fig.9.47. Velocity effects of a seafloor canyon. (Courtesy Seiscom Delta.) (*a*) Before correction for velocity; (*b*) after empirical velocity correction.

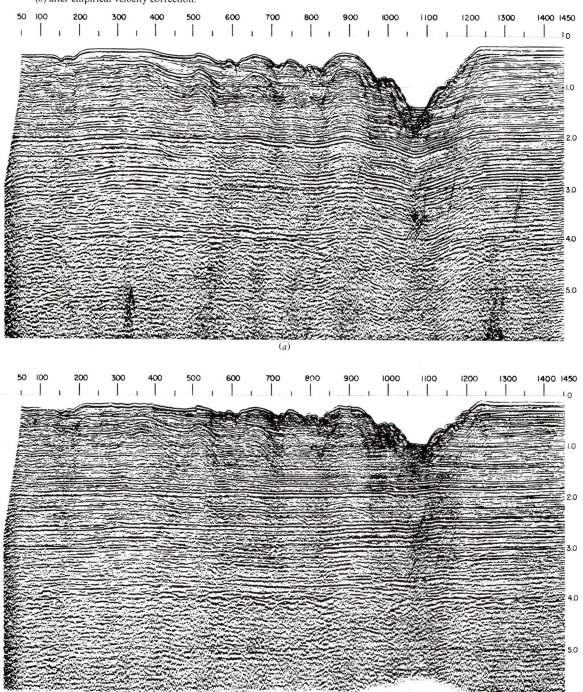

(*a*)

(*b*)

Fig.9.48. Two uplifts in the Mediterranean Sea. The left uplift is a salt pillow with the salt from 5.0 to 5.3 s and the right one is piercement salt. The apparent uplift below the salt is probably all velocity anomaly. (Courtesy CGG.)

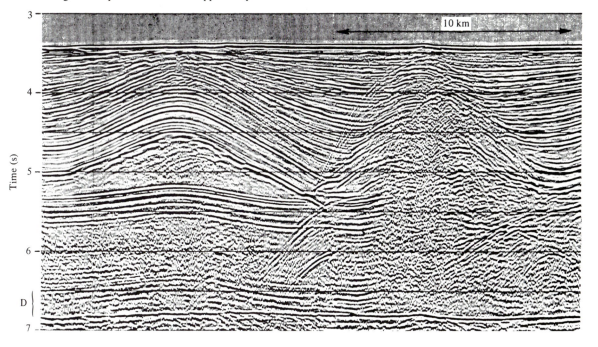

Fig.9.49. Thrust faulting in the Rocky Mountains. High-velocity rocks in an overthrust sheet over the left 40% of the section cause events C and D to arrive earlier than usual. Events C and D are probably continuous and unfaulted across the section. (Courtesy Amoco.)

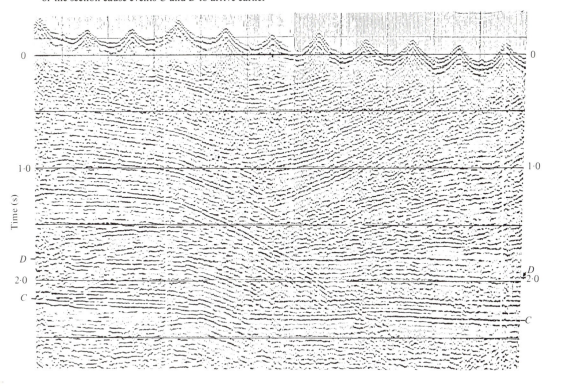

high-velocity sediments in the Rocky Mountain thrust belt. The complications involve not only velocity pull-up but also fictitious faulting evidences, phantom diffractions, etc. The usual solutions to such extreme complications are to trace rays through a model of the section in an effort to achieve reasonable agreement with what is observed (see §9.4.6), but this approach is subject to the uncertainties of modeling, mainly lack of information as to how to construct the model and determine appropriate velocities; such information is apt to be lacking where needed most.

9.6   Three-dimensional interpretation

Reflecting points lie updip from the points where the reflections are observed. Migration accommodates the components of dip in the in-line direction (although not always correctly) so that most of the problems in three-dimensional mapping result from the fact that the component of dip perpendicular to the line is often unknown or not taken into account properly. The *subsurface trace* (line of reflecting points) lies in the direction updip from the seismic line and this should be allowed for in mapping data oriented along seismic lines (see Sheriff, 1978, chapter 21), that is, data ought to be posted on maps at the reflecting points on the reflector being mapped rather than at the shotpoints on the surface. Where data are limited to a grid of a few lines, the first step in mapping is to determine reflecting points where cross-information permits this to be done, as at line intersections. Reflecting points in between the points where such determinations can be made then often can be inferred with adequate accuracy. An alternative sometimes feasible is to map unmigrated data and then migrate the maps to produce a correct structural map (see fig. 8.33).

The techniques of acquiring data specifically for three-dimensional analysis have been discussed in §5.3.7, of 3-D display in §5.4.7, of data-processing in §8.3.6. As of this writing, work with 3-D techniques is one of the fastest-growing areas of geophysics. Most of this work has been devoted to detailing fields once oil has been discovered in order to optimize the field development and exploitation (Dahm and Graebner, 1982; Galbraith and Brown, 1982), and 3-D techniques have proven very cost-effective in this regard despite their present high cost (sometimes $2 million) and the long time delays involved in interpreting 3-D surveys (often exceeding a year). The problems with the processing and interpretation of 3-D surveys mainly involve the huge amounts of data involved. However, there seems to be unanimous agreement that 3-D surveys result in clearer and more accurate pictures of geological detail and that their costs are more

than repaid by the elimination of unnecessary development holes and by the increase in recoverable reserves through the discovery of isolated reservoir pools which otherwise might be missed.

Sets of time-slice maps and sections, including sections which slice through 3-D data sets in arbitrary directions, exhibit the data from different points of view; a new point of view often suggests features otherwise missed (see, for example, fig. 9.36).

9.7   Stratigraphic interpretation
9.7.1   *Introduction*

Extracting non-structural information from seismic data is called *seismic stratigraphy* or *seismic-facies analysis*. *Facies* refers to the sum total of features which characterize the environment in which a sediment was deposited. Facies involves, among other things, sedimentary structure, the form of bedding, original attitude, and the shape, thickness, thickness variations and continuity of sedimentary units. Our interest here is in inferring stratigraphy rather than in locating stratigraphic traps (discussed in §9.3.6), though the implications for stratigraphic traps are obvious. There are three books on seismic stratigraphy: Payton (1977), Sheriff (1980), and Brown and Fisher (1980).

Depositional patterns such as progradation and pinchouts can sometimes be seen in seismic data (for example in fig. 9.50) although many stratigraphic features are too small to be resolvable (Sheriff, 1977*b*) or too gradual to see. Depositional patterns are sometimes associated with depositional energy (which determines the degree of separation of fine from coarse particles), lithology, porosity and other physical properties which are important in hydrocarbon reservoirs.

Seismic stratigraphy is often divided into several sub-areas:

(1) Seismic-sequence analysis, separating out time-depositional units based on detecting unconformities or changes in seismic patterns;

(2) Seismic-facies analysis, determining depositional environment from seismic reflection characteristics;

(3) Reflection-character analysis, examining the lateral variation of individual reflection events, or series of events, to locate where stratigraphic changes occur and identify their nature; the primary tool for this is modeling, by both synthetic seismograms and seismic logs, and this has already been discussed in §9.4;

(4) Detection of hydrocarbon indicators, which will be discussed in §9.8.

9.7.2 *Seismic-sequence analysis*

The concept underlying seismic-sequence analysis is that the boundaries of time-stratigraphic units can be recognized in seismic data. A time-stratigraphic unit is a three-dimensional set of facies deposited contemporaneously as parts of the same system, genetically linked by depositional processes and environments. The key to the definition of a unit is that the top and base represent unconformities. The technique for mapping units is therefore to locate the angularities which mark these unconformities (fig. 9.51) and continue mapping the unconformities through the regions where they are not evident by such angularities.

Vail *et al.* (1977*a*) use diagrams such as those of fig. 9.52 to relate seismic sequences to changes in relative sea level. A relative rise in sea level can be produced by either an absolute rise in sea level or by subsidence of land level. The primary evidence for a relative sea-level rise seen on seismic data is coastal onlap, the progressive termination of reflections in the landward direction. Relative rise of sea level is usually associated with a transgression over an unconformity but it can also be associated with a regression if the influx of sediments is sufficiently rapid. A gradual fall of sea level produces offlap of reflections at an overlying erosional unconformity, whereas a rapid fall of sea level produces a marked seaward shift in the location of onlap.

Vail *et al.* interpret the patterns actually seen in seismic data as showing a sequence of relatively long periods of relative sea-level rise interrupted by short times of rapid sea-level fall (fig. 9.52*e*). They believe the sea-level changes to have been contemporaneous world-wide and have developed a eustatic-level chart showing the world-wide pattern. By correlating local charts of relative change of sea level with a master eustatic-level chart, one can sometimes age-date reflection events with rather high precision.

The seismic-sequence analysis procedure continues on with mapping a sequence over a grid of lines, constructing structure and isopach (thickness) maps of each unit, subdividing these maps according to seismic-facies evidences, relating them to adjacent units, and finally attributing stratigraphic significance to them. This procedure is illustrated by the example shown in fig. 9.53.

Fig. 9.50. Section showing progradation (AA') with sand pinchouts at the top of the unit. (Courtesy Chevron.)

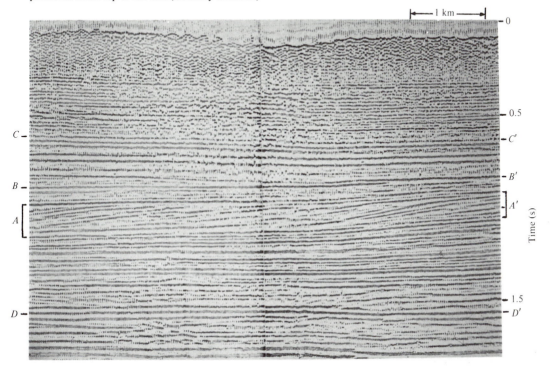

Implied in seismic-sequence analysis is the concept that the attitude of seismic reflections is that of depositional time lines rather than facies lines. A time line indicates a surface that at one time was the surface of the solid Earth. The passage of major storms, floods and other short-term events redistribute sediments within very short periods of time along time lines whereas the long periods between such events often do not leave a record because the new sediments brought in are rearranged by the next major event. Stratal surfaces thus tend to follow time lines. Since the thickness of the stratal units are generally very small, much smaller than the seismic resolving power, they only produce very minor reflection contributions but these tend to interfere in essentially the same way over a widespread area because the stratal surfaces are generally parallel over a wide area and change very slowly laterally. The interference produces the coherent line-ups of reflection events. The fact that seismic reflections parallel time lines is well established by many observations but it is somewhat contradictory to intuitive feelings that reflections should be due to changes in rock nature, such as from sand to shale along facies lines. Facies lines are often based on fairly wide-spread control (for example, on well control) so that the detailed information as to how to draw the facies line is not available. The major portions of correctly-drawn facies lines parallel time lines (fig. 9.54).

9.7.3 *Seismic-facies analysis*

Seismic facies concerns the distinctive characteristics which make one group of reflections look different from adjacent reflections; inferences as to the depositional environment are drawn from seismic facies. The most comprehensive analysis and classification scheme is given by Sangree and Widmier (1979) but Roksandic (1978) has also published on classification. Mitchum *et al.* (1977) and Sangree and Widmier (1979) classify seismic facies according to reflection terminations at seismic-sequence boundaries (fig. 9.51), configurations within sequences (fig. 9.55) and the external shapes of sequences (fig. 9.56).

Fig.9.51. Reflections at boundaries of seismic sequences. (From Mitchum *et al.*, 1977.) (*a*) Relations at top of sequence unit and (*b*) at base of unit; (*c*) relations within an idealized unit.

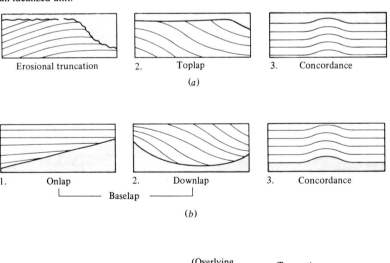

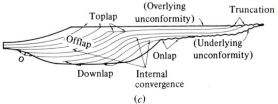

Fig.9.52. Patterns associated with relative sea-level changes. (After Vail *et al.*, 1977a.) (*a*) Relative sea-level rise produces a transgression if terrigenous influx is low, (*b*) a regression if terrigenous influx overwhelms effects of the rise; (*c*) progradation associated with stillstand of sea-level; (*d*) gradual sea-level fall produces downward shift in pattern but tops of patterns are usually eroded off; (*e*) sudden sea-level fall produces major seaward shift in the locale of coastal onlap; pattern indicates gradual rise, then sudden fall (between units 5 and 6), followed by another gradual rise.

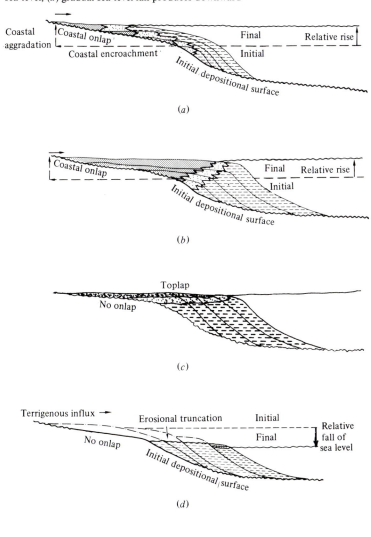

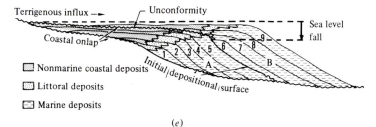

Fig.9.53. Stratigraphic interpretation of seismic data in East Texas. (After Ramsayer, 1979.) (*a*) Portion of seismic section; unconformities bounding the seismic sequence are mapped from the seismic lines; portions of synthetic seismograms made from logs in the wells shown have been superimposed on the section. (*b*) Characteristics of reflections within the seismic sequence are mapped; Top = toplap at top of sequence, C = concordance at top of unit, On = onlap at base of unit, dwn = downlap at base of unit, Thin = unit not thick enough to see internal patterns; Ob = oblique pattern in body of unit, P = parallel reflections in interior of unit; solid arrows indicate direction of onlap, open arrows the direction of downlap. (*c*) Facies interpretation.

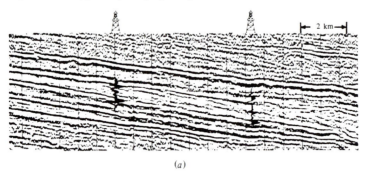

(*a*)

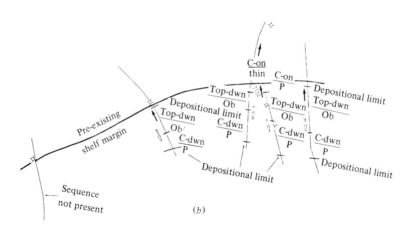

(*b*)

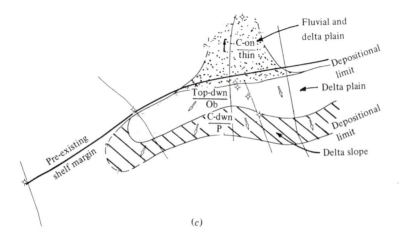

(*c*)

Fig.9.54. The nature of facies lines. (Data from Váil *et al.*, 1977*b*.) (*a*) Facies line as it might be drawn based on data from two wells about 17 km apart; the log curves are SP-curves which distinguish the sand from the surrounding shale. (*b*) Redrawing of the facies line based on many intervening control points; the major portions parallel stratal or time lines. A seismic line shows reflections parallel with the time lines onlapping the unconformity.

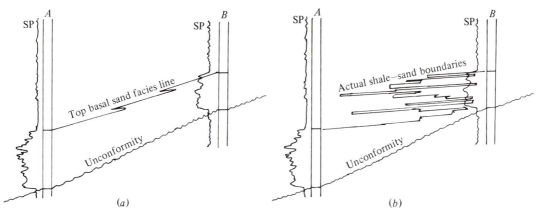

(*a*)   (*b*)

Fig.9.55. Seismic facies classification. (From Sangree and Widmier, 1979.)

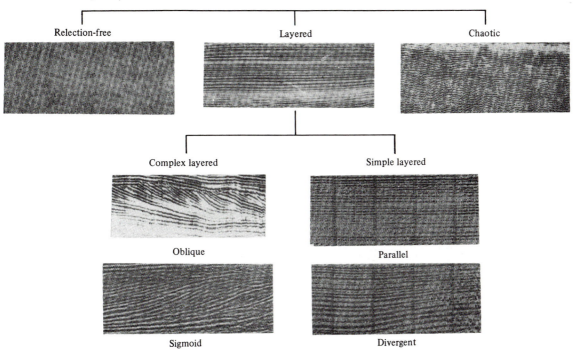

Parallel reflections suggest uniform deposition on a stable or uniformly-subsiding surface whereas divergent reflections indicate variation in the rate of deposition from one area to another or else gradual tilting. Chaotic reflections suggest either relatively high depositional energy, variability of conditions during deposition, or disruption after deposition, such as can be produced by slumping or sliding or turbidity-current flow. A reflection-free interval suggests uniform lithology such as a relatively homogeneous marine shale, salt or massive carbonates; however, distinguishing reflection-free patterns from excessive noise which obscures reflections may be difficult.

Onlap (fig. 9.51) refers to reflection terminations which indicate initially horizontal strata or strata which dip away from the termination, whereas downlap (sometimes called offlap) refers to reflection terminations which indicate strata which dip toward the terminations. The terms are sometimes given a genetic context: onlap is the landward edge of a unit whereas downlap results from inadequate sediment supply (starvation) and thus is the seaward edge of a unit; this occasionally produces a contradiction as at *O* in fig. 9.51*c*, which is geometrically onlap but genetically downlap.

Oblique progradational patterns (fig. 9.55) are characterized by toplap angularity (also sometimes called offlap) and reflection-character variability. The tops of oblique patterns indicate periods during which sea level was not changing markedly (stillstands) and deposition near the wave-base, with consequent high depositional energy. Thus, the tops of oblique patterns often contain relatively clean sands. Sigmoid progradational patterns, on the other hand, are characterized by gentle S-shaped reflections of rather uniform character, the tops of the reflections exhibiting concordance with the top of the sequence unit. These indicate relative sea-level rise and usually consist of fine-grained sediments, sometimes calcareous.

The three-dimensional shape of units provides the principal basis for classification in basin settings (fig. 9.56). Units which drape over preexisting topography are generally low-energy fine-grained pelagic units. Those with mounded tops or chaotic reflections are generally variable-to-high-energy turbidite deposits.

Table 9.3 shows a classification according to seismic-facies characteristics.

High reflection continuity suggests continuous strata deposited in an environment that was relatively quiet and uniform over a widespread area, such as marine shales interbedded with silts or calcareous shales. Fluvial sediments with interbedded clays and coals sometimes produce strong reflections.

Fig.9.56. Seismic facies classification of basinal sequences. (From Sangree and Widmier, 1979.)

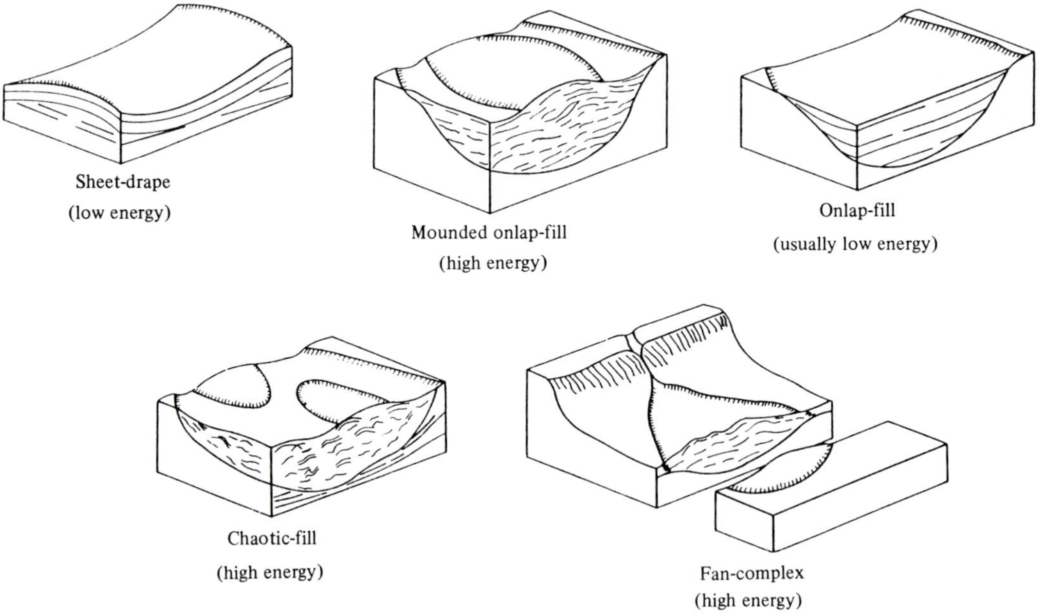

Sheet-drape
(low energy)

Mounded onlap-fill
(high energy)

Onlap-fill
(usually low energy)

Chaotic-fill
(high energy)

Fan-complex
(high energy)

Table 9.3. *Seismic facies classification (after Sangree and Widmier, 1979).*

Regional Setting	Basis of Distinction	Subdivisions	Interpretation	Other Characteristics
Shelf	Reflection character Unit shape: widespread sheet or gentle wedge Reflections generally parallel or divergent	High continuity, high amplitude (Generally marine)	Shelf deposits— alternating neritic shale/limestone, interbedded high/low energy deposits, or shallow marine clastics transported mainly by wave action	Possibly cut by submarine canyons Distinguish on basis of location compared to other facies
		Variable continuity, low amplitude, occasional high amplitude	Fluvial or nearshore clastics, fluvial/wave-transport processes (delta platform), or low-energy turbidity current or wave transport	Distinguish on basis of location compared to other facies Shale-prone if seaward of unit above, Sand-prone if seaward of unit below
		Low continuity, variable amplitude	Nonmarine clastics, fluvial or marginal-marine	Occasional high amplitude and high continuity from coal members
	Mounded shape	Variable continuity and amplitude	Delta complex	Internal reflections gently sigmoid to divergent Occasional high amplitudes
		Local reflection void	Reef	See fig. 9.31
Shelf margin— prograded slope	Internal reflection pattern	Oblique, fan-shaped or overlapping fans	Adequate sediment supply Shelf margin—deltaic High energy deposits in updip portions Occasionally due to strong currents in deep water	Moderate continuity and amplitude, reflections variable. Foreset (clinoform) dips to 10° (averaging 4–5°), steeper dips are calcareous Often fan-shaped (including multiple fans)
		Sigmoid, elongate lens/fan	Low sediment supply Low depositional energy	High continuity, high to moderate amplitude, uniform reflections

Table 9.3 (*Continued*)

Regional Setting	Basis of Distinction	Subdivisions			Interpretation	Other Characteristics
Basin slope, basin floor	Overall unit shape	Drape	Sheet drape		Deep marine hemipelagic; mainly clay Low energy	High continuity, low amplitude Drapes over preexisting topography
		Mounded	Contourite		Deep Low energy	Variable continuity and amplitude
				Fan-shaped	Variable energy, slump/turbidity currents	Discontinuous, variable amplitude At mouth of submarine canyons Composition depends on what was eroded up above
		Fill	Slope front fill		Low energy Deep marine clay and silt	Variable continuity and amplitude Fan-shaped to extensive along slope
			Onlapping fill		Low-velocity turbidity currents	High continuity, variable amplitude
			Mounded onlap fill or chaotic fill		High or variable-energy turbidites	Overall mound in a topographic low, gouge common at base Discontinuous, variable amplitude
			Canyon fill		Variable superimposed strata Coarse turbidites to hemipelagic	Variable continuity and amplitude

The lateral equivalents of units sometimes provide the key to identification. Thus, a low reflection-amplitude facies representing prodelta shales may grade landward into a facies of high continuity and amplitude resulting from interbedded silts and/or sands, whereas a low reflection-amplitude sand facies may grade landward into a non-marine, low-continuity, variable-amplitude facies. The prodelta shale may grade basinward into a prograded-slope facies whereas the sand may grade basinward into high-continuity, high-amplitude marine facies.

9.7.4 *Reflection-character analysis*

Reflection-character analysis involves study of the trace-to-trace changes in the waveshape of one or more reflections with the objective of locating and determining the nature of changes in the stratigraphy or fluid in the pore spaces. Special displays may be used to make it easier to see the changes, such as enlarged displays of the portion of the section being studied, displays of attribute measurements (Taner and Sheriff, 1977; Taner et al., 1979) such as amplitude, instantaneous frequency, etc. (§8.4.2), or seismic-log displays (§9.4.5), often involving color.

Synthetic seismograms (§9.4.4) are often used to determine the nature of the stratigraphic change which a waveshape change indicates. The various stratigraphic changes which are regarded as reasonable possibilities are modeled (Harms and Tackenberg, 1972; Neidell and Poggiagliolmi, 1977) and matched with the observed waveforms. Clement (1977) describes the use of reflection-character analysis in mapping a sand associated with channels on an unconformity surface in Oklahoma. A distinctive reflection was present (fig. 9.57) where the sands were more than six meters in thickness, in which situation it usually was also porous. Several successful wells were drilled on the basis of the predictions from reflection character, but one well encountered tight indurated interbedded sandstone and shale which gave very similar reflection character. This study thus il-lustrates both successful application of these techniques and also the ambiguity of conclusions based on reflection character.

Hun (1978) describes mapping reflection amplitude over an oil field. By using the amplitude of a reflection which is not involved with the local feature as a normalizing factor, uncertainties caused by non-geologic amplitude factors have less effect and the correlation with the accumulation and with well productivity is improved (fig. 9.58).

Maureau and van Wijhe (1979) successfully used seismic logs to predict high-porosity zones in Permian carbonates in the Netherlands. Lindseth (1979) reports mapping porosity in Devonian carbonates in Alberta and other reflection-character analysis studies using seismic logs.

Fig.9.57. Evidences of a channel sand. (From Clement, 1977.) (a) Portion of a seismic line across a channel showing development of an event where channel sand is more than 6 m thick; the well at A just missed the channel. (b) Models of logs for various sand thicknesses and synthetic traces at locations A, B, C; the logs at B and C are modifications of log A; the first of each pair of traces is a synthetic trace, the second is the actual seismic trace.

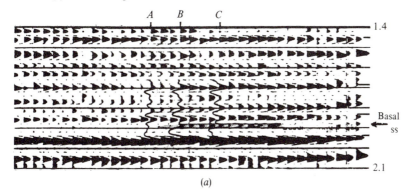

(a)

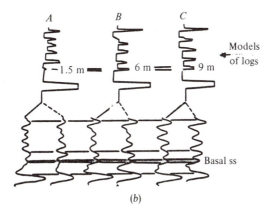

(b)

9.8 Hydrocarbon indicators

The velocity and density of sedimentary rocks depend on porosity and on the properties of the fluids filling the pore space, as was discussed in §7.1.4 and §7.1.7. The dependence of porosity on density is the straightforward relation given by (7.3) but the dependence of porosity on velocity where there is a mixture of fluids in the pore space is not as simple. The change in velocity resulting from a change in the interstitial fluid is often marked, as shown in fig. 7.15, resulting in amplitude anomalies associated with accumulations. The extensive use of automatic gain control obscured these amplitude effects until about 1970 when recognition of their usefulness in locating accumulations of hydrocarbons became widespread. Since the anomaly is most often one of locally-increased amplitude (as in fig. 9.59), they were

Fig.9.58. Ratio of the amplitudes of two events over an oilfield. Outline of the field is shown dashed. A producibility index has values of 160 at A, 480 at B, 95 at C, 32 at D, 250 at E, 300 at F, 40 at H and 260 at J; well G was dry. (After Hun, 1978.)

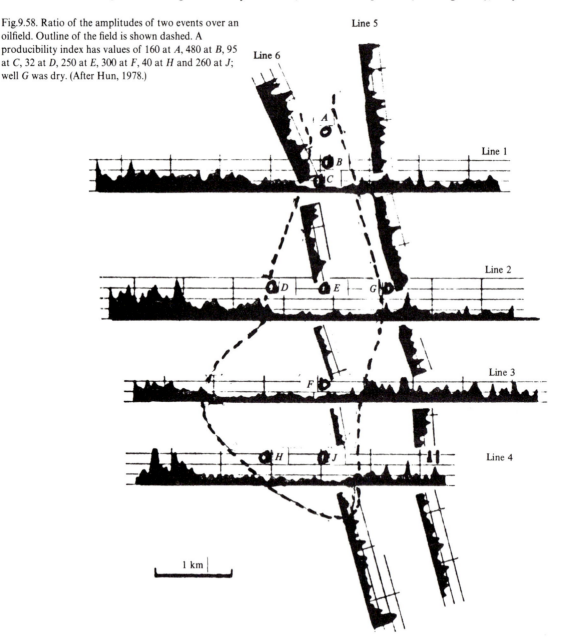

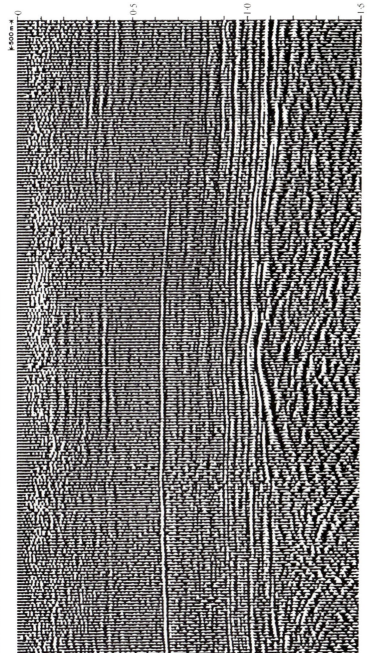

Fig.9.59. Amplitude anomaly caused by gas accumulation. The gas field is just below 0.6 s, basement at about 1.1 s; the accumulation is large despite its shallow depth. (Courtesy Chevron.)

called 'bright spots'. Soon other types of anomalies were found to be associated with hydrocarbon accumulations under some conditions. The relation between hydrocarbon indicators and hydrocarbon accumulations is not simple and universal, however, and many bright spots turned out to result from changes other than commercial hydrocarbon accumulations.

The effect of interstitial fluids on velocity depends on the structure of the rock and is generally greater and simpler for relatively unconsolidated clastic rocks. Thus, the effects are generally greater for young rocks than for older and bright-spot technology is especially applicable to Tertiary clastic basins which are mostly offshore around the periphery of the continents, but being offshore is irrelevant except that marine data are often of better quality than land data and hence anomalies easier to see. The effect on velocity is generally greater (and more complex) for gaseous than for liquid hydrocarbons, as indicated in fig. 7.15.

Replacement of brine by hydrocarbons as the interstitial fluid almost always results in a lowering of velocity but the effect on reflections depends also on the properties of the rock overlying (and underlying) the reservoir rock. If the overlying rock has higher velocity than brine-filled reservoir rock, lowering the reservoir rock velocity by filling it with hydrocarbons increases the contrast and hence increases the amplitude of the reflection from the top of the reservoir. This is common in many Tertiary clastic basins and is the reason for the bright-spot name. If, on the other hand, the overlying rock has a velocity appreciably lower than the reservoir, the effect of hydrocarbons is to decrease the contrast producing a 'dim spot'. This is sometimes observed where carbonate reservoir rocks are capped by shales. Where the overlying rock has a velocity slightly smaller than that of the reservoir rock, lowering the reservoir rock velocity by hydrocarbons may invert the sign of the reflection producing a 'polarity reversal' over the reservoir.

Where a well-defined fluid contact, especially a gas – oil or gas – water contact is present, the contrast may be great enough to give a fairly strong reflection which may stand out on seismic records because of its flat attitude in contrast to the dipping attitudes of other reflections. This is a 'flat spot' and, where seen, it is usually the most definitive and informative of the hydrocarbon indicators. However, reservoir thicknesses are usually small compared to the resolvable limit so that the reflections from reservoir cap, fluid contact and base of the reservoir generally interfere with one another to provide a composite reflection which may show various phase and amplitude changes as the component reflections interfere in different ways. Thus 'phasing' can also be regarded as a hydrocarbon indicator.

The lowering of velocity in a hydrocarbon accumulation will also affect reflections from deeper reflectors by increasing arrival times to cause a 'sag' in reflections seen through the reservoir, but the magnitude of the sag is usually small because most reservoirs are not very thick.

The lowering of velocity can also bend raypaths passing through the reservoir (as indicated in fig. 4.20a) resulting in distortion of deeper reflection events; sometimes the effect is simply a degradation of reflection quality under the reservoir, a 'shadow zone'. The high amplitude associated with a bright spot often results in a lowering of the amplitudes of entire traces by processing procedures which make the mean energy of all traces the same (amplitude normalizing), in order to remove near-surface or recording-system amplitude effects. A consequence is that a lower-amplitude shadow zone may exist above a bright spot as well as below it.

The high amplitude associated with a bright spot also affects multiples involving the reservoir reflectors. Occasionally, sections are made to emphasize multiples by using lower stacking velocities which attenuate primary reflections; an increase in amplitude of multiples seen on such a section can also be used as a hydrocarbon indicator.

A lowering of instantaneous frequency (§8.4.2) is often observed immediately under hydrocarbon accumulations. Such 'low-frequency shadows' seem to be confined to a couple of cycles below (not at) accumulations. No adequate explanation is available (see Taner *et al.*, 1979); proposed explanations generally involve either the removal of higher frequencies because of absorption or other mechanisms, or improper stacking because of erroneous velocity assumptions or raypath distortion, but they cannot account for the magnitudes of the changes which are observed.

Special displays are often used to enhance hydrocarbon indicators and help in their detection and analysis. The most common is a low-amplitude display in which special efforts are made to preserve amplitude relations; reflection events are subdued on such displays so that the amplitude build-ups associated with bright spots stand out more clearly (fig. 9.6 is a low-amplitude display of part of fig. 9.24*d*). Sections are also often displayed in variable-area mode with both normal and inverted polarity, because hydrocarbon-indicator effects are sometimes more evident on one than the other. Measurements of amplitude, envelope amplitude, amplitude ratios, phase, frequency, velocity, etc., may be displayed so as to make lateral changes along reflectors more evident. Complex-trace analysis (§8.4.2) may be used to make such measurements, the results often being displayed as color overlays on seismic sections to facilitate correlation with structural evidences.

Hydrocarbon indicators are also used quantitatively to predict reservoir thicknesses and hydrocarbon

Fig.9.60. Response of a thin bed. (*a*) Peak-to-trough time is nearly independent of thickness below $\frac{1}{4}\lambda$ but nearly linear with thickness above $\frac{1}{4}\lambda$ (after Schramm *et al.*, 1977). (*b*) Maximum peak-to-trough amplitude is nearly linear with thickness below $\frac{1}{8}\lambda$, goes through a maximum (or minimum) at $\frac{1}{4}\lambda$ and is nearly independent of thickness above $\frac{1}{2}\lambda$. Amplitude measurements require a calibration, however, which is often not possible (after Sheriff, 1980).

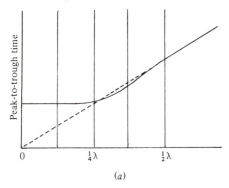

(a)

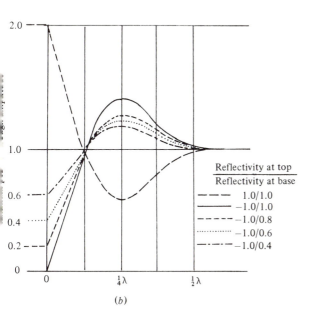

(b)

volumes. Amplitude and dominant-period measurements based on model studies of a wedge pinchout (such as that in fig. 4.22*b*, *c*) are shown in fig. 9.60. Dominant-period or peak-to-trough time measurements can be used to indicate thicknesses when greater than a quarter wavelength and amplitude measurements when smaller than a quarter-wavelength. The use of amplitude measurements, however, requires an amplitude calibration (that is, what would the amplitude be if the reservoir were very thick) which often is not possible. Quantitative analysis also requires extremely careful processing, good data and assumptions about the nature of the rock sequence which may not always be true. Modeling is almost always involved in quantitative analysis.

9.9   Crustal studies

The field and processing techniques used in petroleum seismic exploration – common-depth-point recording, vertical stacking, statics corrections, deconvolution, velocity analysis, the use of long streamers, precise navigation at sea, large Vibroseis sources and long sweeps on land, etc. – are beginning to be applied to studies of the Earth's crust.

Marine seismic lines reveal details of the crustal structure, such as subduction-zone faults (fig. 9.61) and, possibly, the base of the oceanic crustal plate. In the United States the Consortium for Continental Reflection Profiling (COCORP) has been running a series of lines to study the structure of the Earth's crust, and similar studies are beginning in other countries. The preliminary results have been exciting. Many complexities have been found in the crust, including regions of layered reflections which suggest sedimentary rather than igneous origin, although undoubtedly metamorphosed. A line across the Southern Appalachians (Cook *et al.*, 1979, 1980) indicates that the crystalline and other rocks seen on the surface have been thrust from the east a long distance and that sedimentary rocks may underlie portions of the low-angle thrusts. The ideas of 'thin-skin tectonics' are beginning to change concepts of how continents were formed.

Problems

9.1.   Four lines forming a grid are shown in fig. 9.24. (*a*) Map the three horizons encountered at 1.355, 1.830 and 2.660 s at the intersections of lines *B* and *C*. A velocity analysis at this location gives time–velocity (stacking velocity) pairs as follows: 0.100 s, 1520 m/s; 0.600, 1830; 0.800, 1900; 1.200, 2050; 1.400, 2100; 1.600, 2140; 2.000, 2280; 2.700, 2440; 3.000, 2470. (*b*) Map the curved fault plane. (*c*) Estimate and map the throw on the fault.

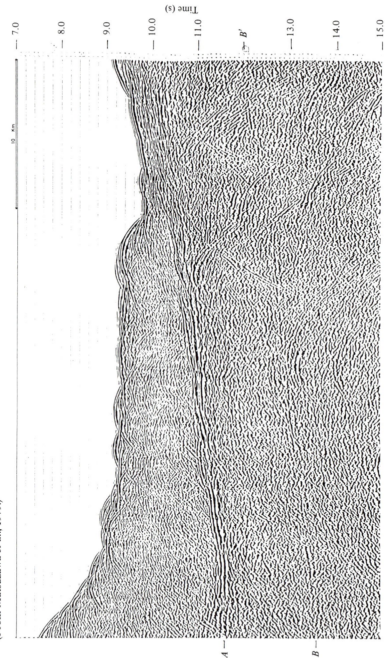

Fig. 9.61. CDP section across the Japan Trench. The strong reflection at *A* is interpreted as the contact along which the Pacific plate is sliding under the Japanese plate (subduction). The discontinuous reflection *BB'* is interpreted as being from the Mohorovičić discontinuity. The interval velocity between the *A* and *B* reflections calculated from stacking velocities is 6.0–6.4 km/s so the 2 s time interval between them represents a thickness of about 6 km for the thickness of the oceanic crust. (From Matsuzawa *et al.*, 1979.)

9.2.   The line in fig. 9.6 shows part of line B of fig. 9.24 (the well is located at W in fig. 9.24), but with relative amplitude information preserved and plotted so that the largest amplitudes are not clipped. What conclusions can be drawn from fig. 9.6 that are less evident from fig. 9.24?

9.3.   (a) Well B is 500 m due east of well A and well C is 600 m due north of A. A fault cuts A, B, C at depths of 800, 1000 and 600 m respectively. Assuming all wells are vertical and the fault surface is a plane, find the surface trace and strike of the fault [see the derivation of (3.17)]. (b) At what depth would you look for this fault in well D located 500 m N30°W from well C? (c) Another fault cuts wells A and C at depths of 1300 and 1000 m respectively and is known to strike N20°W. Where does it cut well B?

9.4.   Interpret the sections shown in figs. (a) 8.26b; (b) 8.28b; (c) 8.29d; (d) 9.27; (e) 9.34. Assume that out-of-the-plane data are not important. [Pick events which involve angularities between primary reflections in order to identify unconformities and/or seismic sequence boundaries.] Deduce the geologic history.

9.5.   Try to match the section shown in fig. 8.12b with the structural styles shown in Table 9.1. With which might it be compatible? [Note that fig. 8.12b is unmigrated but assume that it is nearly perpendicular to the strike.] Do the velocity data from problem 7.20 help?

9.6.   Figure 9.29 shows a salt uplift at a shelf edge. (a) How could one tell that this feature is not caused by reef growth instead? (b) Could it have been caused by shale flowage? (c) Does the relief above the unconformity U_1 indicate post-unconformity salt movement, renewed activity at the shelf edge (downdrop along faulting at the shelf edge), or differential compaction because of the weight of the post-unconformity section?

9.7.   If the nature of a flow structure (such as shown in figs. 9.28 or 9.29) should not be clear, how might gravity, magnetic or refraction measurements be used to distinguish between salt, shale or igneous flow? Between these and a reef?

9.8.   Figure 8.33b maps two separate high closures on a northeast-plunging anticlinal nose. (a) Assuming that the only existing control is that shown by the lines marked //////////// in fig. 8.33a (which is on a different horizon), what additional program would you recommend to check out weaknesses in the interpretation before recommending a well to test for hydrocarbon accumulation? (b) Can you find a fault for which the indicated direction of throw is clearly wrong?

9.9.   How would evidences of thickening/thinning around a saltdome or in a folded structure be distorted on an unmigrated time section? On a migrated time section?

9.10.   In fig. 9.46, $V_1 = 2.00$ km/s, $V_2 = 4.00$ km/s, the horizon dips 10° and the vertical depth of the diffracting point P is 1000 m, the interface between V_1 and V_2 being 350 m vertically above P. Compare the diffraction curve with that which would have been observed if $V_1 = V_2 = 3.00$ km/s. [Hint: ray-trace enough rays to roughly define the diffraction curves.]

9.11.   (a) Using the minimum-phase wavelet of problem 8.11, determine the waveshapes for a sand enclosed in shale where the two-way traveltime through the sand is 12 ms when the upper part contains gas, the two-way traveltime through the gas–sand portion being successively 0, 2, 4, 6, 8, 10 and 12 ms. Plot the traces side-by-side shifted successively by 2 ms as would be the case where the gas–water contact were horizontal. This illustrates a bright-spot, flat-spot situation. Take the reflection coefficients for shale to gas–sand as -0.1, gas–sand to water–sand as $+0.15$, and water–sand to shale as -0.05. (b) Repeat using the zero-phase wavelet of problem 8.11.

9.12.   Attempt a stratigraphic interpretation of fig. 9.50. CC' divides non-marine from marine sediments. Does the surface channel create fictitious deep effects?

10

Background mathematics

Overview

This chapter serves as an appendix to this book rather than a portion of the main text. It can be omitted by those already familiar with mathematics or by those who wish to take the mathematics on faith. More extensive treatments can be found in Wylie (1966), Pipes and Harvill (1970), Robinson (1967*a*, *b*, *c*), Cassand *et al.* (1971), Kanasewich (1973). Båth (1974), Kulhánek (1976), Claerbout (1976), Silvia and Robinson (1979), Robinson and Treitel (1973, 1980).

We begin with short summaries of determinants, vector analysis, matrix analysis, complex numbers, the methods of least squares, finite differences and partial fractions (§10.1). Most of the chapter involves the mathematics of data-processing, especially Fourier (§10.2, 10.3), Laplace (§10.4) and *z*-transforms (§10.6) and almost all deal with linear systems (§10.5). The cepstrum is discussed (§10.7) and a final section deals with filtering. In contrast to chapter 8 which dealt almost entirely with digital data, much of this chapter deals with continuous functions, although some elaborates on digital considerations, especially §10.6.

10.1 Summaries of basic concepts

10.1.1 *Determinants*

A determinant, det (a), is a square array of $(n \times n)$ numbers, a_{ij}, called *elements*, i and j designating the row and column respectively:

$$\det(a) = \begin{vmatrix} a_{11} & a_{12} & \cdots & a_{1n} \\ a_{21} & a_{22} & \cdots & a_{2n} \\ \cdots & & & \\ a_{n1} & a_{n2} & \cdots & a_{nn} \end{vmatrix}. \qquad (10.1)$$

The *minor* M_{ij} of the element a_{ij} is the determinant of order $(n-1)$ formed by deleting the ith row and the jth column of $\det(a)$. The product $(-1)^{i+j}M_{ij}$ is the *cofactor* of a_{ij}. The value of a determinant is defined as

$$\det(a) = \sum_j (-1)^{i+j} a_{ij} M_{ij} = \sum_i (-1)^{i+j} a_{ij} M_{ij} \qquad (10.2)$$

where the summation is taken along one row (*row expansion*, hence $i = $ constant) or along one column (column expansion, $j = $ constant); the result is the same in both cases (see rule (5) below). As an example we expand by the first row:

$$\det(a) = \begin{vmatrix} 1 & 2 & 3 \\ 4 & 5 & 6 \\ 8 & 9 & 0 \end{vmatrix} = (-1)^{1+1} 1 \begin{vmatrix} 5 & 6 \\ 9 & 0 \end{vmatrix}$$

$$+ (-1)^{1+2} 2 \begin{vmatrix} 4 & 6 \\ 8 & 0 \end{vmatrix} + (-1)^{1+3} 3 \begin{vmatrix} 4 & 5 \\ 8 & 9 \end{vmatrix}$$

$$= 1(5 \times 0 - 6 \times 9) - 2(4 \times 0 - 6 \times 8)$$

$$+ 3(4 \times 9 - 5 \times 8) = 30.$$

A determinant is thus a single number.

Equations (10.1) and (10.2) can be used to derive the following rules (see Wylie, 1966, p. 403 – 10 for proofs):

(1) if all the elements of one row are zero, or if the elements of one row are proportional to the corresponding elements of another row, then the determinant equals zero;

(2) multiplying all the elements of a row by a constant multiplies the determinant by the same constant;

(3) interchanging any two rows changes the sign of the determinant;

(4) interchanging rows with columns does not change the value;

(5) (10.2) gives the same value regardless of the row selected;

(6) any row can be multiplied by a constant and added to another row without changing the value of the determinant;

(7) 'row' can be replaced with 'column' in any of the above rules.

Determinants can be used to solve a system of linear equations:

$$a_{11}x_1 + a_{12}x_2 + \ldots + a_{1n}x_n = b_1,$$

$$a_{21}x_1 + a_{22}x_2 + \ldots + a_{2n}x_n = b_2,$$

$$\ldots$$

$$a_{n1}x_1 + a_{n2}x_2 + \ldots + a_{nn}x_n = b_n;$$

Cramer's rule states that

$$x_r = \{\det(a_r)\} / \{\det(a)\} \qquad (10.3)$$

where

$$\det(a) = \begin{vmatrix} a_{11} & a_{12} & \cdots & a_{1n} \\ a_{21} & a_{22} & \cdots & a_{2n} \\ \cdots & & & \\ a_{n1} & a_{n2} & \cdots & a_{nn} \end{vmatrix}$$

and $\det(a_r)$ is $\det(a)$ with the rth column replaced by $b_1, b_2, \ldots, b_n$ (Pipes and Harvill, 1970, p. 101).

10.1.2 *Vector analysis*

(a) *Basic definitions.* A *scalar* quantity, such as temperature, has magnitude only while a *vector* quantity, such as force, has both magnitude and direction. Vectors (represented by bold-face type) can be added to give a resultant as shown in fig. 10.1a. Subtraction is equivalent

Fig.10.1. Operations on vectors. (a) Addition and subtraction; (b) resolution into components; (c) cross-product of vectors; (d) scalar function of position; (e) curvilinear coordinates.

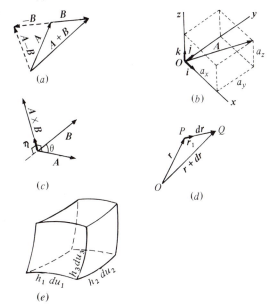

to reversing one of the vectors and then adding. Multiplication of a vector by a scalar changes the magnitude of the vector but not its direction (except that multiplication by negative numbers reverses the direction).

A vector can be resolved into components along coordinate axes (fig. 10.1*b*) and expressed as a vector sum of its components:

$$\mathbf{A} = a_x\mathbf{i} + a_y\mathbf{j} + a_z\mathbf{k} \tag{10.4}$$

where **i**, **j**, **k** are unit vectors along the *x*-, *y*-, *z*-axes. (Vectors can also be expressed in other coordinate systems such as cylindrical or spherical.)

Vectors can be added by adding corresponding components. Thus, if

$$\mathbf{A} = a_x\mathbf{i} + a_y\mathbf{j} + a_z\mathbf{k}, \text{etc.,}$$

$$\mathbf{A} + 2\mathbf{B} - 3\mathbf{C} = (a_x + 2b_x - 3c_x)\mathbf{i} + (a_y + 2b_y - 3c_y)\mathbf{j}$$
$$+ (a_z + 2b_z - 3c_z)\mathbf{k}.$$

The magnitude of a vector, written |**A**|, is given by

$$|\mathbf{A}| = (a_x^2 + a_y^2 + a_z^2)^{\frac{1}{2}}. \tag{10.5}$$

If **A** has direction cosines (l, m, n), then

$$\mathbf{A} = |\mathbf{A}|(l\mathbf{i} + m\mathbf{j} + n\mathbf{k}). \tag{10.6}$$

(b) Vector products. The *scalar* or *dot product* of two vectors, **A** and **B**, is written $\mathbf{A} \cdot \mathbf{B}$; it is a scalar:

$$\mathbf{A} \cdot \mathbf{B} = |\mathbf{A}||\mathbf{B}| \cos\theta \tag{10.7}$$

when θ is the smaller angle between the vectors (fig. 10.1*c*). Thus, the dot product equals the magnitude of one of the vectors times the projection of the second vector onto the first. If **A** is a force acting on a point mass which suffers displacement **B**, $\mathbf{A} \cdot \mathbf{B}$ gives the work done on the mass. Obviously

$$\mathbf{A} \cdot \mathbf{B} = \mathbf{B} \cdot \mathbf{A},$$

$$= 0 \quad \text{when} \quad \mathbf{A} \perp \mathbf{B},$$

$$= |\mathbf{A}||\mathbf{B}| \quad \text{when} \quad \mathbf{A} \parallel \mathbf{B}.$$

Also,

$$\mathbf{i} \cdot \mathbf{i} = \mathbf{j} \cdot \mathbf{j} = \mathbf{k} \cdot \mathbf{k} = 1, \quad \mathbf{i} \cdot \mathbf{j} = \mathbf{j} \cdot \mathbf{k} = \mathbf{k} \cdot \mathbf{i} = 0,$$

hence

$$\mathbf{A} \cdot \mathbf{B} = \mathbf{B} \cdot \mathbf{A} = a_x b_x + a_y b_y + a_z b_z \tag{10.8}$$

and

$$\mathbf{A}^2 = \mathbf{A} \cdot \mathbf{A} = a_x^2 + a_y^2 + a_z^2.$$

The *vector*, or *cross*, *product* of **A** and **B**, $\mathbf{A} \times \mathbf{B}$, is a vector defined by the relation

$$\mathbf{A} \times \mathbf{B} = (|\mathbf{A}||\mathbf{B}|\sin\theta)\mathbf{\eta} \tag{10.9}$$

where $\mathbf{\eta}$ is a unit vector perpendicular to the plane containing **A** and **B** and in the direction of advance of a right-handed screw rotated from **A** to **B** through the angle θ (fig. 10.1*c*). Because the direction of $\mathbf{\eta}$ depends on the sequence,

$$\mathbf{A} \times \mathbf{B} = -\mathbf{B} \times \mathbf{A},$$

$$= 0 \quad \text{when} \quad \mathbf{A} \parallel \mathbf{B},$$

$$= |\mathbf{A}||\mathbf{B}|\mathbf{\eta} \quad \text{when} \quad \mathbf{A} \perp \mathbf{B}.$$

The magnitude of $\mathbf{A} \times \mathbf{B}$ equals the area of the parallelogram defined by **A** and **B**. The torque about an axis of rotation is given by a vector product (see problem 10.1). Applying (10.9) to the unit vectors $\mathbf{i}, \mathbf{j}, \mathbf{k}$ gives

$$\mathbf{i} \times \mathbf{j} = \mathbf{k}, \quad \mathbf{j} \times \mathbf{k} = \mathbf{i}, \quad \mathbf{k} \times \mathbf{i} = \mathbf{j};$$
$$\mathbf{i} \times \mathbf{i} = \mathbf{j} \times \mathbf{j} = \mathbf{k} \times \mathbf{k} = 0.$$

$\mathbf{A} \times \mathbf{B}$ can be expressed as a determinant (problem 10.2),

$$\mathbf{A} \times \mathbf{B} = \begin{vmatrix} \mathbf{i} & \mathbf{j} & \mathbf{k} \\ a_x & a_y & a_z \\ b_x & b_y & b_z \end{vmatrix}. \tag{10.10}$$

Products of more than two vectors can be formed in various ways (see problem 10.3).

(c) Vector operators. Let $\psi(x, y, z)$ be a scalar function of position (for example, temperature). If the value is ψ at $P(x, y, z)$ in fig. 10.1*d*, then the value at a nearby point Q is $\psi + d\psi$ where

$$d\psi = \frac{\partial\psi}{\partial x}dx + \frac{\partial\psi}{\partial y}dy + \frac{\partial\psi}{\partial z}dz$$

$$= \left(\frac{\partial\psi}{\partial x}\mathbf{i} + \frac{\partial\psi}{\partial y}\mathbf{j} + \frac{\partial\psi}{\partial z}\mathbf{k}\right) \cdot (\mathbf{i}\,dx + \mathbf{j}\,dy + \mathbf{k}\,dz)$$

$$= \nabla\psi \cdot d\mathbf{r}$$

thus

$$d\psi/dr = \nabla\psi \cdot \mathbf{r}_1 \tag{10.11}$$

where $\mathbf{r}_1$ is a unit vector along $d\mathbf{r}$. The vector $\nabla\psi$ (pronounced 'del ψ') is the *gradient* of ψ (grad ψ); it is a vector in the direction of the maximum rate of increase of ψ and its magnitude equals this maximum rate of increase (see problem 10.4*b*). The rate of increase of ψ in an arbitrary direction **p**, where **p** is a unit vector, is given by $\nabla\psi \cdot \mathbf{p}$ (see problem 10.4*c*).

The vector operator, *del*, $\mathbf{V} = \mathbf{i}(\partial/\partial x) + \mathbf{j}(\partial/\partial y) + \mathbf{k}(\partial/\partial z)$, is often used as if it were a vector. Thus we can take the dot product of $\mathbf{V}$ and a vector $\mathbf{A}$, called the *divergence* of $\mathbf{A}$ or *div* $\mathbf{A}$:

$$\text{div } \mathbf{A} = \mathbf{V} \cdot \mathbf{A} = \frac{\partial a_x}{\partial x} + \frac{\partial a_y}{\partial y} + \frac{\partial a_z}{\partial z}. \quad (10.12)$$

We can also take the vector product of $\mathbf{V}$ and $\mathbf{A}$, called the *curl*:

$$\text{curl } \mathbf{A} = \mathbf{V} \times \mathbf{A} = \begin{vmatrix} \mathbf{i} & \mathbf{j} & \mathbf{k} \\ \dfrac{\partial}{\partial x} & \dfrac{\partial}{\partial y} & \dfrac{\partial}{\partial z} \\ a_x & a_y & a_z \end{vmatrix} \quad (10.13)$$

on using (10.10). The operator $\mathbf{V}$ can be applied more than once. For example,

$$\text{div grad } \psi = \mathbf{V} \cdot \mathbf{V}\psi = \mathbf{V}^2\psi = \left[\frac{\partial^2}{\partial x^2} + \frac{\partial^2}{\partial y^2} + \frac{\partial^2}{\partial z^2} \right] \psi \quad (10.14)$$

$$= \text{\textit{Laplacian} of } \psi.$$

In the same way we can form products such as $\mathbf{V} \times \mathbf{V}\psi$, $\mathbf{V} \cdot \mathbf{V} \times \mathbf{A}$, $\mathbf{V} \times \mathbf{V} \times \mathbf{A}$, $\mathbf{V}(\mathbf{V} \cdot \mathbf{A})$, etc. (see problem 10.5).

(d) Orthogonal curvilinear coordinates. While Cartesian coordinates are usually the most convenient, at times cylindrical, spherical or other orthogonal curvilinear coordinates lead to simpler results. We write u_1, u_2, u_3 for such coordinates, the surfaces $u_1 = c_1, u_2 = c_2, u_3 = c_3$ (c_i = constant) being orthogonal. When u_i changes by $\mathrm{d}u_i$, the element of length, $\mathrm{d}s$, is given by

$$\mathrm{d}s^2 = (h_1 \mathrm{d}u_1)^2 + (h_2 \mathrm{d}u_2)^2 + (h_3 \mathrm{d}u_3)^2,$$

where $h_i = h_i(u_1, u_2, u_3)$ is a variable scale factor.

To express $\mathbf{V}\psi$ in curvilinear coordinates, we note that $\partial\psi/\partial x$ corresponds to $\partial\psi/(h_1 \partial u_1)$, hence

$$\mathbf{V}\psi = \sum_{i=1}^{3} \frac{1}{h_i} \frac{\partial\psi}{\partial u_i} \mathbf{u}_i \quad (10.15)$$

where $\mathbf{u}_i$ are unit vectors along the curvilinear axes. To obtain $\mathbf{V} \cdot \mathbf{A}$ we use Gauss' theorem (Pipes and Harvill, 1970, p. 909) which states that

$$\iiint_{\mathscr{V}} \mathbf{V} \cdot \mathbf{A} \, \mathrm{d}\mathscr{V} = \iint_{\mathscr{S}} \mathbf{A} \cdot \mathrm{d}\mathscr{S}$$

where the surface $\mathscr{S}$ encloses the volume $\mathscr{V}$ and the outward-drawn normal is positive. Applying this to an element of volume $\mathrm{d}\mathscr{V}$ (fig. 10.1e), we get

$$\mathbf{V} \cdot \mathbf{A} \, \mathrm{d}\mathscr{V} = \text{surface integral over the 6 faces,}$$
$$= -A_1(h_2 \mathrm{d}u_2 h_3 \mathrm{d}u_3) + A_1(h_2 \mathrm{d}u_2 h_3 \mathrm{d}u_3)$$
$$+ \left\{ \frac{\partial}{\partial u_1} (A_1 h_2 h_3) \, \mathrm{d}u_1 \right\} \mathrm{d}u_2 \, \mathrm{d}u_3$$

+ similar expressions for
the other pairs of faces.

Thus

$$\mathbf{V} \cdot \mathbf{A}(h_1 \mathrm{d}u_1 h_2 \mathrm{d}u_2 h_3 \mathrm{d}u_3) = \left\{ \frac{\partial}{\partial u_1} (A_1 h_2 h_3) + \frac{\partial}{\partial u_2} (A_2 h_3 h_1) \right.$$
$$\left. + \frac{\partial}{\partial u_3} (A_3 h_1 h_2) \right\} \mathrm{d}u_1 \mathrm{d}u_2 \mathrm{d}u_3,$$

$$\mathbf{V} \cdot \mathbf{A} = \frac{1}{h_1 h_2 h_3} \sum \frac{\partial}{\partial u_i} (A_i h_j h_k), \quad (10.16)$$

i, j, k being in cyclic order.

The Laplacian, $\mathbf{V}^2\psi$, is equal to div grad ψ, i.e.

$$\mathbf{V}^2\psi = \frac{1}{h_1 h_2 h_3} \sum \frac{\partial}{\partial u_i} \left(\frac{h_j h_k}{h_i} \frac{\partial\psi}{\partial u_i} \right) \quad (10.17)$$

10.1.3 *Matrix analysis*

(a) Definitions. A *matrix* is a rectangular array of numbers a_{ij} arranged in r rows and s columns; an entire matrix is here indicated by script bold-face type:

$$\mathscr{A} = \begin{Vmatrix} a_{11} & a_{12} & a_{13} & \cdots \\ a_{21} & a_{22} & a_{23} & \cdots \\ a_{31} & a_{32} & a_{33} & \cdots \\ \cdots \end{Vmatrix}. \quad (10.18)$$

The *order* of a matrix is $(r \times s)$. If $r = 1$, we have a *row matrix*; if $s = 1$ a *column matrix*; these are also called *vectors of the first and second kinds*, respectively. A *null matrix* $\mathscr{O}$ has zeros for all elements. The *transpose* of a matrix has rows and columns interchanged; thus

$$\mathscr{A}^{\mathrm{T}} = \begin{Vmatrix} a_{11} & a_{21} & a_{31} & \cdots \\ a_{12} & a_{22} & a_{32} & \cdots \\ a_{13} & a_{23} & a_{33} & \cdots \\ \cdots \end{Vmatrix}. \quad (10.19)$$

A matrix of order $(r \times r)$ is a *square matrix*. The *principal diagonal* of a square matrix has the elements a_{nn}. A *diagonal matrix* is a square matrix with zeros for all elements which are not on the principal diagonal, that is,

$a_{ij} = 0$ if $i \neq j$, and at least one of the $a_{ii} \neq 0$. An *identity matrix* $\mathscr{I}$ is a diagonal matrix where $a_{ii} = 1$ for all i. A matrix with zeros below (above) the principal diagonal is called an upper (lower) *triangular matrix*. A *symmetric matrix* equals its transpose, that is, $\mathscr{A} = \mathscr{A}^{\mathrm{T}}$, and a *skew-symmetric matrix* equals the negative of its transpose, $\mathscr{A} = -\mathscr{A}^{\mathrm{T}}$. A symmetric matrix where all the elements along any diagonal parallel to the principal diagonal are the same is a *Toeplitz* matrix.

The *cofactor* of an element a_{rs} of a square matrix is $(-1)^{r+s}$ times the determinant formed by deleting the rth row and the sth column. The *adjoint* of a square matrix, adj $(\mathscr{A})$, is the transpose of the matrix $\mathscr{A}$ with each element replaced by its cofactor. The *determinant of a square matrix*, det $(\mathscr{A})$, is a single number given by det $(\mathscr{A}) = \sum_i a_{ik} A_{ik} = \sum_k a_{ik} A_{ik}$, where A_{ik} is the cofactor of a_{ik}. The *inverse* of a square matrix may be found by dividing the adjoint by the determinant [if det $(\mathscr{A}) \neq 0$], that is,

$$\mathscr{A}^{-1} = [1/\det(\mathscr{A})]\ \mathrm{adj}\,(\mathscr{A}), \quad \mathscr{A}^{-1}\mathscr{A} = \mathscr{I} \qquad (10.20)$$

(see problem 10.8*b*).

(*b*) *Matrix operations.* Operations performed on matrices change the values of the matrix elements. Corresponding elements can be added, that is, if $\mathscr{C} = \mathscr{A} + \mathscr{B}$, $c_{rs} = a_{rs} + b_{rs}$; matrices must be of the same order to be added. Matrices may be multiplied by scalars, that is, if $\mathscr{D} = k\mathscr{A}$, $d_{rs} = ka_{rs}$. In matrix multiplication, the ith row of the first matrix is multiplied element-by-element with the jth column of the second matrix and the products are summed to give the ijth element of the product matrix, that is, if $\mathscr{E} = \mathscr{A}\mathscr{B}$, $e_{ij} = \sum_k a_{ik} b_{kj}$. The first matrix must have the same number of columns as the second matrix has rows for matrices to be multiplied. In $\mathscr{E} = \mathscr{A}\mathscr{B}$, if the order of $\mathscr{A}$ is $(m \times n)$ and that of $\mathscr{B}$ is $(n \times p)$, the order of $\mathscr{E}$ is $(m \times p)$. When more than two matrices are multiplied, products can be formed in pairs; thus $\mathscr{A}\mathscr{B}\mathscr{C} = (\mathscr{A}\mathscr{B})\mathscr{C} = \mathscr{A}(\mathscr{B}\mathscr{C})$. The transpose of a product is the product of the transposes of the individual matrices in inverse order, that is, $(\mathscr{A}\mathscr{B})^{\mathrm{T}} = \mathscr{B}^{\mathrm{T}}\mathscr{A}^{\mathrm{T}}$ (see problem 10.9*b*).

It is sometimes convenient to *partition a matrix*, that is, to represent it as a matrix whose elements are submatrices of the original matrix. For example,

$$\mathscr{A} = \begin{Vmatrix} 2 & 0 & 0 & 3 & 5 \\ 1 & -1 & 4 & -2 & 1 \\ 3 & 2 & 1 & -5 & 4 \\ 0 & 0 & 1 & 1 & 0 \end{Vmatrix} = \begin{Vmatrix} \mathscr{P} & \mathscr{Q} \\ \mathscr{R} & \mathscr{S} \end{Vmatrix},$$

where

$$\mathscr{P} = \begin{Vmatrix} 2 & 0 & 0 \\ 1 & -1 & 4 \\ 3 & 2 & 1 \end{Vmatrix}, \quad \mathscr{Q} = \begin{Vmatrix} 3 & 5 \\ -2 & 1 \\ -5 & 4 \end{Vmatrix},$$

$$\mathscr{R} = \begin{Vmatrix} 0 & 0 & 1 \end{Vmatrix}, \quad \mathscr{S} = \begin{Vmatrix} 1 & 0 \end{Vmatrix}.$$

To add $\mathscr{A}$ to a similar (4×5) matrix $\mathscr{B}$, $\mathscr{B}$ must be partitioned in the same way, that is,

$$\mathscr{B} = \begin{Vmatrix} \mathscr{T} & \mathscr{U} \\ \mathscr{V} & \mathscr{W} \end{Vmatrix}$$

where $\mathscr{T}$ is (3×3), $\mathscr{U}$ is (3×2), etc. Then

$$\mathscr{A} + \mathscr{B} = \begin{Vmatrix} \mathscr{P} + \mathscr{T} & \mathscr{Q} + \mathscr{U} \\ \mathscr{R} + \mathscr{V} & \mathscr{S} + \mathscr{W} \end{Vmatrix}.$$

When partitioning matrices which are to be multiplied, the submatrices must be conformable. Thus, if

$$\mathscr{C} = \mathscr{A}\mathscr{B}$$

where $\mathscr{A}$ is $(m \times n)$, $\mathscr{B}$ is $(n \times p)$, we can partition $\mathscr{A}$ and $\mathscr{B}$ as follows:

$$\mathscr{A} = \begin{Vmatrix} \mathscr{C} & \mathscr{D} \\ \mathscr{E} & \mathscr{F} \end{Vmatrix}, \quad \mathscr{B} = \begin{Vmatrix} \mathscr{G} & \mathscr{H} \\ \mathscr{J} & \mathscr{K} \end{Vmatrix},$$

$\mathscr{C}$ being $(a \times b)$, $\mathscr{D}(a \times c)$, $\mathscr{E}(d \times b)$, $\mathscr{F}(d \times c)$, $\mathscr{G}(b \times j)$, $\mathscr{H}(b \times k)$, $\mathscr{J}(c \times j)$, $\mathscr{K}(c \times k)$, $a + d = m$, $b + c = n$, $j + k = p$; then

$$\mathscr{A}\mathscr{B} = \begin{Vmatrix} \mathscr{C}\mathscr{G} + \mathscr{D}\mathscr{J} & \mathscr{C}\mathscr{H} + \mathscr{D}\mathscr{K} \\ \mathscr{E}\mathscr{G} + \mathscr{F}\mathscr{J} & \mathscr{E}\mathscr{H} + \mathscr{F}\mathscr{K} \end{Vmatrix}.$$

Matrices can be used to solve simultaneous equations. If we write a set of linear equations as

$$\left. \begin{array}{l} a_{11}x_1 + a_{12}x_2 + a_{13}x_3 + \cdots = c_1, \\ a_{21}x_1 + a_{22}x_2 + a_{23}x_3 + \cdots = c_2, \\ a_{31}x_1 + a_{32}x_2 + a_{33}x_3 + \cdots = c_3, \end{array} \right\} \qquad (10.21)$$

and let $\mathscr{A}$ be the elements a_{rs}, $\mathscr{X}$ be a column matrix with elements $x_{i1} = x_i$, $\mathscr{C}$ be a column matrix with elements c_i, then we can write

$$\mathscr{A}\mathscr{X} = \mathscr{C}$$

and solve for $\mathscr{X}$:

$$\mathscr{A}^{-1}\mathscr{A}\mathscr{X} = \mathscr{A}^{-1}\mathscr{C},$$

$$\mathscr{X} = \mathscr{A}^{-1}\mathscr{C} \quad \text{because} \quad \mathscr{A}^{-1}\mathscr{A} = \mathscr{I}. \qquad (10.22)$$

This solution requires that $\mathscr{A}$ be a square matrix and that the equations are independent, i.e., det $(\mathscr{A}) \neq 0$ (Pipes and Harvill, 1970, pp. 98–102).

Convolution, $a_t * b_t = c_t$, can be performed by the operation $\mathscr{A}\mathscr{B} = \mathscr{C}$ if $\mathscr{A}$ is a matrix of the form indicated below and $\mathscr{B}$ and $\mathscr{C}$ are column matrices. For simplicity we assume that a_t and b_t both have $n + 1$ data points, zeros being added to achieve this. Then,

$$
\begin{Vmatrix}
a_0 & 0 & 0 & \cdots & 0 \\
a_1 & a_0 & 0 & \cdots & 0 \\
\multicolumn{5}{c}{\cdots\cdots\cdots\cdots\cdots\,\,0} \\
a_n & a_{n-1} & \cdots & & a_0 \\
0 & a_n & \cdots & & a_1 \\
\multicolumn{5}{c}{\cdots\cdots\cdots\cdots\cdots} \\
0 & 0 & \cdots & & a_n
\end{Vmatrix}
\begin{Vmatrix}
b_0 \\ \vdots \\ b_n
\end{Vmatrix}
=
\begin{Vmatrix}
c_0 \\ c_1 \\ \vdots \\ c_n \\ c_{n+1} \\ \vdots \\ c_{2n}
\end{Vmatrix}
\qquad (10.23)
$$

where $c_i = \sum_k a_{i-k} b_k = \sum_k a_k b_{i-k}$. Thus (10.23) gives the same result as (8.18). Note that matrix $\mathscr{A}$ is of order $p \times (n + 1)$ where $p = 2n + 1$.

Cross-correlation can be written as $\mathscr{E}^{\mathrm{T}}\mathscr{D} = \phi_{ed}$, that is,

$$
\begin{Vmatrix}
e_0 & e_1 & \cdot & \cdot & e_n & 0 & \cdot & \cdot & 0 \\
0 & e_0 & e_1 & \cdot & \cdot & e_n & \cdot & \cdot & 0 \\
\cdot & \cdot & \cdot & \cdot & & \cdot & \cdot & \cdot & \\
0 & \cdot & \cdot & 0 & e_0 & \cdot & \cdot & \cdot & e_n \\
0 & \cdot & \cdot & \cdot & 0 & e_0 & \cdot & \cdot & e_{n-1} \\
\cdot & \cdot & \cdot & \cdot & & \cdot & \cdot & \cdot & \\
0 & \cdot & \cdot & \cdot & & \cdot & \cdot & 0 & e_0 \\
d_0 & 0 & \cdot & \cdot & 0 & 0 & \cdot & \cdot & 0 \\
d_1 & d_0 & \cdot & \cdot & & \cdot & \cdot & & \cdot \\
\cdot & d_1 & & & 0 & & \cdot & & \cdot \\
d_n & \cdot & \cdot & \cdot & d_0 & 0 & \cdot & \cdot & 0 \\
0 & d_n & \cdot & \cdot & \cdot & d_0 & \cdot & \cdot & 0 \\
\cdot & \cdot & \cdot & \cdot & & \cdot & \cdot & \cdot & \\
0 & \cdot & \cdot & \cdot & d_n & d_{n-1} & \cdot & \cdot & d_0
\end{Vmatrix}
$$

$$
=
\begin{Vmatrix}
\phi_{ed}(0) & \phi_{ed}(-1) & \cdots & \phi_{ed}(-n) & \cdots & & \cdots & 0 \\
\phi_{ed}(+1) & \phi_{ed}(0) & \cdots & \cdots & & \cdots & \cdots \cdots \\
\cdots & \cdots & \cdots & \cdots & \cdots & \cdots & \cdots \cdots \\
\phi_{ed}(+n) & \cdots & & \cdots & \phi_{ed}(0) & \cdots & \cdots & \phi_{ed}(-n) \\
0 & \phi_{ed}(+n) & \cdots & \cdots & & \phi_{ed}(0) & \cdots \cdots \\
\cdots & \cdots & \cdots & \cdots & & \cdots & \cdots \cdots \\
0 & 0 & \cdots & \phi_{ed}(+n) & \cdots & & \cdots & \phi_{ed}(0)
\end{Vmatrix}.
$$

$$(10.24)$$

This gives the same values as (8.36). The cross-correlation matrix is a Toeplitz matrix. Another scheme for cross-correlating e_t with d_t is given by

$$
\begin{Vmatrix}
e_n & 0 & \cdots & 0 \\
e_{n-1} & e_n & \cdots & 0 \\
\cdots & \cdots & \cdots & \cdots \\
e_0 & e_1 & \cdots & e_n \\
0 & e_0 & \cdots & e_{n-1} \\
\cdots & \cdots & \cdots & \cdots \\
0 & 0 & \cdots & e_0
\end{Vmatrix}
\begin{Vmatrix}
d_0 \\ \vdots \\ d_n
\end{Vmatrix}
=
\begin{Vmatrix}
\phi_{ed}(-n) \\ \cdots \\ \phi_{ed}(-1) \\ \phi_{ed}(0) \\ \phi_{ed}(+1) \\ \cdots \\ \phi_{ed}(+n)
\end{Vmatrix}.
\qquad (10.25)
$$

Autocorrelation is given by

$$\phi_{dd} = \mathscr{D}^{\mathrm{T}}\mathscr{D}. \qquad (10.26)$$

The autocorrelation is also a Toeplitz matrix.

The Wiener filter normal equations, (8.58), can be expressed in matrix form as

$$\phi_{gg}\mathscr{F} = \phi_{gh} \qquad (10.27)$$

where

$$
\phi_{gg} =
\begin{Vmatrix}
\phi_{gg}(0) & \phi_{gg}(-1) & \cdots & \phi_{gg}(-n) \\
\phi_{gg}(1) & \phi_{gg}(0) & \cdots & \phi_{gg}(-n+1) \\
\cdots & \cdots & & \cdots \cdots \\
\phi_{gg}(n) & \phi_{gg}(n-1) & \cdots & \phi_{gg}(0)
\end{Vmatrix},
\qquad (10.28)
$$

$$
\mathscr{F} =
\begin{Vmatrix}
f_0 \\ f_1 \\ \vdots \\ f_n
\end{Vmatrix},
\quad
\phi_{gh} =
\begin{Vmatrix}
\phi_{gh}(0) \\ \phi_{gh}(1) \\ \vdots \\ \phi_{gh}(n)
\end{Vmatrix}.
\qquad (10.29)
$$

The filter $\mathscr{F}$ is given by

$$\mathscr{F} = \phi_{gg}^{-1}\phi_{gh}. \qquad (10.30)$$

Sometimes the solution of a matrix equation involves the inversion of a matrix which is not square, as in $\mathscr{A}\mathscr{B} = \mathscr{C}$, where $\mathscr{A}$ is of size $(m \times n)$, $\mathscr{B}$ of size $(n \times p)$ and $\mathscr{C}$ of size $(m \times p)$. To solve for $\mathscr{B}$, we multiply by $\mathscr{A}^{\mathrm{T}}$ (note that $\mathscr{A}^{\mathrm{T}}\mathscr{A}$ is always square), then multiply by $(\mathscr{A}^{\mathrm{T}}\mathscr{A})^{-1}$ to get $\mathscr{B}$:

$$\mathscr{B} = (\mathscr{A}^{\mathrm{T}}\mathscr{A})^{-1}\mathscr{A}^{\mathrm{T}}\mathscr{C}. \qquad (10.31)$$

10.1.4 *Complex numbers*

The square roots of negative numbers are *imaginary numbers* and numbers which are partly real and partly imaginary are *complex numbers*. If we write $j = \sqrt{-1}$, that is, $j^2 = -1$ (some writers use i instead of j), we can write, for example, the imaginary number $\sqrt{-9} = \sqrt{9}\sqrt{-1} = 3j$.

A complex number, $z = a + jb$, can be represented by plotting in the *complex plane* where the direction of imaginary numbers is at right angles to the real direction, as in fig. 10.2. We can also express complex numbers in polar form:

$$z = a + jb = r(\cos\theta + j\sin\theta) = re^{j\theta} \qquad (10.32)$$

(see problem 10.11a) where $r = (a^2 + b^2)^{\frac{1}{2}}$ = modulus of $z = |z|$ and $\theta = \tan^{-1}(b/a)$ = arg (z). The conjugate complex of z, $\bar{z}$, is defined as $\bar{z} = a - jb = r(\cos\theta - j\sin\theta) = re^{-j\theta}$ (see fig. 10.2).

The sum (or difference) of complex numbers is obtained by adding (or subtracting) the real and imaginary parts. If $z_1 = a + jb$, $z_2 = c + jd$, then $(z_1 \pm z_2) = (a \pm c) + j(b \pm d)$. A complex number is zero only if both its real and imaginary parts are zero, hence two complex numbers are equal only if both their real and imaginary parts are equal. Multiplication and division obey the usual algebraic rules. For example (see also problem 10.11b),

$$\left.\begin{aligned}
z_1 z_2 &= (a + jb)(c + jd) = (ac - bd) \\
&\quad + j(ad + bc) \\
&= r_1 r_2 \{\cos(\theta_1 + \theta_2) + j\sin(\theta_1 + \theta_2)\} \\
&= r_1 r_2 \, e^{j(\theta_1 + \theta_2)};
\end{aligned}\right\} \qquad (10.33)$$

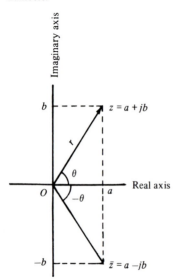

Fig.10.2. Geometrical representation of complex numbers.

$$\left.\begin{aligned}
z_1/z_2 &= \frac{(a + jb)}{(c + jd)} = \frac{(a + jb)(c - jd)}{(c + jd)(c - jd)} \\
&= \frac{(ac + bd) + j(bc - ad)}{c^2 + d^2} \\
&= (r_1/r_2)\{\cos(\theta_1 - \theta_2) + j\sin(\theta_1 - \theta_2)\} \\
&= (r_1/r_2)\,e^{j(\theta_1 - \theta_2)}.
\end{aligned}\right\} \qquad (10.34)$$

The nth root of z, z_0, can be found by writing

$$\begin{aligned}
z = re^{j\theta} = z_0^n = (r_0\,e^{j\theta_0})^n &= \{r_0(\cos\theta_0 + j\sin\theta_0)\}^n \\
&= r_0^n(\cos n\theta_0 + j\sin n\theta_0)
\end{aligned}$$

by de Moivre's theorem (see problem 10.11a). Hence,

$$\begin{aligned}
z^{1/n} = z_0 &= r_0(\cos\theta_0 + j\sin\theta_0), \\
r_0 &= r^{1/n}, \quad \theta_0 = (\theta + 2\pi k)/n, \\
k &= 0, 1, 2, 3, \ldots, n - 1.
\end{aligned} \qquad (10.35)$$

Figure 10.3 shows roots plotted in polar form for the case where $z = r = re^{j2\pi}$ = real, $n = 5$ and 6.

10.1.5 *Method of least squares*

Let us assume that we wish to obtain the 'best-fit' curve of order m,

$$y_i = a_0 + a_1 x_i + a_2 x_i^2 + \cdots + a_m x_i^m \qquad (10.36)$$

to represent a set of n pairs of measured values (x_i, y_i). If $n = m + 1$, the curve will pass through all n points, (x_i, y_i). If $n > m + 1$, the curve will not pass through all n points and we seek the 'best-fit' curve such that the sum of the squares of the 'errors' between the curve and each point (x_i, y_i) is a minimum, the errors e_i being the differences between the measured values y_i and those given by the curve. Thus,

$$\begin{aligned}
e_i &= y_i - (a_0 + a_1 x_i + \cdots + a_m x_i^m), \\
i &= 1, 2, 3, \ldots, n,
\end{aligned}$$

and we wish to minimize E where

$$E = \sum_{i=1}^{n} e_i^2 = \sum_i \{y_i - (a_0 + a_1 x_i + \ldots + a_m x_i^m)\}^2.$$

Fig.10.3. Roots of a complex number.

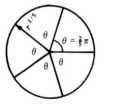
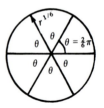

Since E is a function of the parameters a_k only, the minimum is given by

$$\frac{\partial E}{\partial a_k} = 2\sum_i (y_i - a_0 - a_1 x_i - \ldots - a_m x_i^m)(-x_i^k) = 0,$$

$$a_0 \sum_i x_i^k + a_1 \sum_i x_i^{k+1} + \ldots + a_m \sum_i x_i^{k+m} = \sum_i x_i^k y_i,$$

$$k = 0, 1, 2, \ldots, m. \tag{10.37}$$

There are $m + 1$ such *normal equations*, so we can solve for the $m + 1$ unknowns, a_k.

Sometimes we wish to find a least-squares solution subject to a certain condition on the unknown parameters (*constraint*), for example, we could require that $a_1 = a_4$ and/or $a_1 + a_2 + a_3 = 0$. We can write each constraint in the form $C(a_1, a_2, \ldots, a_m) = 0$. Since the a_i s are chosen so that $\partial E / \partial a_i = 0$ (and $\partial C / \partial a_i = 0$), we can write the least-squares condition with constraints in the form

$$\frac{\partial}{\partial a_i}(E + \lambda C) = 0, i = 0, 1, \ldots, m, \tag{10.38}$$

λ having the same significance here as in the Lagrange method of undetermined multipliers (Pipes and Harvill, 1970, p. 968). These $(m + 1)$ equations plus the equation $C(a_1, a_2, \ldots, a_m) = 0$ suffice to solve for λ plus the $(m + 1)$ values of a_i. The extension to several constraints involves solving

$$\frac{\partial}{\partial a_i}(E + \sum_j \lambda_j C_j) = 0.$$

Equation (10.37) can be written in matrix form as

$$\mathcal{Y}^* = \mathcal{X}^* \mathcal{A}^* \tag{10.39}$$

where

$$\mathcal{X}^* = \begin{Vmatrix} \sum_i x_i^0 & \sum_i x_i^1 & \cdots & \sum_i x_i^m \\ \sum_i x_i^1 & \sum_i x_i^2 & \cdots & \sum_i x_i^{m+1} \\ \cdots & \cdots & \cdots & \cdots \\ \sum_i x_i^m & \sum_i x_i^{m+1} & \cdots & \sum_i x_i^{2m} \end{Vmatrix},$$

$$\mathcal{A}^* = \begin{Vmatrix} a_0 \\ a_1 \\ \vdots \\ a_m \end{Vmatrix}, \quad \mathcal{Y}^* = \begin{Vmatrix} \sum_i y_i \\ \sum_i x_i y_i \\ \cdots \\ \sum_i x_i^m y_i \end{Vmatrix}.$$

Since $\mathcal{X}^*$ is square,

$$\mathcal{A}^* = (\mathcal{X}^*)^{-1} \mathcal{Y}^*. \tag{10.40}$$

If we solve the least-squares problem using matrices from the beginning, we can obtain a more general result which is also well adapted to computer calculations. We write (10.36) in the form

$$\mathcal{X} \mathcal{A} = \mathcal{Y}$$

where

$$\mathcal{X} = \begin{Vmatrix} x_{11} & x_{12} & \cdots & x_{1m} \\ x_{21} & x_{22} & \cdots & x_{2m} \\ \cdots & \cdots & \cdots & \cdots \\ x_{n1} & x_{n2} & \cdots & x_{nm} \end{Vmatrix}, \quad \mathcal{A} = \begin{Vmatrix} a_1 \\ a_2 \\ \vdots \\ a_m \end{Vmatrix}, \quad \mathcal{Y} = \begin{Vmatrix} y_1 \\ y_2 \\ \vdots \\ y_n \end{Vmatrix},$$

x_{ij}, y_i being known, a_i unknown and $n > m$. (In general x_{ij} can be any $m \times n$ known quantities, including powers of x_i as in (10.36).)

Since we have more equations than unknowns, the equations cannot hold exactly; writing $\mathcal{E}$ as the column matrix of the errors e_i, $i = 1, 2, \ldots, n$, we obtain

$$\mathcal{X} \mathcal{A} - \mathcal{Y} = \mathcal{E}.$$

This can be simplified by writing (Claerbout, 1976, p. 107)

$$\begin{Vmatrix} -y_1 & x_{11} & \cdots & x_{1m} \\ \cdot & \cdot & & \cdot \\ \cdot & & & \cdot \\ \cdot & & & \cdot \\ \cdot & & & \cdot \\ -y_n & x_{n1} & \cdots & x_{nm} \end{Vmatrix} \begin{Vmatrix} 1 \\ a_1 \\ \cdot \\ \cdot \\ \cdot \\ a_m \end{Vmatrix} = \begin{Vmatrix} e_1 \\ \cdot \\ \cdot \\ \cdot \\ \cdot \\ e_n \end{Vmatrix}.$$

This can be partitioned thus:

$$\left\| -\mathcal{Y} \mid \mathcal{X} \right\| \begin{Vmatrix} 1 \\ --- \\ \mathcal{A} \end{Vmatrix} = \mathcal{E} = \mathcal{B} \begin{Vmatrix} 1 \\ --- \\ \mathcal{A} \end{Vmatrix}$$

where the first column of $\mathcal{B}$ is $-\mathcal{Y}$ and the rest is $\mathcal{X}$. We can accommodate this by taking the first column as the 0th column, that is, $b_{i0} = -y_i$ and $b_{ij} = x_{ij}, j > 0$.

İndividual errors are given by

$$e_i = \| 1\, a_1 \ldots a_m \| \begin{Vmatrix} -y_i \\ x_{i1} \\ \vdots \\ x_{im} \end{Vmatrix} = \| -y_i \quad x_{i1} \ldots x_{im} \| \begin{Vmatrix} 1 \\ a_1 \\ \vdots \\ a_m \end{Vmatrix}.$$

Then,

$$e_i^2 = \| 1\, a_1 \ldots a_m \| \begin{Vmatrix} -y_i \\ x_{i1} \\ \vdots \\ x_{im} \end{Vmatrix} \| -y_i \quad x_{i1} \ldots x_{im} \| \begin{Vmatrix} 1 \\ a_1 \\ \vdots \\ a_m \end{Vmatrix},$$

and

$$E = \sum_i e_i^2 = \|1\, a_1 \ldots a_m\| \sum_i \left(\left\| \begin{matrix} -y_i \\ x_{i1} \\ \vdots \\ x_{im} \end{matrix} \right\| \times \| -y_i \quad x_{i1} \ldots x_{im} \| \right) \left\| \begin{matrix} 1 \\ a_1 \\ \vdots \\ a_m \end{matrix} \right\|$$

$$= \left\| 1 \vdots \mathscr{A}^{\mathrm{T}} \right\| \mathscr{R} \left\| \begin{matrix} 1 \\ --- \\ \mathscr{A} \end{matrix} \right\|$$

where

$$\mathscr{R} = \sum_i \left\| \begin{matrix} -y_i \\ x_{i1} \\ \vdots \\ x_{im} \end{matrix} \right\| \| -y_i \quad x_{i1} \ldots x_{im} \| = \mathscr{B}^{\mathrm{T}} \mathscr{B}. \quad (10.41)$$

Setting derivatives of E with respect to a_i equal to zero gives

$$\frac{\partial E}{\partial a_i} = 0 = \|0 \ldots 0\,1\,0 \ldots 0\| \mathscr{R} \left\| \begin{matrix} 1 \\ --- \\ \mathscr{A} \end{matrix} \right\| + \|1 \vdots \mathscr{A}^{\mathrm{T}}\| \mathscr{R} \left\| \begin{matrix} 0 \\ \vdots \\ 0 \\ 1 \\ 0 \\ \vdots \\ 0 \end{matrix} \right\|$$

$$= 2\|0 \ldots 0\,1\,0 \ldots 0\| \mathscr{R} \left\| \begin{matrix} 1 \\ --- \\ \mathscr{A} \end{matrix} \right\|$$

since $\mathscr{R} = \mathscr{B}^{\mathrm{T}} \mathscr{B}$ is symmetrical. This result can be written

$$\| r_{i1} \ldots r_{im} \| \left\| \begin{matrix} 1 \\ --- \\ \mathscr{A} \end{matrix} \right\| = 0, \quad i = 1, 2, 3, \ldots, m.$$

If we combine the m equations, we get

$$\mathscr{R} \left\| \begin{matrix} 1 \\ --- \\ \mathscr{A} \end{matrix} \right\| = \mathcal{O}$$

except that we lack the 0th row of $\mathscr{R}$. We define a quantity v by the relation

$$\| r_{00} r_{01} \ldots r_{0m} \| \left\| \begin{matrix} 1 \\ --- \\ \mathscr{A} \end{matrix} \right\| = v; \quad (10.42)$$

we now have

$$\mathscr{R} \left\| \begin{matrix} 1 \\ --- \\ \mathscr{A} \end{matrix} \right\| = \left\| \begin{matrix} v \\ 0 \\ \vdots \\ 0 \end{matrix} \right\| = \mathscr{B}^{\mathrm{T}} \mathscr{B} \left\| \begin{matrix} 1 \\ --- \\ \mathscr{A} \end{matrix} \right\|$$

$$= \left\| \begin{matrix} -\mathscr{Y}^{\mathrm{T}} \\ ------ \\ \mathscr{X}^{\mathrm{T}} \end{matrix} \right\| \| -\mathscr{Y} \vdots \mathscr{X} \| \left\| \begin{matrix} 1 \\ --- \\ \mathscr{A} \end{matrix} \right\| \quad (10.43)$$

$$= \left\| \begin{matrix} \mathscr{Y}^{\mathrm{T}} \mathscr{Y} & -\mathscr{Y}^{\mathrm{T}} \mathscr{X} \\ -\mathscr{X}^{\mathrm{T}} \mathscr{Y} & \mathscr{X}^{\mathrm{T}} \mathscr{X} \end{matrix} \right\| \left\| \begin{matrix} 1 \\ --- \\ \mathscr{A} \end{matrix} \right\|,$$

that is,

$$\left\| \begin{matrix} v \\ 0 \\ \vdots \\ 0 \end{matrix} \right\| = \left\| \begin{matrix} \mathscr{Y}^{\mathrm{T}} \mathscr{Y} - \mathscr{Y}^{\mathrm{T}} \mathscr{X} \mathscr{A} \\ -\mathscr{X}^{\mathrm{T}} \mathscr{Y} + \mathscr{X}^{\mathrm{T}} \mathscr{X} \mathscr{A} \end{matrix} \right\|.$$

Thus, we have

$$v = \mathscr{Y}^{\mathrm{T}} \mathscr{Y} - \mathscr{Y}^{\mathrm{T}} \mathscr{X} \mathscr{A},$$

and

$$\left\| \begin{matrix} 0 \\ 0 \\ \vdots \\ 0 \end{matrix} \right\| = -\mathscr{X}^{\mathrm{T}} \mathscr{Y} + \mathscr{X}^{\mathrm{T}} \mathscr{X} \mathscr{A},$$

hence

$$\mathscr{A} = (\mathscr{X}^{\mathrm{T}} \mathscr{X})^{-1} \mathscr{X}^{\mathrm{T}} \mathscr{Y}. \quad (10.44)$$

At times we wish to give extra weight to one or more sets of observations $(y_i, x_{i1}, x_{i2}, \ldots, x_{im})$. We can do this by multiplying the error e_i and the ith row of $\mathscr{B}$ by a weighting factor $\sqrt{w_i}$. Then,

$$E = \sum w_i e_i^2 = \|1 \vdots \mathscr{A}^{\mathrm{T}}\| \sum \left(w_i \left\| \begin{matrix} -y_i \\ x_{i1} \\ \vdots \\ x_{im} \end{matrix} \right\| \times \| -y_i \quad x_{i1} \ldots x_{im} \| \right) \left\| \begin{matrix} 1 \\ --- \\ \mathscr{A} \end{matrix} \right\| = \|1 \vdots \mathscr{A}^{\mathrm{T}}\| \mathscr{R}^* \left\| \begin{matrix} 1 \\ --- \\ \mathscr{A} \end{matrix} \right\|$$

$$(10.45)$$

where $\mathscr{R}^*$ is the weighted form of $\mathscr{R} = \mathscr{B}^{\mathrm{T}} \mathscr{B}$, i.e. the product of $\mathscr{B}^{\mathrm{T}}$ and $\mathscr{B}$ when the ith row of $\mathscr{B}$ and the ith column of $\mathscr{B}^{\mathrm{T}}$ are multiplied by $\sqrt{w_i}$. Equation (10.44) is still valid except that the ith row of $\mathscr{X}$ and $\mathscr{Y}$ and the ith column of $\mathscr{X}^{\mathrm{T}}$ are multiplied by $\sqrt{w_i}$.

Constraints can be considered as additional equations in the unknowns a_i which must be satisfied exactly, that is, they have infinite weights in comparison with the error equations. We can write k linear constraint equations, $k < m$, in the form

$$\mathscr{C}\left\|\begin{array}{c}1\\---\\\mathscr{A}\end{array}\right\| = \left\|\begin{array}{c}0\\0\\0\\\vdots\\0\end{array}\right\| = \mathscr{O} \qquad (10.46)$$

where

$$\mathscr{C} = \left\|\begin{array}{cccc}c_{10} & c_{11} & \cdots & c_{1m}\\c_{20} & c_{21} & \cdots & c_{2m}\\\cdots & \cdots & \cdots & \cdots\\c_{k0} & c_{k1} & \cdots & c_{km}\end{array}\right\|.$$

Following Claerbout (1976, pp. 112–3), we assign the weight $\sqrt{w}$ to each constraint equation, insert the weighted left-hand side of (10.46) in the error equations, derive the result, then let w approach infinity.

Without constraints we had the result (see (10.41) and (10.43))

$$\mathscr{R}\left\|\begin{array}{c}1\\\mathscr{A}\end{array}\right\| = \sum_{i=1}^{n}\left(\left\|\begin{array}{c}-y_i\\x_{i1}\\\vdots\\x_{im}\end{array}\right\|\|-y_i\ x_{i1}\dots x_{im}\|\right)\left\|\begin{array}{c}1\\\mathscr{A}\end{array}\right\| = \left\|\begin{array}{c}v\\0\\\vdots\\0\end{array}\right\|;$$

with constraints, this becomes

$$\left\{\sum_{i=1}^{n}\left(\left\|\begin{array}{c}-y_i\\x_{i1}\\\vdots\\x_{im}\end{array}\right\|\|-y_i\ \ x_{i1}\dots x_{im}\|\right)\right.$$

$$\left.+\sum_{i=1}^{k}w\left(\left\|\begin{array}{c}c_{io}\\c_{i1}\\\vdots\\c_{im}\end{array}\right\|\|c_{io}\,c_{i1}\dots c_{im}\|\right)\right\}\left\|\begin{array}{c}1\\---\\\mathscr{A}\end{array}\right\| = \left\|\begin{array}{c}v^*\\0\\\vdots\\0\end{array}\right\| \qquad (10.47)$$

that is,

$$(\mathscr{B}^{\mathrm{T}}\mathscr{B} + w\mathscr{C}^{\mathrm{T}}\mathscr{C})\left\|\begin{array}{c}1\\---\\\mathscr{A}\end{array}\right\| = \left\|\begin{array}{c}v^*\\0\\\vdots\\0\end{array}\right\| = \mathscr{V}^*$$

We write

$$w = 1/\varepsilon, \qquad \left\|\begin{array}{c}1\\---\\\mathscr{A}\end{array}\right\| = \mathscr{A}_0^* + \varepsilon\mathscr{A}_1^* + \varepsilon^2\mathscr{A}_2^* + \dots,$$

where $\mathscr{A}_0^*$ is the matrix which gives the desired solution, $\mathscr{A}_1^*$ is a similar matrix with different unknowns, etc. Substituting, we get

$$(\mathscr{B}^{\mathrm{T}}\mathscr{B} + \tfrac{1}{\varepsilon}\mathscr{C}^{\mathrm{T}}\mathscr{C})(\mathscr{A}_0^* + \varepsilon\mathscr{A}_1^* + \dots) = \mathscr{V}^*.$$

Equating powers of ε gives:

$$\left.\begin{array}{l}\varepsilon^{-1}:\ \mathscr{C}^{\mathrm{T}}\mathscr{C}\mathscr{A}_0^* = \mathscr{O}\\[4pt]\varepsilon^{0}:\ \mathscr{B}^{\mathrm{T}}\mathscr{B}\mathscr{A}_0^* + \mathscr{C}^{\mathrm{T}}\mathscr{C}\mathscr{A}_1^* = \mathscr{V}^*.\end{array}\right\} \qquad (10.48)$$

Since

$$\mathscr{C}\mathscr{A}_0^* = \mathscr{C}\left\|\begin{array}{c}1\\---\\\mathscr{A}\end{array}\right\| = \mathscr{O}$$

from (10.46), the first equation is satisfied automatically, hence provides no new information. In the second equation, we substitute $\mathscr{C}\mathscr{A}_1^* = \mathscr{L}$, where $\mathscr{L}$ is a $(k \times 1)$ matrix whose elements are the equivalents of Lagrangian undetermined multipliers. Then

$$\mathscr{B}^{\mathrm{T}}\mathscr{B}\mathscr{A}_0^* + \mathscr{C}^{\mathrm{T}}\mathscr{L} = \mathscr{V}^* \qquad (10.49)$$

Equations (10.46) and (10.49), which give the solutions for the m unknowns a_i and the k unknowns λ_i, can be combined in the form

$$\left\|\begin{array}{c:c}\mathscr{B}^{\mathrm{T}}\mathscr{B} & \mathscr{C}^{\mathrm{T}}\\\hdashline\mathscr{C} & \mathscr{O}\end{array}\right\|\left\|\begin{array}{c}\mathscr{A}_0^*\\\mathscr{L}\end{array}\right\| = \left\|\begin{array}{c}\mathscr{V}^*\\---\\\mathscr{O}\end{array}\right\|,$$

or

$$\left\|\begin{array}{c}\mathscr{B}^{\mathrm{T}}\mathscr{B}\mathscr{A}_0^* \ + \ \mathscr{C}^{\mathrm{T}}\mathscr{L}\\\hdashline\mathscr{C}\mathscr{A}_0^* \ + \ \mathscr{O}\end{array}\right\| = \left\|\begin{array}{c}\mathscr{V}^*\\---\\\mathscr{O}\end{array}\right\|. \qquad (10.50)$$

10.1.6 *Finite differences*

The calculus of finite differences has wide application in many practical situations, for example, calculations with data in digital form, interpolation, numerical differentiation and integration, numerical solution of differential equations. We shall discuss the basic relations assuming a function $f(x)$ which is given at equal intervals Δ, that is, we discuss the discrete function $f(x + n\Delta)$, $n = 0, \pm 1, \pm 2, \dots$ We shall use the notation $f_n = f(x_0 + n\Delta)$ (note that $f_0 = f(x_0)$).

We define the following operators:

$$3\{f_0\} = f(x_0 + \Delta) = f_1,$$

that is, $3\{f_{n-1}\} = f_n,$

$$(10.51)$$

$$\mathfrak{D}\{f_n\} = f_{n+1} - f_n, \tag{10.52}$$

$$\mathfrak{R}\{f_n\} = \frac{df(x)}{dx}\bigg|_{x=x_0+n\Delta}, \tag{10.53}$$

$$\mathfrak{I}\{f_0\} = \int_{x_0}^{x_0+\Delta} f(x)\,dx. \tag{10.54}$$

These are respectively the delay, difference, derivative and integration operators.

The first and second differences are given by

$$\mathfrak{D}\{f_n\} = f_{n+1} - f_n,$$
$$\mathfrak{D}^2\{f_n\} = \mathfrak{D}\{\mathfrak{D}\{f_n\}\} = \mathfrak{D}\{f_{n+1}\} - \mathfrak{D}\{f_n\}.$$

By successive applications of the above we find that

$$\mathfrak{D}^r\{f_n\} = f_{n+r} - r f_{n+r-1} + \frac{r(r-1)}{2!} f_{n+r-2}$$
$$+ \ldots + (-1)^{r-1} \frac{r(r-1)\ldots 3.2}{(r-1)!} f_{n+1}$$
$$+ (-1)^r f_n. \tag{10.55}$$

Also,

$$\mathfrak{D}\{f_n\} = f_{n+1} - f_n = \mathfrak{J}\{f_n\} - f_n,$$

hence

$$\mathfrak{D} = (\mathfrak{J} - 1). \tag{10.56}$$

[Sometimes the difference is referenced to $\{x + (n + \frac{1}{2})\Delta\}$ instead of $(x + n\Delta)$ in which case $\mathfrak{D}\{f_{n+\frac{1}{2}}\}$ is called the *central difference*.]

Combinations of operators are often used, for example,

$$\mathfrak{J}\{\mathfrak{R}\{f_0\}\} = \mathfrak{J}\left\{\left(\frac{df}{dx}\right)\bigg|_{x=x_0}\right\} = \left(\frac{df}{dx}\right)\bigg|_{x=x_0+\Delta} = f'_1;$$

$$\mathfrak{D}\{\mathfrak{R}\{f_n\}\} = f'_{n+1} - f'_n.$$

The operators obey the basic laws of algebra such as association, distribution and commutation.

We can apply Taylor's series to $\mathfrak{J}\{f_0\}$ to obtain

$$\mathfrak{J}\{f_0\} = f(x_0 + \Delta) = f(x_0) + \Delta f'(x_0)$$
$$+ (\Delta^2/2!)f''(x_0) + \ldots,$$
$$= \{1 + (\Delta\mathfrak{R}) + (1/2!)(\Delta\mathfrak{R})^2 + \ldots\}f(x_0) \bigg\} \tag{10.57}$$
$$= e^{\Delta\mathfrak{R}}\{f_0\}.$$

To obtain an expression for interpolation between $f(x_0)$ and $f(x_0 + \Delta)$, we use (10.56) to write

$$\mathfrak{J}^r = (1 + \mathfrak{D})^r = 1 + r\mathfrak{D} + \frac{r(r-1)}{2!}\mathfrak{D}^2 + \ldots \tag{10.58}$$

Although r is normally a positive integer in this expression, the equation is still valid when $0 < r < +1$ (Wylie, 1966), which enables us to interpolate between f_0 and f_1; in this case the result is known as the *Forward Gregory–Newton formula* (a Backward Gregory–Newton formula exists for use near the upper end of the tabulated values – see Wylie, *loc. cit.*).

An expression for the derivative, $\mathfrak{R}\{f_0\}$, can be found as follows:

$$f(x_0 + r\Delta) = \mathfrak{J}^r\{f_0\} = (1 + \mathfrak{D})^r f_0,$$
$$= \{1 + r\mathfrak{D} + \frac{r(r-1)}{2!}\mathfrak{D}^2 + \ldots\}f_0.$$

Differentiating with respect to r (that is, treating r as if it were a continuous variable) gives

$$\frac{df(x_0 + r\Delta)}{\Delta\,dr} = \frac{1}{\Delta}\left(\mathfrak{D} + \frac{2r-1}{2!}\mathfrak{D}^2\right.$$
$$\left. + \frac{3r^2 - 6r + 2}{3!}\mathfrak{D}^3 + \ldots\right)f_0. \tag{10.59}$$

We let r go to zero and obtain

$$\mathfrak{R}\{f_0\} = \frac{df}{dx}\bigg|_{x=x_0} = \lim_{r\to 0}\frac{df(x_0 + r\Delta)}{\Delta\,dr}$$
$$= \frac{1}{\Delta}(\mathfrak{D} - \frac{1}{2}\mathfrak{D}^2 + \frac{1}{3}\mathfrak{D}^3 - \frac{1}{4}\mathfrak{D}^4 + \ldots)f_0. \tag{10.60}$$

Repeating the process, we find that

$$\mathfrak{R}^2\{f_0\} = \frac{d^2f}{dx^2} = \frac{1}{\Delta^2}(\mathfrak{D} - \mathfrak{D}^2 + \frac{11}{12}\mathfrak{D}^3 - \ldots)f_0. \tag{10.61}$$

The integral, $\mathfrak{I}\{f_0\}$, can be found as follows:

$$\mathfrak{I}\{\mathfrak{R}\{f_0\}\} = \int_{x_0}^{x_0+\Delta} f'(x)\,dx = f(x_0 + \Delta) - f(x_0)$$
$$= \mathfrak{D}\{f_0\},$$

so that

$$\mathfrak{I}\mathfrak{R} = \mathfrak{D}. \tag{10.62}$$

From (10.56) and (10.57), we have

$$\mathfrak{J} = 1 + \mathfrak{D} = \exp\{\Delta\mathfrak{R}\},$$

or

$$\mathfrak{R} = \frac{1}{\Delta}\ln(1 + \mathfrak{D}) = \frac{1}{\Delta}(\mathfrak{D} - \frac{1}{2}\mathfrak{D}^2 + \frac{1}{3}\mathfrak{D}^3 - \ldots).$$

Thus

$$\mathfrak{I} = \mathfrak{D}/\mathfrak{R} = \Delta(1 - \frac{1}{2}\mathfrak{D} + \frac{1}{3}\mathfrak{D}^2 - \frac{1}{4}\mathfrak{D}^3 + \ldots)^{-1}.$$

Therefore,

$$\Im\{f_0\} = \Delta(1 + \tfrac{1}{2}\mathfrak{D} - \tfrac{1}{12}\mathfrak{D}^2 + \tfrac{1}{24}\mathfrak{D}^3 - \ldots)f_0. \qquad (10.63)$$

If we multiply together the operators in (10.60) and (10.63), we find that the product is $\mathfrak{D}$ in agreement with (10.62).

Often one must solve differential equations by numerical methods, generally because the equation is too complex to be solved analytically or the data are in digital form. The basic problem is that of finding $y(x)$ where $dy/dx = y' = f(x, y)$, $y = y_0$ when $x = x_0$, $f(x, y)$ being given either in functional form or as a table of values of $f(x, y)$ at equal intervals Δ of x (if the values are at unequal intervals, it is usually necessary to interpolate to get evenly-spaced values).

We use Taylor's series (see (10.57)) to write

$$y_1 = y(x_0 + \Delta) \approx y(x_0) + \Delta\frac{dy}{dx} \approx y_0 + \Delta f(x_0, y_0),$$

$$y_2 = y(x_0 + 2\Delta) \approx y_1 + \Delta f(x_0 + \Delta, y_1),$$

$$\ldots$$

$$y_{n+1} \approx y_n + \Delta f(x_0 + n\Delta, y_n). \qquad (10.64)$$

This method, called the *Euler–Cauchy method*, involves calculating successive values of the derivative, $f(x_0 + r\Delta, y_r)$, and using Taylor's series to find the approximate value of y_{r+1}. The accuracy is low unless Δ is small but in this case the computing time can become excessive. Many methods have been devised, most of which involve higher-order terms in the Taylor series, to increase the accuracy and decrease the computing time. We shall describe briefly two of these.

Milne's method starts from (10.59) with

$$r = 1, 2, 3, 4.$$

Writing y in place of f, $y_r' = \dfrac{dy}{dx}\bigg|_{x = x_0 + r\Delta}$, we get

$$y_1' = \frac{1}{\Delta}(\mathfrak{D} + \tfrac{1}{2}\mathfrak{D}^2 - \tfrac{1}{6}\mathfrak{D}^3 + \tfrac{1}{12}\mathfrak{D}^4)y_0,$$

$$y_2' = \frac{1}{\Delta}(\mathfrak{D} + \tfrac{3}{2}\mathfrak{D}^2 + \tfrac{1}{3}\mathfrak{D}^3 - \tfrac{1}{12}\mathfrak{D}^4)y_0,$$

$$y_3' = \frac{1}{\Delta}(\mathfrak{D} + \tfrac{5}{2}\mathfrak{D}^2 + \tfrac{11}{6}\mathfrak{D}^3 + \tfrac{1}{4}\mathfrak{D}^4)y_0,$$

$$y_4' = \frac{1}{\Delta}(\mathfrak{D} + \tfrac{7}{2}\mathfrak{D}^2 + \tfrac{13}{3}\mathfrak{D}^3 + \tfrac{25}{12}\mathfrak{D}^4)y_0.$$

Using (10.55) these become

$$\left.\begin{array}{l} y_1' = (-3y_0 - 10y_1 + 18y_2 - 6y_3 + y_4)/12\Delta, \\ y_2' = (y_0 - 8y_1 + 8y_3 - y_4)/12\Delta, \\ y_3' = (-y_0 + 6y_1 - 18y_2 + 10y_3 + 3y_4)/12\Delta, \\ y_4' = (3y_0 - 16y_1 + 36y_2 - 48y_3 + 25y_4)/12\Delta. \end{array}\right\} \quad (10.65)$$

Twice the sum of the first and third equations minus the second gives

$$y_4 = y_0 + (4\Delta/3)(2y_1' - y_2' + 2y_3');$$

since (x_0, y_0) can be any one of the set of values, this relation applies to any four successive values, that is,

$$y_{n+1} = y_{n-3} + (4\Delta/3)(2y_{n-2}' - y_{n-1}' + 2y_n'). \qquad (10.66)$$

If we know y_r for four consecutive values, $r = n, n-1, n-2, n-3$, we can find y_r' for these values, hence obtain y_{n+1} from (10.66). To get started we need the values of y_0, y_1, y_2, y_3; the derivatives y_1', y_2', y_3', can then be found by substitution in the differential equation after which we can calculate y_4. Since y_0 is known, we use Taylor's series to find y_1, y_2, y_3; to do this, we must know y', y'', y''', etc. at $x = x_0$. We have

$$y' = f(x, y), \quad y'' = \frac{\partial f}{\partial x} + \frac{\partial f}{\partial y}y', \text{ etc.}$$

The derivatives can be evaluated in the neighborhood of (x_0, y_0) and so y_1, y_2, y_3 can be found to any desired accuracy after which the differential equation gives y_1', y_2', y_3', and (10.66) gives y_4.

Like most methods, Milne's method is subject to errors which at times are cumulative. A check is furnished by the equation obtained by adding the second and fourth equations in (10.65) to four times the fourth, the general result being

$$y_{n+1} = y_{n-1} + \tfrac{1}{3}\Delta(y_{n-1}' + 4y_n' + y_{n+1}'). \qquad (10.67)$$

Once y_{n+1} is found by (10.66), we find y_{n+1}', then substitute in (10.67) to get a check value of y_{n+1}.

Up to this point we have used derivatives evaluated at $x_0 + r\Delta$, r integral. Higher accuracy will result in general if we use derivatives evaluated at intermediate points such as $x + (r + \tfrac{1}{2})\Delta$. This concept is basic in the *Runge–Kutta method* which uses the equation

$$\left.\begin{array}{l} y_{n+1} = y_n + \tfrac{1}{6}\Delta(k_0 + 2k_1 + 2k_2 + k_3) \\ \text{where} \\ \qquad k_0 = f(x_n, y_n), \\ \qquad k_1 = f(x_n + \tfrac{1}{2}\Delta, y_n + \tfrac{1}{2}k_0\Delta), \\ \qquad k_2 = f(x_n + \tfrac{1}{2}\Delta, y_n + \tfrac{1}{2}k_1\Delta), \\ \qquad k_3 = f(x_n + \Delta, y_n + k_2\Delta) \end{array}\right\} \quad (10.68)$$

(for the derivation of these equations, see Wylie, 1966, pp. 111–5, or Kuo, 1965, chapter 7).

Higher-order differential equations can be solved by reducing them to simultaneous first-order equations. Thus, the equation,

$$y'' = f(x, y), \quad y = y_0 \text{ and } y' = y_0' \text{ at } x = x_0,$$

is equivalent to

$$y' = z(x, y), \quad y = y_0 \text{ at } x = x_0,$$
$$z' = f(x, y), \quad z = y_0' \text{ at } x_0.$$

We solve the second equation for $z(x, y)$, then solve the first equation to get $y = y(x)$.

10.1.7 *Partial fractions*

It is often convenient to express a function of the form $N(x)/D(x)$ in *partial fractions*, that is,

$$\frac{N(x)}{D(x)} = \frac{A_1}{x - a_1} + \frac{A_2}{x - a_2} + \ldots + \frac{A_m}{x - a_m}$$

$$+ \frac{B_n}{(x - b_1)^n} + \frac{B_{n-1}}{(x - b_1)^{n-1}} + \ldots + \frac{B_1}{(x - b_1)}$$

where the a_is are single roots of $D(x)$ and b_1 is a multiple root of order n. Obviously $N(x), D(x)$ are polynomials in x, and we take the order of $N(x)$ less than that of $D(x)$ (if this is not so, we carry out long division and the remainder will be a fraction of this type). To find the values of A_i, B_i, we note that we can write

$$\frac{N(x)}{D(x)} = \frac{N(x)}{k(x - a_1)(x - a_2) \ldots (x - a_m)(a - b_1)^n}$$

$$= \frac{A_1}{x - a_1} + \ldots + \frac{A_m}{x - a_m} + \frac{B_n}{(x - b_1)^n} + \ldots$$

$$+ \frac{B_1}{(x - b_1)}$$

where k is the coefficient of the highest power in $D(x)$.

To find A_1, we multiply both sides of the above expression by $(x - a_1)$ and then set $x = a_1$; since the factor cancels on the left and in the first term on the right, and appears in all other terms on the right side, we find

$$A_1 = \frac{N(x)}{D^*(x)}\bigg|_{x = a_1} \tag{10.69}$$

where the asterisk means that the factor $(x - a_1)$ is deleted from $D(x)$. We can get all of the coefficients A_i in the same way. To get B_n we multiply both sides by $(x - b_1)^n$ and

then set $x = b_1$; thus,

$$B_n = \frac{N(x)}{D^{**}(x)}\bigg|_{x = b_1}$$

where the double asterisk means that the factor $(x - b_1)^n$ has been deleted. To get B_{r-1} we differentiate once before setting $x = b_1$

$$B_{n-1} = \frac{d}{dx}\left\{\frac{N(x)}{D^{**}(x)}\right\}\bigg|_{x = b_1}$$

In general,

$$B_{n-s} = \frac{1}{s!}\frac{d^s}{dx^s}\left\{\frac{N(x)}{D^{**}(x)}\right\}\bigg|_{x = b_1} \tag{10.70}$$

As an example of the above, let it be required to find the inverse Laplace transform (see §10.4) of

$$\frac{s^2 - 2}{s(s^2 - 5s + 6)(s - 1)^2} = \frac{s^2 - 2}{s(s - 2)(s - 3)(s - 1)^2}.$$

Then,

$$\frac{s^2 - 2}{s(s - 2)(s - 3)(s - 1)^2} = \frac{A_1}{s} + \frac{A_2}{s - 2} + \frac{A_3}{s - 3}$$

$$+ \frac{B_2}{(s - 1)^2} + \frac{B_1}{s - 1}$$

and

$$A_1 = \frac{s^2 - 2}{(s - 2)(s - 3)(s - 1)^2}\bigg|_{s = 0} = -\frac{1}{3},$$

$$A_2 = \frac{s^2 - 2}{s(s - 3)(s - 1)^2}\bigg|_{s = 2} = -1,$$

$$A_3 = \frac{s^2 - 2}{s(s - 2)(s - 1)^2}\bigg|_{s = 3} = \frac{7}{12},$$

$$B_2 = \frac{s^2 - 2}{s(s - 2)(s - 3)}\bigg|_{s = 1} = -\frac{1}{2},$$

$$B_1 = \frac{d}{ds}\left\{\frac{s^2 - 2}{s(s - 2)(s - 3)}\right\}\bigg|_{s = 1}$$

$$= \frac{s(s - 2)(s - 3)2s - (s^2 - 2)\{(s - 2)}{\{s(s - 2)(s - 3)\}^2}\bigg|_{s = 1} = \frac{3}{4}.$$

$$\times (s - 3) + s(s - 3) + s(s - 2)\}$$

10.2 Fourier series and Fourier integral
10.2.1 *Fourier series*

Let $g(t)$ be a periodic function with period T, that is,

$$g(t \pm nT) = g(t), \quad n = 0, 1, 2, \ldots$$

Provided that $g(t)$ is reasonably well-behaved, that is, provided $g(t)$ obeys the Dirichlet conditions: (1) it has at most a finite number of maxima, minima and discontinuities in an interval T and (2)

$$\int_{-\frac{1}{2}T}^{+\frac{1}{2}T} |g(t)|\,\mathrm{d}t$$

is finite, then $g(t)$ can be expanded in a Fourier series:

$$g(t) = \tfrac{1}{2}a_0 + \sum_{n=1}^{\infty} (a_n \cos n\omega_0 t + b_n \sin n\omega_0 t), \qquad (10.71)$$

where $\omega_0 = 2\pi\nu_0 = 2\pi/T$. We have written $\tfrac{1}{2}a_0$ instead of a_0 so that all values of a_n are given by the same formula, (10.77). To obtain the coefficients a_n and b_n we use the fact that, for any value d and for m and n integral,

$$\left.\begin{array}{l} \displaystyle\int_{d}^{d+2\pi} \sin m\theta \sin n\theta\,\mathrm{d}\theta = 0, \qquad (10.72) \\[2.5em] \displaystyle\int_{d}^{d+2\pi} \cos m\theta \cos n\theta\,\mathrm{d}\theta = 0, \qquad (10.73) \end{array}\right\} m \neq n$$

$$\int_{d}^{d+2\pi} \sin m\theta \cos n\theta\,\mathrm{d}\theta = 0, \qquad (10.74)$$

$$\int_{d}^{d+2\pi} \sin^2 n\theta\,\mathrm{d}\theta = \pi, \qquad (10.75)$$

$$\int_{d}^{d+2\pi} \cos^2 n\theta\,\mathrm{d}\theta = \pi. \qquad (10.76)$$

If we multiply both sides of equation (10.71) by $\cos n\omega_0 t$ and integrate over the period T, we get

$$a_n = (2/T)\int_{-\frac{1}{2}T}^{\frac{1}{2}T} g(t)\cos n\omega_0 t\,\mathrm{d}t; \qquad (10.77)$$

likewise if we multiply both sides of (10.71) by $\sin n\omega_0 t$ and integrate over the period T, we get

$$b_n = (2/T)\int_{-\frac{1}{2}T}^{\frac{1}{2}T} g(t)\sin n\omega_0 t\,\mathrm{d}t. \qquad (10.78)$$

In particular, for $n = 0$,

$$\tfrac{1}{2}a_0 = (1/T)\int_{-\frac{1}{2}T}^{\frac{1}{2}T} g(t)\,\mathrm{d}t = \text{average value of } g(t); \qquad (10.79)$$

hence $a_0 = 0$ whenever $g(t)$ is an odd function.

The sine and cosine series can be combined into one series by introducing phase angles, γ_n:

$$g(t) = \tfrac{1}{2}c_0 + \sum_{n=1}^{\infty} c_n \cos(n\omega_0 t - \gamma_n) \qquad (10.80)$$

where

$$c_n^2 = a_n^2 + b_n^2; \quad c_0 = a_0; \quad \gamma_0 = 0;$$

$$\gamma_n = \tan^{-1}(b_n/a_n), \quad n > 0, \qquad (10.81)$$

Equation (10.80) shows that $g(t)$ can be expressed as an infinite series of harmonics of the fundamental frequency ω_0. The constants c_n and γ_n give the amplitudes and phase angles of the harmonics and are referred to as the *amplitude spectrum* and *phase spectrum* of $g(t)$. [*Frequency spectrum* is used for both the amplitude spectrum and the combined amplitude and phase spectra.] For very large n, the amplitudes must get smaller, i.e., $\lim_{n \to \infty} c_n = 0$, since otherwise

$$\int_{-\frac{1}{2}T}^{\frac{1}{2}T} |g(t)|\,\mathrm{d}t \text{ would not be finite.}$$

Equation (10.71) can be written

$$g(t) = (1/T)\int_{-\frac{1}{2}T}^{\frac{1}{2}T} g(y)\,\mathrm{d}y + (2/T)\sum_{n=1}^{\infty}\int_{-\frac{1}{2}T}^{\frac{1}{2}T}$$

$$g(y)\cos\{n\omega_0(t-y)\}\,\mathrm{d}y. \qquad (10.82)$$

where the variable t in (10.77)–(10.81) has been replaced by the dummy variable y.

Often in practical work the function $g(t)$ is given only at equal intervals, Δ, for example $g(t) = g(n\Delta)$, $n = 0, 1, \ldots, m$. Equations (10.71), (10.77) to (10.81) still hold except that the sums in (10.71) and (10.80) are finite sums and the integrals are replaced by sums. For further details the reader can consult Wylie (1966). An interesting aspect of Fourier series is that the finite series obtained by discarding all terms in (10.71) above a certain value of n is the best fit for $g(t)$ in the least-squares sense (see problem 10.14a).

The two infinite series in (10.71) can be combined using Euler's formula (see problem 10.11a) to give an exponential form of the series:

$$g(t) = \sum_{n=-\infty}^{\infty} \alpha_n\,\mathrm{e}^{jn\omega_0 t}; \qquad (10.83)$$

$$\left.\begin{array}{l} \alpha_0 = \tfrac{1}{2}a_0, \\[1em] \alpha_{\pm n} = \tfrac{1}{2}(a_n \mp jb_n) \\[1.5em] \text{or} \\[1em] \alpha_{\pm n} = (1/T)\displaystyle\int_{-\frac{1}{2}T}^{\frac{1}{2}T} g(t)\,\mathrm{e}^{\mp jn\omega_0 t}\,\mathrm{d}t, \\[1em] n = 0, 1, 2, \ldots, \infty. \end{array}\right\} \qquad (10.84)$$

At times we wish to represent a function $g(t)$ by a Fourier series in an interval such as $(-\tfrac{1}{2}T, +\tfrac{1}{2}T)$, regardless of the values outside this interval (for example, see §10.3.10 and §10.6.1); the Fourier series then repeats the same portion of $g(t)$ each time that t increases or decreases by T.

10.2.2 *Fourier integral*

When $g(t)$ is periodic, the Fourier coefficients constitute a discrete frequency spectrum with components at intervals of ω_0. If T increases, $g(t)$ repeats at longer intervals while the frequency components occur at smaller and smaller intervals. When T becomes infinite, the frequency spectrum becomes continuous and the sum in (10.83) becomes the Fourier integral.

We shall give a heuristic demonstration of the transition from the Fourier series to the Fourier integral as T approaches infinity; a rigorous derivation can be found in Churchill (1963). We can substitute (10.84) in (10.83), obtaining

$$g(t) = \sum_{n=-\infty}^{+\infty} \left\{ \int_{-\frac{1}{2}T}^{\frac{1}{2}T} g(t) e^{-jn\omega_0 t} \, dt \right\} e^{jn\omega_0 t} (1/T).$$

As T approaches infinity, $1/T$ becomes infinitesimal; hence

$$1/T \to dv_0 = d\omega_0/2\pi.$$

The difference between adjacent harmonics, $n\omega_0$ and $(n+1)\omega_0$, becomes infinitesimal, that is, $n\omega_0$ becomes a continuous variable ω. Thus, the discrete spectrum $n\omega_0$ becomes a continuous spectrum in the limit. (We can drop the subscript on v and ω because in the limit it will have no significance.) At the same time the summation in (10.83) becomes an integral:

$$g(t) = \int_{-\infty}^{+\infty} \left\{ \int_{-\infty}^{\infty} g(t) e^{-j\omega t} \, dt \right\} e^{j\omega t} \, d\omega/2\pi. \tag{10.85}$$

The integral with respect to t is calculated first, the result being a function of ω; then in the second integration, ω disappears and we have again a function of t. The factor $1/2\pi$ can be combined with $d\omega$ to give dv but then ω in the exponential terms should be replaced by $2\pi v$.

10.3 **Fourier transforms**
10.3.1 *Introduction*

If we write $G(\omega) = \displaystyle\int_{-\infty}^{\infty} g(t) e^{-j\omega t} \, dt$ (10.86)

$= $ *Fourier transform of $g(t)$,*

then

$$g(t) = (1/2\pi) \int_{-\infty}^{\infty} G(\omega) e^{j\omega t} \, d\omega \tag{10.87}$$

$= $ *inverse Fourier transform of $G(\omega)$.*

(Some authors distribute the $1/2\pi$ factor differently, such as putting $(1/2\pi)^{\frac{1}{2}}$ in front of both the Fourier transform and the inverse Fourier transform.) The relation between $g(t)$ and $G(\omega)$ is often written

$$g(t) \leftrightarrow G(\omega). \tag{10.88}$$

A sufficient condition for the existence of the Fourier transform of a function $g(t)$ is that $\int_{-\infty}^{+\infty} |g(t)| \, dt$ be finite. However, this condition is not necessary and a second, somewhat more complicated, condition can be stated (see Papoulis, 1962, p. 9).

The Fourier integral can be written in several ways. Assuming $g(t)$ is real, as is usually the case, and noting that t inside the braces in (10.85) is a dummy variable, we can write

$$g(t) = (1/2\pi) \int_{-\infty}^{\infty} e^{j\omega t} \left\{ \int_{-\infty}^{\infty} g(y) e^{-j\omega y} \, dy \right\} d\omega$$

$$= (1/2\pi) \int_{-\infty}^{\infty} \int_{-\infty}^{\infty} g(y) e^{+j\omega(t-y)} \, dy \, d\omega$$

$$= (1/2\pi) \int_{-\infty}^{\infty} \int_{-\infty}^{\infty} g(y) \{\cos \omega(t-y)$$
$$+ j \sin \omega(t-y)\} \, dy \, d\omega$$

$$= (1/2\pi) \int_{-\infty}^{\infty} \int_{-\infty}^{\infty} g(y) \cos \omega(t-y) \, dy \, d\omega$$

(since $g(t)$ is real)

$$= (1/2\pi) \int_{-\infty}^{\infty} \int_{-\infty}^{\infty} g(y) \{\cos \omega t \cos \omega y$$
$$+ \sin \omega t \sin \omega y\} \, dy \, d\omega$$

$$g(t) = (1/2\pi) \int_{-\infty}^{\infty} R(\omega) \cos \omega t \, d\omega$$

$$- (1/2\pi) \int_{-\infty}^{\infty} X(\omega) \sin \omega t \, d\omega, \tag{10.89}$$

where

$$R(\omega) = \int_{-\infty}^{\infty} g(y) \cos \omega y \, dy$$

$$= \textit{cosine transform of } g(t)$$

and   (10.90)

$$-X(\omega) = \int_{-\infty}^{\infty} g(y) \sin \omega y \, dy$$

$$= \textit{sine transform of } g(t).$$

If we express the exponential term in (10.86) in terms of $\cos \omega t$ and $\sin \omega t$ and compare with the above we find that

$$G(\omega) = R(\omega) + jX(\omega)$$

$$= A(\omega) e^{j\gamma(\omega)}$$

where

$$A(\omega) = [R^2(\omega) + X^2(\omega)]^{\frac{1}{2}}$$

$$= \textit{amplitude spectrum of } g(t)$$

and   (10.91)

$$\gamma(\omega) = \tan^{-1}[X(\omega)/R(\omega)]$$

$$= \textit{phase spectrum of } g(t).$$

Although the independent variables are usually time and frequency in (10.86) and (10.87), this is not necessarily so. We can, for example, calculate the Fourier transform with respect to x so that ω has inverse-length dimensions instead of inverse-time, in which case we call ω *spatial frequency* or *wavenumber*, often indicated by the symbol κ.

10.3.2 *Multidimensional Fourier series and transforms*

The Fourier series in (10.71) was defined for $g(t)$, a function of one variable only. We can expand $g(x, t)$ in a Fourier series if we assume that $g(x, t)$ is also periodic in x with 'period' equal to $\kappa_0 = 2\pi/\lambda$ (equivalent to $\omega_0 = 2\pi/T$). Then

$$g(x, t) = \sum_{m=0}^{\infty} \sum_{n=0}^{\infty} (a_{mn} \cos m\kappa_0 x \cos n\omega_0 t$$
$$+ b_{mn} \cos m\kappa_0 x \sin n\omega_0 t$$
$$+ c_{mn} \sin m\kappa_0 x \cos n\omega_0 t$$
$$+ d_{mn} \sin m\kappa_0 x \sin n\omega_0 t),$$
      (10.92)

the coefficients being given by equations similar to (10.77) and (10.78), for example,

$$b_{mn} = (4/\lambda T) \int_{-\frac{1}{2}\lambda}^{\frac{1}{2}\lambda} \int_{-\frac{1}{2}T}^{\frac{1}{2}T}$$

$$g(x, t) \cos m\kappa_0 x \sin n\omega_0 t \, dx \, dt$$

$$(10.93)$$

Similarly the Fourier transform equations (10.86) and (10.87) become

$$G(\kappa, \omega) = \int_{-\infty}^{\infty} \int_{-\infty}^{\infty} g(x, t) e^{-j(\kappa x + \omega t)} \, dx \, dt \quad (10.94)$$

$$g(x, t) = [1/(2\pi)^2] \int_{-\infty}^{\infty} \int_{-\infty}^{\infty} G(\kappa, \omega) e^{j(\kappa x + \omega t)} \, d\kappa \, d\omega.$$

$$(10.95)$$

The extension to any number of dimensions r is obvious (the factor $1/(2\pi)^2$ becomes $1/(2\pi)^r$). For further details see Fail and Grau (1963) and Treitel *et al* (1967).

10.3.3 *Special functions*

We shall have many occasions to use certain special functions, especially step(t), the *unit step*, defined by

$$\text{step}(t) = \quad 0, \quad t \leq 0;$$
$$= +1, \quad t \geq 0,$$
      (10.96)

Obviously step(t) has a discontinuity at $t = 0$. Multiplying a function $g(t)$ by step(t) 'wipes out' the function for negative values of t and leaves it unchanged for positive values of t. A unit step shifted t_0 units to the right can be written step$(t - t_0)$; multiplication by step$(t - t_0)$ wipes out a function for all values of t less that t_0. Moreover, k step(t) is a step of strength k.

To get the transform of step(t), we define

$$\text{sgn}(t) = -1, \quad t \leq 0;$$
$$= +1, \quad t \geq 0.$$
      (10.97)

The simplest way of proving the relation,

$$\text{sgn}(t) \leftrightarrow 2/j\omega, \quad\quad\quad (10.98)$$

is to calculate the inverse transform:

$$\frac{2}{j\omega} \leftrightarrow \frac{1}{2\pi} \int_{-\infty}^{\infty} \frac{2}{j\omega} e^{j\omega t} \, d\omega = \frac{1}{j\pi} \int_{-\infty}^{\infty} \left(\frac{\cos \omega t + j \sin \omega t}{\omega} \right) d\omega$$

$$= \frac{2}{\pi} \int_{0}^{\infty} \frac{\sin \omega t}{\omega} \, d\omega,$$

since $(\cos \omega t)/\omega$ is odd and the portions for negative and positive ω will cancel, whereas $(\sin \omega t)/\omega$ is even. The definite integral has the values 0, $\frac{1}{2}\pi$ or $-\frac{1}{2}\pi$ according as $t = 0$, $t > 0$ or $t < 0$ (*Handbook of Chemistry and Physics*, 1975, integral 621); hence the right side equals $\text{sgn}(t)$. Thus

$$\text{step}(t) = \{1 + \text{sgn}(t)\}/2 \leftrightarrow \pi\delta(\omega) + 1/j\omega \quad (10.99)$$

(using the transform pair $1 \leftrightarrow 2\pi\delta(\omega)$; see (10.102) and (10.106)).

The unit step, $\text{step}(t)$, is useful in dealing with discontinuous functions such as $g(t)$ in fig. 10.4, which has values $g(a+)$ and $g(a-)$ at the discontinuity $t = a$ as we approach from the right and from the left, respectively. We write $g_c(t)$ for the continuous function obtained by using $g(t)$ to the left of $t = a$ and the dashed curve to the right of $t = a$ which is merely $g(t)$ displaced parallel to the vertical axis by the amount of the discontinuity. Then, $g(t) = g_c(t) + \{g(a+) - g(a-)\}\ \text{step}(t - a)$. The derivative of $g(t)$ is the same as that of $g_c(t)$ except for an impulse (see (10.102) and (10.107)) of strength $g(a+) - g(a-)$ at $t = a$.

The *gate* or *boxcar*, $\text{box}_a(t)$, is defined by

$$\begin{aligned}\text{box}_a(t) &= 0, \quad t \le -\tfrac{1}{2}a; \\ &= 1, \quad -\tfrac{1}{2}a \le t \le \tfrac{1}{2}a; \quad (10.100) \\ &= 0, \quad t \ge \tfrac{1}{2}a.\end{aligned}$$

The boxcar thus has a width a and a height 1, with area a. It can be expressed as the difference of two steps (see

Fig.10.4. Application of step (t) to describe discontinuous functions.

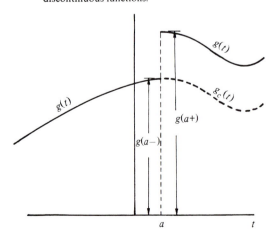

problem 10.17a). The transform of $\text{box}_a(t)$ is

$$\text{box}_a(t) \leftrightarrow \int_{-\frac{1}{2}a}^{\frac{1}{2}a} e^{-j\omega t}\, dt = \frac{e^{+\frac{1}{2}j\omega a} - e^{-\frac{1}{2}j\omega a}}{j\omega}$$

$$= \frac{2\sin(\frac{1}{2}\omega a)}{\omega} = a\,\text{sinc}(\tfrac{1}{2}\omega a),$$

$$(10.101)$$

where $\text{sinc}\,\theta = (\sin\theta)/\theta$. The transform of $\text{box}_a(t)$ is shown in fig. 10.5a.

The *unit impulse* or *Dirac delta*, $\delta(t)$, is defined by the relation

$$\int_{-\infty}^{\infty} \delta(t)g(t)\, dt = g(0), \quad (10.102)$$

that is, $\delta(t)$ is an operator which sets the argument equal to zero.

[The unit impulse, $\delta(t)$, is an example of a distribution (a readily understood account of distributions is given in Papoulis, 1962, Appendix I). A *distribution*, $\alpha(t)$, is an operator which causes $g(t)$ to be replaced by some function of $g(t)$, $\phi\{g(t)\}$, that is,

$$\int_{-\infty}^{\infty} \alpha(t)g(t)\, dt = \phi\{g(t)\}.$$

The integral sign does not have its usual significance here; it is used only because of certain analogies between the above definition and real integrals. The *derivative of a distribution* is defined by the relation,

$$\int_{-\infty}^{\infty} (d\alpha/dt)g(t)\, dt = -\int_{-\infty}^{\infty} \alpha(t)(dg/dt)\, dt = -\phi\{dg/dt\};$$

in other words, $d\alpha/dt$ is a distribution which attributes to dg/dt the same functional value that $\alpha(t)$ does to $g(t)$, except for the minus sign. Thus

$$\int_{-\infty}^{\infty} (d\delta/dt)g(t)\, dt = -dg/dt|_{t=0}.$$

To give a physical concept of the impulse (rather than a rigorous mathematical analysis), we start with a unit boxcar, $\text{box}_a(t)/a$, and let a approach zero while holding the area at unity, so that the height approaches infinity as the width approaches zero:

$$\lim_{a \to 0} \{\text{box}_a(t)/a\} = \delta(t).$$

In the limit $\delta(t)$ is an impulse of infinite height but of zero duration, the 'strength' of the impulse being unity. Furthermore, $\delta(t - t_0)$ is a unit impulse occurring at $t = t_0$. If we multiply $g(t)$ by $\delta(t - t_0)$, the result is the

Fig.10.5. Examples of functions and their Fourier
transforms. (a) $\text{box}_a(t)$; (b) sinc $(\tfrac{1}{2}at)$; (c) $\delta(t)$;
(d) $\delta(t - t_0)$; (e) $e^{-kt}\text{step}(t)$; (f) $e^{-|kt|}$;
(g) $e^{-k(t-t_0)}\text{step}(t - t_0)$; (h) $te^{-kt}\text{step}(t)$
[Note: a and k are positive constants.]

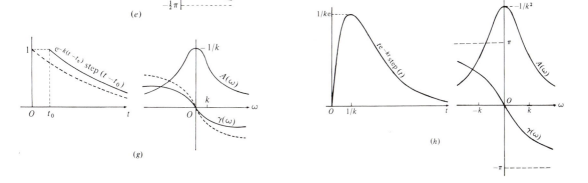

value of $g(t)$ at the time the impulse occurs, that is, $g(t_0)$. Papoulis (1962, p.280) shows that

$$\delta(t) = \lim_{\omega \to \infty} \frac{\sin \omega t}{\pi t}. \qquad (10.103)$$

We can derive another expression for $\delta(t)$ from this result:

$$\int_{-\infty}^{\infty} \cos \omega t \, d\omega = \lim_{a \to \infty} \int_{-a}^{a} \cos \omega t \, d\omega = 2 \lim_{a \to \infty} \int_{0}^{a} \cos \omega t \, d\omega$$

$$= 2 \lim_{a \to \infty} \frac{\sin \omega t}{t}\Big|_{0}^{a} = 2\pi \lim_{a \to \infty} \frac{\sin at}{\pi t} = 2\pi \, \delta(t). \qquad (10.104)$$

The Fourier transform of $\delta(t)$ is

$$\delta(t) \leftrightarrow \int_{-\infty}^{\infty} \delta(t) e^{-j\omega t} \, dt = e^{-j\omega t}\Big|_{t=0} = +1. \qquad (10.105)$$

Thus the spectrum of the unit impulse is flat, that is, all frequencies are present and have the same amplitude and zero phase (see fig. 10.5c). Conversely, the transform of a constant in the time domain is an impulse in the frequency domain:

$$1 \leftrightarrow \int_{-\infty}^{\infty} e^{-j\omega t} \, dt = \int_{-\infty}^{\infty} \cos \omega t \, dt - j \int_{-\infty}^{\infty} \sin \omega t \, dt$$

$$= 2\pi \, \delta(\omega), \qquad (10.106)$$

using (10.104) and noting that the sine integral vanishes because the sine is odd.

We now consider the relation between $\text{step}(t)$ and $\delta(t)$. The derivative of $\text{step}(t)$ is everywhere zero except at the origin, where we may think of it as a distribution:

$$\int_{-\infty}^{\infty} \frac{d(\text{step}(t))}{dt} g(t) \, dt = -\int_{-\infty}^{\infty} \text{step}(t) \frac{dg}{dt} \, dt$$

$$= -\int_{0}^{\infty} \frac{dg}{dt} \, dt = -g(t)\big|_{0}^{\infty} = g(0),$$

since $g(+\infty) = 0$. Thus,

$$(d/dt) \, \text{step}(t) = \delta(t). \qquad (10.107)$$

A *comb* is a series of equally spaced unit impulses, the series usually being considered infinite:

$$\text{comb}(t) = \sum_{n=-\infty}^{+\infty} \delta(t - n\Delta) \qquad (10.108)$$

where n is integral and Δ is a fixed time interval. Multiplication of $g(t)$ by $\text{comb}(t)$ replaces the continuous function with a digitized function sampled at intervals of

Δ. The transform of the comb is derived in the next section.

A *linear ramp* is

$$\text{ramp}(t) = t, \quad 0 \le t \le b,$$
$$= 0 \quad \text{for all other values of } t;$$

$$\text{ramp}(t) \leftrightarrow \int_{0}^{b} t e^{-j\omega t} \, dt = (1/\omega^2)\{e^{-j\omega b}(1 + j\omega b) - 1\}. \qquad (10.109)$$

An *exponential decay*, e^{-kt}, for $t \ge 0$ and k positive, has the transform

$$e^{-kt} \, \text{step}(t) \leftrightarrow \int_{-\infty}^{\infty} e^{-kt} \, \text{step}(t) e^{-j\omega t} \, dt$$

$$\leftrightarrow \int_{0}^{\infty} e^{-(k+j\omega)t} \, dt = 1/(k + j\omega) \qquad (10.110)$$

(see fig. 10.5e).

A *double-sided exponential decay*, $e^{-k|t|}$ for k positive, has the transform

$$e^{-k|t|} \leftrightarrow \int_{-\infty}^{0} e^{kt} e^{-j\omega t} \, dt + \int_{0}^{\infty} e^{-kt} e^{-j\omega t} \, dt$$

$$\leftrightarrow 2k/(k^2 + \omega^2) \qquad (10.111)$$

(see fig. 10.5g).

10.3.4 *Theorems on Fourier transforms*

Many theorems exist regarding the properties of Fourier transforms; the most important are listed below:

$$g(t) \leftrightarrow G(\omega),$$
$$k \, g(t) \leftrightarrow k \, G(\omega); \qquad (10.112)$$

$$k_1 g_1(t) + k_2 g_2(t) \leftrightarrow k_1 G_1(\omega) + k_2 G_2(\omega); \qquad (10.113)$$

Shift theorems $\begin{cases} g(t - a) \leftrightarrow e^{-j\omega a} G(\omega); & (10.114) \\ e^{-jat} g(t) \leftrightarrow G(\omega + a); & (10.115) \end{cases}$

Scaling theorems $\begin{cases} g(at) \leftrightarrow (1/|a|)G(\omega/a); & (10.116) \\ (1/|a|)g(t/a) \leftrightarrow G(a\omega); & (10.117) \end{cases}$

Symmetry theorem $G(t) \leftrightarrow 2\pi g(-\omega);$ \qquad (10.118)

Derivative theorems $\begin{cases} \dfrac{d^n g(t)}{dt^n} \leftrightarrow (j\omega)^n G(\omega); & (10.119) \\[2mm] (-jt)^n g(t) \leftrightarrow \dfrac{d^n G(\omega)}{d\omega^n}; & (10.120) \end{cases}$

$$\text{Integral theorems} \begin{cases} \displaystyle\int_{-\infty}^{t} g(t)\,dt \leftrightarrow (1/j\omega)G(\omega); & (10.121) \\ \\ -(1/jt)g(t) \leftrightarrow \displaystyle\int_{-\infty}^{\omega} G(\omega)\,d\omega; \\ & (10.122) \end{cases}$$

Convolution theorems

$$\begin{cases} g_1(t) * g_2(t) = \displaystyle\int_{-\infty}^{\infty} g_1(\tau)g_2(t-\tau)\,d\tau \\ \\ \qquad \leftrightarrow G_1(\omega)G_2(\omega); & (10.123) \\ \\ 2\pi\, g_1(t)g_2(t) \leftrightarrow G_1(\omega) * G_2(\omega); & (10.124) \end{cases}$$

Cross-correlation theorem

$$\phi_{12}(\tau) = \int_{-\infty}^{\infty} g_1(t)g_2(\tau+t)\,dt \leftrightarrow \overline{G_1(\omega)}G_2(\omega) \qquad (10.125)$$

where the superscribed bar means the complex conjugate. The convolution and cross-correlation theorems are discussed in §10.3.6 and §10.3.7 respectively.

To prove (10.114), we write

$$g(t-a) \leftrightarrow \int_{-\infty}^{\infty} g(t-a)\,e^{-j\omega t}\,dt$$

$$= \int_{-\infty}^{\infty} g(y)\,e^{-j\omega(y+a)}\,dy, \quad (y = t-a),$$

$$= e^{-j\omega a} \int_{-\infty}^{\infty} g(y)\,e^{-j\omega y}\,dy = e^{-j\omega a}\,G(\omega).$$

The proof of (10.115) is similar. As examples of the application of (10.114) and (10.115) we start with (10.99), (10.101) and (10.109) and obtain:

(a) shifted step: $\text{step}\,(t-t_0) \leftrightarrow e^{-j\omega t_0}\{\pi\delta(\omega) + 1/j\omega\};$

$$(10.126)$$

(b) windowed exponential:

$$e^{-jkt}\,\text{box}_a(t) \leftrightarrow a\,\text{sinc}\,\{\tfrac{1}{2}a(\omega+k)\}; \qquad (10.127)$$

(c) shifted ramp:

$$g(t-t_0) \leftrightarrow (1/\omega^2)e^{-j\omega t_0}\{e^{-j\omega b}(1+j\omega b) - 1\},$$

$$(10.128)$$

where

$$g(t-t_0) = (t-t_0), \quad t_0 \leqslant t \leqslant t_0 + b;$$

$$= 0 \text{ for all other } t.$$

Equation (10.114) shows that the effect of a time shift is to leave the magnitude of the transform, $A(\omega)$, unchanged and to add a linear shift to the phase, $\gamma(\omega)$. This is illustrated in fig. 10.5 by comparing (c) with (d) or (e) with (g).

The proof of (10.116) is straightforward if we take a positive at first, then consider the necessary changes when a is negative. As an example, if

$$g(t) = \pm 2t, \quad 0 \leqslant t \leqslant b,$$

$$= 0 \text{ for all other } t,$$

then from (10.109),

$$g(t) \leftrightarrow G(\omega) = \tfrac{1}{2}G(\pm\tfrac{1}{2}\omega)$$

$$= (1/2\omega^2)\{e^{\pm\frac{1}{2}j\omega b}(1 \pm \tfrac{1}{2}j\omega b) - 1\}.$$

An important corollary of (10.116) is obtained by setting $a = -1$, resulting in

$$g(-t) \leftrightarrow G(-\omega). \qquad (10.129)$$

Equation (10.116) can be used to illustrate an important relation between a time function and its transform, namely, that the more the time function is concentrated, the more spread out is the transform and *vice versa*. Thus, in fig. 10.5e, f, g, h, the larger k is, the more closely the time function approaches a spike but at the same time the magnitude of the transform, $A(\omega)$, broadens out. In the limiting case of $\delta(t)$, the time function is concentrated entirely at one instant while the transform is spread uniformly from $-\infty$ to $+\infty$.

Equation (10.118) can be proved by substituting $-t$ for t in (10.87), giving

$$2\pi g(-t) = \int_{-\infty}^{\infty} G(\omega)\,e^{-j\omega t}\,d\omega,$$

and then interchanging the symbols t and ω (since this does not affect the equation); thus

$$2\pi\, g(-\omega) = \int_{-\infty}^{\infty} G(t)\,e^{-j\omega t}\,dt.$$

As an example, we have from (10.110)

$$e^{-kt}\,\text{step}\,(t) \leftrightarrow 1/(k+j\omega),$$

and hence

$$1/(k+jt) \leftrightarrow 2\pi\,e^{k\omega}\,\text{step}\,(-\omega)$$

(note that $\text{step}\,(-\omega)$ equals $+1$ or 0 according as ω is negative or positive). Again, (10.101) gives

$$\text{box}_a(t) \leftrightarrow a\,\text{sinc}\,(\tfrac{1}{2}\omega a)$$

hence

$$a \operatorname{sinc}(\tfrac{1}{2}at) \leftrightarrow 2\pi \operatorname{box}_a(-\omega) = 2\pi \operatorname{box}_a(\omega), \qquad (10.130)$$

since $\operatorname{box}_a(\omega)$ is even (see fig. 10.5b).

The derivative theorems are of fundamental importance. The proof follows directly from (10.87) if we differentiate with respect to t and ω. Thus

$$dg/dt = (1/2\pi)\int_{-\infty}^{\infty} G(\omega)\frac{d}{dt}(e^{j\omega t})\,d\omega$$

$$= (1/2\pi)\int_{-\infty}^{\infty}\{j\omega G(\omega)\}e^{j\omega t}\,d\omega,$$

hence

$$dg/dt \leftrightarrow j\omega G(\omega).$$

Differentiating n times, we get (10.119):

$$\frac{d^n g(t)}{dt^n} \leftrightarrow (j\omega)^n G(\omega).$$

The most important application of (10.119) is in the solution of differential equations where it permits us to replace derivatives with a function of ω. It also enables us to obtain new transform pairs by differentiation of a known pair. An interesting application of this theorem is

$$\delta(t) = (d/dt)\operatorname{step}(t) \leftrightarrow j\omega\{\pi\delta(\omega) + 1/j\omega\} = 1$$

(using (10.107) and (10.99) and the fact that $\omega\delta(\omega) = 0$).

Equations (10.121) are easily proved using (10.119). We write

$$g_1(t) = \int_{-\infty}^{t} g(t)\,dt = \int_{-\infty}^{t} g(x)\,dx \leftrightarrow G_1(\omega),$$

$$dg_1(t)/dt = g(t) \leftrightarrow j\omega G_1(\omega) = G(\omega),$$

where we have used Leibnitz' rule to differentiate the integral on the right. [Leibnitz' rule states that, given $F(t) = \int_{a(t)}^{b(t)}\phi(t,\lambda)\,d\lambda$, then the derivative of $F(t)$ with respect to t is given by

$$\frac{dF}{dt} = \int_{a(t)}^{b(t)}\frac{\partial\phi}{\partial t}\,d\lambda + \phi(t,b)\frac{db}{dt} - \phi(t,a)\frac{da}{dt} \qquad (10.131)$$

(see Kaplan, 1952, p. 220)]. Therefore

$$\int_{-\infty}^{t} g(t)\,dt \leftrightarrow (1/j\omega)\,G(\omega).$$

Equation (10.122) can be proved in the same way, starting from (10.120). Equations (10.123) and (10.125) will be discussed later.

Using (10.105), (10.114), (10.115) and (10.118), we can write the following transform pairs:

$$\left.\begin{aligned}
\delta(t) &\leftrightarrow +1;\\
\delta(t \pm t_0) &\leftrightarrow e^{\pm j\omega t_0};\\
(1/2)\{\delta(t+t_0) + \delta(t-t_0)\} &\leftrightarrow \cos\omega t_0;\\
(1/2j)\{\delta(t+t_0) - \delta(t-t_0)\} &\leftrightarrow \sin\omega t_0;\\
1 &\leftrightarrow 2\pi\delta(\omega);\\
e^{\pm j\omega_0 t} &\leftrightarrow 2\pi\delta(\omega \mp \omega_0);\\
\cos\omega_0 t &\leftrightarrow \pi\{\delta(\omega+\omega_0) + \delta(\omega-\omega_0)\};\\
\sin\omega_0 t &\leftrightarrow j\pi\{\delta(\omega+\omega_0) - \delta(\omega-\omega_0)\}.
\end{aligned}\right\} \qquad (10.132)$$

The transforms of $\sin\omega_0 t$ and $\cos\omega_0 t$ are shown in fig. 10.6.

To obtain the transform of the comb, we find its Fourier series expansion using (10.82) and then find the transform of the series; replacing $g(y)$ in (10.82) with $\delta(y)$ and writing $T = \Delta$, we get

$$\operatorname{comb}(t) = \sum_{n=-\infty}^{\infty}\delta(t-n\Delta) = 1/\Delta + (2/\Delta)\sum_{n=1}^{\infty}\cos n\omega_0 t,$$

$$\omega_0 = 2\pi/\Delta.$$

Using (10.132), we obtain

$$\sum_{n=-\infty}^{\infty}\delta(t-n\Delta) \leftrightarrow \omega_0\sum_{m=-\infty}^{\infty}\delta(\omega - m\omega_0),$$

that is,

$$\operatorname{comb}(t) \leftrightarrow \omega_0\operatorname{comb}(\omega). \qquad (10.133)$$

10.3.5 *Gibbs' phenomenon*

At a discontinuity the Fourier integral (and the Fourier series) gives the average value, $\tfrac{1}{2}\{g(a-) + g(a+)\}$ in fig. 10.4. However, $g(a \pm \varepsilon)$, ε infinitesimal, does not approach the average value smoothly as ε goes to zero. For simplicity, we take $a = 0$ and write

$$g(t) = g_c(t) + \{g(0+) - g(0-)\}\operatorname{step}(t).$$

Fig.10.6. Fourier transforms of (a) $\cos\omega_0 t$ and (b) $\sin\omega_0 t$.

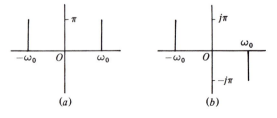

Equation (10.87) gives

$$g(t) = \lim_{\lambda \to \infty} (1/2\pi) \int_{-\lambda}^{+\lambda} G(\omega) e^{j\omega t} \, d\omega$$

$$= \lim_{\lambda \to \infty} (1/2\pi) \int_{-\lambda}^{+\lambda} \left\{ \int_{-\infty}^{+\infty} g(y) e^{-j\omega y} \, dy \right\} e^{j\omega t} \, d\omega.$$

Interchanging the order of integration (see §10.3.6) we have

$$g(t) = \lim_{\lambda \to \infty} (1/2\pi) \int_{-\infty}^{+\infty} g(y) \left\{ \int_{-\lambda}^{+\lambda} e^{j\omega(t-y)} \, d\omega \right\} dy$$

$$= \lim_{\lambda \to \infty} (1/2\pi) \int_{-\infty}^{+\infty} g(y) \frac{e^{j\omega(t-y)}}{j(t-y)} \bigg|_{-\lambda}^{+\lambda} dy$$

$$= \lim_{\lambda \to \infty} \int_{-\infty}^{+\infty} g(y) \frac{\sin \lambda(t-y)}{\pi(t-y)} \, dy$$

$$= \lim_{\lambda \to \infty} \int_{-\infty}^{+\infty} [g_c(y) + \{g(0+)$$

$$- g(0-)\} \, \text{step}(y)] \frac{\sin \lambda(t-y)}{\pi(t-y)} \, dy$$

$$= \int_{-\infty}^{+\infty} g_c(y) \lim_{\lambda \to \infty} \frac{\sin \lambda(t-y)}{\pi(t-y)} \, dy$$

$$+ \{g(0+) - g(0-)\} \lim_{\lambda \to \infty} \int_{0}^{+\infty} \frac{\sin \lambda(t-y)}{\pi(t-y)} \, dy$$

$$= \int_{-\infty}^{+\infty} g_c(t)\delta(t-y) \, dy + (1/\pi)\{g(0+)$$

$$- g(0-)\} \lim_{\lambda \to \infty} \int_{-\infty}^{\lambda t} \text{sinc} \, x \, dx$$

using (10.103) and taking $\lambda(t-y) = x$ so that $dy = -dx/\lambda$ (we do not go to the limit in the right-hand integral because we wish to study the behaviour of this term as λ goes to infinity). The right-hand term can be written

$$g'(t) = (1/\pi)\{g(0+) - g(0-)\} \left\{ \int_{-\infty}^{0} \text{sinc} \, x \, dx \right.$$

$$\left. + \lim_{\lambda \to \infty} \int_{0}^{\lambda t} \text{sinc} \, x \, dx \right\}.$$

Since $\int_{-\infty}^{0} \text{sinc} \, x \, dx = \frac{1}{2}\pi$ (see the derivation of (10.99)), we have finally

$$g'(t) = \{g(0+) - g(0-)\} \left\{ \frac{1}{2} + (1/\pi) \lim_{\lambda \to \infty} \int_{0}^{\lambda t} \text{sinc} \, x \, dx \right\}.$$

The graph of the second factor is shown in fig. 10.7. As $\lambda \to \infty$ the peak values do not change but the ripple moves toward the discontinuity from both sides. At the discontinuity, $t = 0$, we have for $g(t)$

$$g(0) = g_c(0) + \frac{1}{2}\{g(0+) - g(0-)\};$$

however, an infinitesimal distance away on either side, we have an 'overshoot' of about 18% (9% at top and 9% at bottom).

Gibbs' phenomenon is important whenever discontinuities are present, for example, in the application of filters and windows. If we multiply by a boxcar in the time-domain, the discontinuities at the edges of the boxcar will produce 'ringing'. The objective of 'window carpentry', that is, of shaping windows, is to remove discontinuities (and discontinuities in derivatives) so as to minimize ringing. See also §10.8.5.

10.3.6 *Convolution theorem* (see also §8.1.2a)

The *convolution* of two functions, $g_1(t)$ and $g_2(t)$, usually written in the form $g_1(t) * g_2(t)$, eq. (10.123), is defined as

$$g_1(t) * g_2(t) = \int_{-\infty}^{+\infty} g_1(\tau)g_2(t-\tau) \, d\tau.$$

We see that $g_2(\tau - t)$ denotes $g_2(\tau)$ displaced t units to the right while $g_2(-\tau + t)$ is $g_2(\tau - t)$ reflected in the vertical axis. Thus, convolution involves reflecting one of the curves in the vertical axis (often called 'folding'), shifting it by t units, multiplying corresponding coordinates of the two curves and summing from $-\infty$ to $+\infty$ (see fig. 10.8a). The value depends only on the time shift t and is

Fig.10.7.Gibbs' phenomenon at a discontinuity.

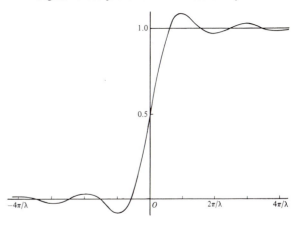

independent of which curve is shifted and reflected, that is,

$$g_1(t) * g_2(t) = g_2(t) * g_1(t). \tag{10.134}$$

This is illustrated in fig. 10.8*b*.

The *convolution theorem*, (10.123), can be proven as follows:

$$g_1(t) * g_2(t) \leftrightarrow \int_{-\infty}^{\infty} \left\{ \int_{-\infty}^{\infty} g_1(\tau) g_2(t - \tau) \, d\tau \right\} e^{-j\omega t} \, dt.$$

We assume that $g_1(t)$ and $g_2(t)$ have transforms, that is,

$$\int_{-\infty}^{\infty} |g_i(t)| \, dt < \infty, \quad i = 1, 2.$$

We shall assume that this relation means that $\int_{-\infty}^{\infty} |g_i(t)|^2 \, dt$ is also finite. In this case the order of integration can be interchanged (Papoulis, 1962, p.27), giving

$$g_1(t) * g_2(t) \leftrightarrow \int_{-\infty}^{\infty} g_1(\tau) \left\{ \int_{-\infty}^{\infty} g_2(t - \tau) e^{-j\omega t} \, dt \right\} d\tau,$$

$$\leftrightarrow \int_{-\infty}^{\infty} g_1(\tau) \left\{ \int_{-\infty}^{\infty} g_2(y) e^{-j\omega(y + \tau)} \, dy \right\} d\tau$$

(where $y = t - \tau$),

$$g_1(t) * g_2(t) \leftrightarrow \int_{-\infty}^{\infty} g_1(\tau) e^{-j\omega\tau} \, d\tau \int_{-\infty}^{\infty} g_2(y) e^{-j\omega y} \, dy,$$

$$\leftrightarrow G_1(\omega) G_2(\omega).$$

Fig.10.8. Convolution in the time-domain.
(*a*) Illustrating the reflecting and shifting of $g_2(\tau)$;
(*b*) geometrical demonstration that $g_1(t) * g_2(t) = g_2(t) * g_1(t)$; the overlap is the same regardless of which is reflected and shifted.

The inverse relation, (10.124), can be proven as follows:

$$G_1(\omega) * G_2(\omega) \leftrightarrow (1/2\pi) \int_{-\infty}^{\infty} \left\{ \int_{-\infty}^{\infty} \right.$$

$$\left. \times G_1(y) G_2(\omega - y) \, dy \right\} e^{j\omega t} \, d\omega,$$

$$= (1/2\pi) \int_{-\infty}^{\infty} G_1(y) \left\{ \int_{-\infty}^{\infty} \right.$$

$$\left. \times G_2(x) e^{j(x + y)t} \, dx \right\} dy, \quad x = \omega - y$$

$$= (1/2\pi) \int_{-\infty}^{\infty} G_1(y) e^{jyt} \, dy \int_{-\infty}^{\infty} G_2(x) e^{jxt} \, dx,$$

$$= (1/2\pi)\{2\pi g_1(t)\}\{2\pi g_2(t)\} = 2\pi g_1(t) g_2(t).$$

10.3.7 *Cross-correlation theorem* (see also §8.1.3*a*)

The *cross-correlation function*, $\phi_{12}(t)$, defined by (10.125),

$$\phi_{12}(t) = \int_{-\infty}^{\infty} g_1(\tau) g_2(t + \tau) \, d\tau,$$

is closely related to the convolution function. Obviously $\phi_{12}(t)$ is the result of displacing $g_2(\tau)$ t units to the left and summing the products of the ordinates. It is evident that

$$\phi_{12}(t) = \phi_{21}(-t). \tag{10.135}$$

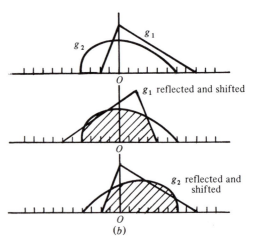

(*a*)

(*b*)

The cross-correlation function can be regarded as a convolution of two functions, the second of which has been reversed in time:

$$\left.\begin{aligned} \phi_{12}(t) &= g_1(t) * g_2(-t); \\ \phi_{21}(t) &= g_1(-t) * g_2(t); \\ g_1(t) * g_2(t) &= \text{cross-correlation of} \\ g_1(t) &\text{ with } g_2(-t), \\ &= \text{cross-correlation of} \\ g_2(t) &\text{ with } g_1(-t). \end{aligned}\right\} \quad (10.136)$$

These relations can be verified by drawing curves of $g_1(t), g_2(t), g_1(-t), g_2(-t)$, and checking geometrically the above relations. They can also be verified by substituting in the integrals; for example,

$$g_1(t) * g_2(-t) = \int_{-\infty}^{\infty} g_1(\tau)g_2(t+\tau)\,d\tau = \phi_{12}(t)$$

(note that for zero shift the argument of the functions in the integral is τ, not t, hence replacing $g_2(t)$ with $g_2(-t)$ on the left-hand side changes the sign of τ in $g_2(\tau)$).

The cross-correlation theorem (10.125) states that

$$\phi_{12}(t) \leftrightarrow \overline{G_1(\omega)}\,G_2(\omega) = \Phi_{12}(\omega)$$

where $\overline{G_1(\omega)}$ is the conjugate complex of $G_1(\omega)$. The proof is straightforward if we note the first relation of (10.136) and change the sign of τ in $g_2(t-\tau)$ in the first line of the derivation of (10.123). This changes the first exponential in the next to the last line of the derivation to $e^{j\omega\tau}$ and (10.125) follows immediately. Writing $\Phi_{12}(\omega)$ for the transform of $\phi_{12}(t)$, we have

$$\Phi_{12}(\omega) = \overline{G_1(\omega)}\,G_2(\omega), \quad \Phi_{21}(\omega) = G_1(\omega)\,\overline{G_2(\omega)}. \quad (10.137)$$

The transform $\Phi_{12}(\omega)$ is known as the *cross-energy spectrum* for reasons which will become clearer in the following section. When $t = 0$, (10.125), (10.135) and (10.137) give

$$\phi_{12}(0) = \phi_{21}(0) = \int_{-\infty}^{\infty} g_1(\tau)g_2(\tau)\,d\tau$$

$$= (1/2\pi)\int_{-\infty}^{\infty} \overline{G_1(\omega)}\,G_2(\omega)\,d\omega$$

$$= (1/2\pi)\int_{-\infty}^{\infty} G_1(\omega)\,\overline{G_2(\omega)}\,d\omega. \quad (10.138)$$

This is referred to as *Parseval's theorem*.

10.3.8 *Autocorrelation*

When $g_2(t) = g_1(t)$, the cross-correlation function becomes the autocorrelation of $g_1(t)$, that is,

$$\phi_{11}(t) = \int_{-\infty}^{\infty} g_1(\tau)g_1(t+\tau)\,d\tau \leftrightarrow |G_1(\omega)|^2. \quad (10.139)$$

Clearly $\phi_{11}(t) = \phi_{11}(-t)$, so that ϕ_{11} is an even function. When $t = 0$, Parseval's theorem (10.138) becomes

$$\phi_{11}(0) = \int_{-\infty}^{\infty} |g_1(\tau)|^2\,d\tau = (1/2\pi)\int_{-\infty}^{\infty} |G_1(\omega)|^2\,d\omega \quad (10.140)$$

In most cases $g_1(t)$ is a voltage, current, velocity, displacement, etc., so that $|g_1(t)|^2$ is proportional to the energy; adopting this point of view and choosing the units properly, we can say that $\phi_{11}(0)$ is the total energy of $g_1(t)$. The quantity $(1/2\pi)|G_1(\omega)|^2$ is thus the energy in the spectrum between the frequencies ω and $\omega + d\omega$; it is called the *energy density* or *spectral density*. We note that the integration is from $-\infty$ to $+\infty$; if we recall that we replaced sines and cosines by the exponential terms $e^{\pm j\omega t}$, it is clear that we must recombine terms for $\pm\omega$ to get the result, that is,

energy density of frequency $\omega = |G_1(-\omega)|^2 + |G_1(+\omega)|^2$.

When $g_1(t)$ is a real function, $G_1(-\omega) = \overline{G_1(\omega)}$ (see problem 10.16b) so that $2|G_1(+\omega)|^2$ gives the total energy density for the frequency ω (if we are interested only in relative energies, as is usually the case, we can forget doubling).

The autocorrelation function assumes its greatest value for zero shift, that is,

$$\phi_{11}(0) > \phi_{11}(t), \quad t \neq 0 \quad (10.141)$$

(see problem 10.21a).

10.3.9 *Multidimensional convolution*

Equation (10.123) can be extended to more than one dimension. In two dimensions we define the convolution by the expression

$$g_1(x,t) * g_2(x,t) = \int_{-\infty}^{\infty}\int_{-\infty}^{\infty} g_1(\sigma,\tau)g_2(x-\sigma,t-\tau)\,d\sigma\,d\tau. \quad (10.142)$$

We can derive the two-dimensional convolution theorem as follows. Using (10.94), we have

$$g_1(x,t) * g_2(x,t) \leftrightarrow \int_{-\infty}^{\infty}\int_{-\infty}^{\infty} \left\{ \int_{-\infty}^{\infty}\int_{-\infty}^{\infty} g_1(\sigma,\tau) \right.$$

$$\left. g_2(x-\sigma, t-\tau)\,d\sigma\,d\tau \right\} e^{-j(\kappa x + \omega t)}\,dx\,dt.$$

Interchanging the order of integration gives

$$g_1(x, t) * g_2(x, t) \leftrightarrow \int_{-\infty}^{\infty} \int_{-\infty}^{\infty} g_1(\sigma, \tau)$$

$$\times \left\{ \int_{-\infty}^{\infty} \int_{-\infty}^{\infty} g_2(x - \sigma, t - \tau) \right.$$

$$\times \left. e^{-j(\kappa x + \omega t)} dx \, dt \right\} d\sigma \, d\tau,$$

$$\leftrightarrow \int_{-\infty}^{\infty} \int_{-\infty}^{\infty} g_1(\sigma, \tau) e^{-j(\kappa \sigma + \omega \tau)}$$

$$\times \left\{ \int_{-\infty}^{\infty} \int_{-\infty}^{\infty} g_2(x - \sigma, t - \tau) \right.$$

$$\times e^{-j(\kappa(x - \sigma) + \omega(t - \tau))} d(x - \sigma)$$

$$\times \left. d(t - \tau) \right\} d\sigma \, d\tau,$$

$$\leftrightarrow \left\{ \int_{-\infty}^{\infty} \int_{-\infty}^{\infty} g_1(\sigma, \tau) e^{-j(\kappa \sigma + \omega \tau)} d\sigma \, d\tau \right\}$$

$$\times G_2(\kappa, \omega),$$

$$\leftrightarrow G_1(\kappa, \omega) G_2(\kappa, \omega) \tag{10.143}$$

which is the equivalent of (10.123).

10.3.10 *Random functions*

The periodic and aperiodic functions which we have been considering hitherto have one property in common: if the process which generates one of these functions is repeated exactly, the same function is generated. However, in many instances repetition of the process gives a different result each time, for example, measurements of microseisms give a function $g(t)$ which never repeats no matter how often we repeat the measurement. Functions of this kind which cannot be predicted exactly no matter how often we repeat the measurements are called *random functions*.

Since random functions cannot be predicted, we use probability theory to deduce their properties. The set of functions obtained if an experiment were repeated an infinite number of times is called an *ensemble*. If we arrive at the same value for some property of the ensemble (for example, the average power density for the frequency ω) whether we average the values for each of the traces at a certain instant in time or average all values for one trace,

then the ensemble is said to be *ergodic* and we can determine the statistical properties using a sufficiently long portion of one function rather than having to make measurements on many functions.

A *stationary* time series is one whose statistical properties are independent of the location of the origin $t = 0$. It can be shown that ergodic ensembles must also be stationary (see Lee, 1960, p.208–9; Bendat and Piersol, 1966, p.11–12). We assume that the random time series with which we deal are ergodic and stationary.

The *autocorrelation of a random function* is defined by the equation,

$$\phi_{11}(t) = \lim_{T \to \infty} (1/2T) \int_{-T}^{T} g_1(\tau) g_1(t + \tau) \, d\tau. \tag{10.144}$$

Note that this definition differs in form from that in (10.139) by the factor $1/2T$ as well as the limiting process. Since a random function does not approach zero as t approaches $\pm \infty$, the integral for the Fourier transform (10.86) does not converge and $G(\omega)$ does not exist. Thus there is no equivalent of (10.139) for random functions. When $t = 0$,

$$\phi_{11}(0) = \lim_{T \to \infty} (1/2T) \int_{-T}^{T} g_1^2(\tau) \, d\tau$$

$$= \text{mean square value of } g_1(t).$$

The portion of one measurement of a random function, $g_1(t)$, included in the interval $(-T, +T)$ will be written $g_1'(t)$. If we express $g_1'(t)$ as a Fourier series with period $2T$, the series represents $g_1(t)$ exactly in the interval $(-T, +T)$ but not elsewhere since it repeats $g_1'(t)$ in each interval of length $2T$. In the limit as $T \to \infty$, $g_1'(t)$ becomes $g_1(t)$. We wrote $\phi_{11}'(t)$ for the autocorrelation function of $g_1'(t)$:

$$\phi_{11}'(t) = \int_{-T}^{T} g_1'(\tau) g_1'(t + \tau) \, d\tau.$$

We can expand $\phi_{11}'(t)$ in a Fourier series:

$$\phi_{11}'(t) = \sum_{n=-\infty}^{\infty} \alpha_n e^{jn\omega_0 t} = \sum_{n=-\infty}^{\infty} e^{jn\omega_0 t}$$

$$\times \left\{ \frac{1}{2T} \int_{-T}^{T} \phi_{11}'(t) e^{-jn\omega_0 t} \, dt \right\}$$

using (10.84) where $\omega_0 = \pi/T$. If we divide $\phi_{11}'(t)$ by $2T$ and let $T \to \infty$, $\phi_{11}'(t)$ approaches $\phi_{11}(t)$ as defined in

(10.144). Thus

$$\phi_{11}(t) = \lim_{T \to \infty} \sum_{n=-\infty}^{\infty} e^{jn\omega_0 t}(1/2T) \int_{-T}^{T} \phi'_{11}(t) e^{-jn\omega_0 t} dt,$$

$$= (1/2\pi) \int_{-\infty}^{\infty} e^{j\omega t} d\omega \int_{-\infty}^{\infty} \phi_{11}(t) e^{-j\omega t} dt,$$

$$= (1/2\pi) \int_{-\infty}^{\infty} \Phi_{11}(\omega) e^{j\omega t} d\omega,$$

where $\Phi_{11}(\omega) = \int_{-\infty}^{\infty} \phi_{11}(t) e^{-j\omega t} dt.$ \hfill (10.145)

Thus, although a random function does not have a Fourier transform, its autocorrelation function has a Fourier transform: $\phi_{11}(t) \leftrightarrow \Phi_{11}(\omega)$.

Also,

$$\phi_{11}(0) = \lim_{T \to \infty} (1/2T) \int_{-T}^{T} g_1^2(t) dt$$

$$= (1/2\pi) \int_{-\infty}^{\infty} \Phi_{11}(\omega) d\omega. \hfill (10.146)$$

Because of the factor $1/2T$, $\Phi_{11}(\omega)$ gives the power density, not energy density as with aperiodic functions. Equations (10.145), and (10.146) express the *Wiener autocorrelation theorem*, that the Fourier transform of the autocorrelation of a random function exists and gives the power density spectrum. The autocorrelation function and the power density spectrum are real, even functions and $\phi_{11}(t)$ has its greatest value at the origin (see problem 10.21b).

The *cross-correlation of random functions* is defined in a manner similar to (10.144). Let $g_1(t)$ and $g_2(t)$ be random functions from different ensembles, for example, the input and output noise of an amplifier; then

$$\phi_{12}(t) = \lim_{T \to \infty} \frac{1}{2T} \int_{-T}^{T} g_1(\tau) g_2(t + \tau) d\tau. \hfill (10.147)$$

Equation (10.135), $\phi_{21}(t) = \phi_{12}(-t)$, holds for both random and aperiodic functions.

10.3.11 *Hilbert transforms*

The Hilbert is a special form of Fourier transform. Let $g(t)$ be any real function and

$$g(t) \leftrightarrow G(\omega) = R(\omega) + jX(\omega).$$

Then $g(t)$ can always be divided into even and odd parts,

$g_e(t)$ and $g_o(t)$ (see fig. 10.9) where

$$\left.\begin{array}{l} g_e(t) = \tfrac{1}{2}\{g(t) + g(-t)\}, \\ g_o(t) = \tfrac{1}{2}\{g(t) - g(-t)\} \end{array}\right\} \hfill (10.148)$$

Since $g_e(t)$ is even,

$$g_e(t) \leftrightarrow \int_0^{\infty} g(t) \cos \omega t \, dt = R(\omega) \hfill (10.149)$$

(see (10.90)). Likewise,

$$g_o(t) \leftrightarrow jX(\omega). \hfill (10.150)$$

If $g(t)$ is also causal (see §10.6.6a), $g_e(t) = g_o(t)$ for $t > 0$ and $g_e(t) = -g_o(t)$ for $t < 0$, that is

$$g_o(t) = g_e(t) \operatorname{sgn}(t),$$

$$g_e(t) = g_o(t) \operatorname{sgn}(t)$$

(see (10.97)). Recalling that $\operatorname{sgn}(t) \leftrightarrow 2/j\omega$ (see (10.98)), we obtain

$$jX(\omega) = (1/2\pi)R(\omega) * (2/j\omega)$$

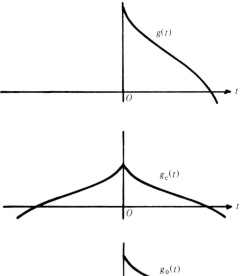

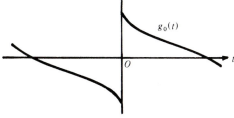

Fig.10.9. Relation between a real causal function $g(t)$ and the derived even and odd functions, $g_e(t)$ and $g_o(t)$.

on using (10.124). Thus

$$X(\omega) = -\frac{1}{\pi}\mathfrak{P}\int_{-\infty}^{\infty}\{R(y)/(\omega - y)\}\,dy$$

$$= -R(\omega)*(1/\pi\omega) \qquad (10.151)$$

where $\mathfrak{P}$ means that we take the Cauchy principle value at $\omega = y$ (see Papoulis, 1962, pp. 9–10 and footnote on p. 10). Likewise,

$$R(\omega) = \frac{1}{\pi}\mathfrak{P}\int_{-\infty}^{\infty}\{X(y)/(\omega - y)\}\,dy$$

$$= X(\omega)*(1/\pi\omega). \qquad (10.152)$$

Equations (10.151) and (10.152) define the Hilbert transform. Given either $R(\omega)$ or $X(\omega)$, the other can be calculated.

Writing $G(\omega) = A(\omega)\,e^{j\gamma(\omega)}$ and taking logarithms gives

$$\ln\{G(\omega)\} = \ln\{A(\omega)\} + j\gamma(\omega);$$

in this case, $R(\omega) = \ln\{A(\omega)\}$ and $X(\omega) = \gamma(\omega)$. Since $A(\omega)$ is the amplitude of the frequency ω, $A(\omega)$ is the square root of the Fourier transform of the autocorrelation function (see (10.139)); knowing $A(\omega)$, we can calculate $\gamma(\omega)$ by (10.151). The Hilbert transform thus allows the calculation of the phase from the autocorrelation function:

$$\gamma(\omega) = -\frac{1}{\pi}\mathfrak{P}\int_{-\infty}^{\infty}[\ln\{A(y)\}/(\omega - y)]\,dy. \qquad (10.153)$$

Given the real part of the Fourier transform of a real, causal time function, the Hilbert transform enables us to find the corresponding time function. A similar problem is: given the real part of a complex time function whose imaginary part has a transform $\pm 90°$ out-of-phase with that of the real part, find the complex time function. Let $f(t)$ be a complex time function where

$$f(t) = x(t) + jy(t); \quad F(\omega) = X(\omega) + jY(\omega)$$
$$= X(\omega)\{1 + jQ(\omega)\},$$

$Q(\omega)$ being a filter which changes the phase by $\pm 90°$ but has no effect on the amplitude spectrum; $Q(\omega)$ is referred to as a *quadrature filter* and $y(t)$ as the *quadrature trace*.

Since $e^{\pm\frac{1}{2}j\pi} = \pm j$, we can select $Q(\omega)$ equal to $+j$, $-j$ or $\pm j\,\mathrm{sgn}(\omega)$. Selection of the first two choices gives $q(t) = \pm j\delta(t)$ from (10.105), which is not a useful result. We thus select $Q(\omega) = -j\,\mathrm{sgn}(\omega)$. Then,

$$y(t) \leftrightarrow X(\omega)\,Q(\omega) = -jX(\omega)\,\mathrm{sgn}(\omega)$$
$$= jX(\omega)\,\mathrm{sgn}(-\omega).$$

Using (10.98), (10.118) and (10.123) we have

$$y(t) = x(t)*(1/\pi t), \qquad (10.154)$$

that is, $x(t)$ and $-y(t)$ form a Hilbert transform pair (see (10.151)).

Although $f(t)$ can be found by using (10.154), it is easier to calculate $F(\omega)$ and transform to get $f(t)$. Thus,

$$F(\omega) = X(\omega) + jY(\omega) = X(\omega)\{1 + jQ(\omega)\},$$
$$= X(\omega)\{1 + \mathrm{sgn}(\omega)\},$$
$$= 0, \quad \omega < 0,$$
$$= 2X(\omega), \quad \omega > 0,$$
$$= 2X(\omega)\,\mathrm{step}(\omega). \qquad (10.155)$$

Equations (10.154) and (10.155) find application in complex-trace analysis (§8.4.2).

10.4 Laplace transforms
10.4.1 *Introduction*

The Laplace transform is closely related to the Fourier transform. If we do not use distributions, many functions such as $\sin at$ and $\cos at$ do not have Fourier transforms because the integral giving the transform does not converge. However, if we multiply the function by a convergence factor $e^{-\sigma|t|}$, σ being real and positive and large enough that $\lim_{t\to\pm\infty}\{e^{-\sigma|t|}g(t)\} = 0$, then $\{[e^{-\sigma|t|}g(t)]\}$ has a Fourier transform which is called the *Laplace transform* of $g(t)$. If $g(t) = 0$ for $t < 0$, we get the *one-sided Laplace transform*:

$$g(t) \leftrightarrow \int_0^{\infty} e^{-\sigma t}g(t)e^{-j\omega t}\,dt$$

$$= \int_0^{\infty} g(t)e^{-st}\,dt = G(s) \qquad (10.156)$$

where $s = \sigma + j\omega$, the real part of s being large enough that $\lim_{t\to\infty}\{e^{-st}g(t)\} = 0$. (The Laplace transform, $G(s)$, is distinguished from the Fourier transform, $G(\omega)$, by the variable s instead of ω.)

The *inverse Laplace transformation* becomes

$$e^{-\sigma t}g(t) = (1/2\pi)\int_0^{\infty} G(s)e^{j\omega t}\,d\omega$$

or

$$g(t) = (1/2\pi)\int_{-\infty}^{\infty} G(s)e^{(\sigma + j\omega)t}\,d\omega$$

$$= (1/2\pi j)\int_{\sigma - j\infty}^{\sigma + j\infty} G(s)e^{st}\,ds \qquad (10.157)$$

where the path of integration is a line to the right of the origin parallel to the imaginary axis such that the integral converges. The calculation of Laplace transforms is usually relatively simple in comparison with Fourier transforms but the inverse transformation is generally difficult.

The Fourier transform is more convenient when we wish to discuss properties which depend upon frequency and/or phase. It is very useful in certain areas of probability theory and in solving linear differential equations with boundary conditions which can be expressed in Fourier or Fourier–Bessel series. The Laplace transform is very useful when we investigate the analytical properties of the transform (as in circuit analysis) and in solving linear differential equations with constant coefficients when initial conditions are given.

The Laplace transform of some common functions can be easily derived. Thus,

$$\text{step}(t) \leftrightarrow \int_0^\infty \text{step}(t) e^{-st} \, dt = \frac{e^{-st}}{-s}\Big|_0^\infty = 1/s; \quad (10.158)$$

$$\delta(t) \leftrightarrow \int_0^\infty \delta(t) e^{-st} \, dt = e^{-st}\Big|_{t=0} = +1; \quad (10.159)$$

$$t \leftrightarrow \int_0^\infty t e^{-st} \, dt = \frac{e^{-st}}{s^2}(-st-1)\Big|_0^\infty = 1/s^2;$$
$$\quad (10.160)$$

$$e^{-kt} \leftrightarrow \int_0^\infty e^{-kt} e^{-st} \, dt = \frac{e^{-(k+s)t}}{-(k+s)}\Big|_0^\infty = +1/(s+k)$$
$$\quad (10.161)$$

(note that $Re[s] > Re[k]$ so that $e^{-(k+s)t} = 0$ for $t = +\infty$);

$$\left.\begin{array}{l} \cos at \leftrightarrow s/(s^2 + a^2); \\ \sin at \leftrightarrow a/(s^2 + a^2); \end{array}\right\} \quad (10.162)$$

$$\left.\begin{array}{l} \cosh at \leftrightarrow s/(s^2 - a^2); \\ \sinh at \leftrightarrow a/(s^2 - a^2). \end{array}\right\} \quad (10.163)$$

The last four results can be obtained directly by integration or by substituting $k = \pm ja$ or $k = \pm a$ in (10.161) and combining exponential terms to get $\cos at$, etc. Equations (10.158), (10.159) and (10.161) should be compared with (10.99) (10.105) and (10.110).

10.4.2 *Theorems on Laplace transforms*

Most of the theorems on Fourier transforms have counterparts for Laplace transforms. The most useful theorems are listed below. Note that $a > 0$ in all cases.

$$g(t) \leftrightarrow G(s)$$
$$k\,g(t) \leftrightarrow k\,G(s); \quad (10.164)$$
$$k_1 g_1(t) + k_2 g_2(t) \leftrightarrow k_1 G_1(s) + k_2 G_2(s); \quad (10.165)$$

Shift theorems $\begin{cases} g(t-a)\,\text{step}\,(t-a) \leftrightarrow e^{-as}G(s); & (10.166) \\ e^{-at}g(t) \leftrightarrow G(s+a); & (10.167) \end{cases}$

Scaling theorems $\begin{cases} g(at) \leftrightarrow (1/a)G(s/a) & (10.168) \\ (1/a)g(t/a) \leftrightarrow G(as); & (10.169) \end{cases}$

Derivative theorems

$$\begin{cases} \dfrac{d^n g(t)}{dt^n} \leftrightarrow s^n G(s) - s^{n-1}g(0+) - s^{n-2}g^1(0+1)\ldots \\ \qquad\qquad\qquad - g^{n-1}(0+) \\ (-t)^n g(t) \leftrightarrow d^n G(s)/ds^n \end{cases}$$

$$\quad (10.170)$$
$$\quad (10.171)$$

Integral theorems $\begin{cases} \displaystyle\int_0^t g(t)\,dt \leftrightarrow (1/s)G(s) & (10.172) \\ (1/t)g(t) \leftrightarrow \displaystyle\int_s^{+\infty} G(s)\,ds & (10.173) \end{cases}$

Convolution theorem

$$g_1(t) * g_2(t)$$
$$= \int_0^t g_1(\tau)g_2(t-\tau)\,d\tau \leftrightarrow G_1(s)G_2(s)$$
$$\quad (10.174)$$

In (10.170) the symbols $g(0+)$ and $g'(0+)$ denote the values of $g(t)$ and its rth derivative at $t = 0$ when t approaches zero from the positive side.

The proofs of most of the above theorems are simple. For example, for (10.166) we have

$$\int_0^\infty g(t-a)\,\text{step}\,(t-a)e^{-st}\,dt$$

$$= \int_a^\infty g(t-a)e^{-st}\,dt$$

$$= \int_0^\infty g(y)e^{-s(y+a)}\,dy, \; y = t - a,$$

$$= e^{-as}\int_0^\infty g(y)e^{-sy}\,dy = e^{-as}G(s).$$

For (10.170), we write

$$\frac{dg}{dt} \leftrightarrow \int_0^\infty \frac{dg}{dt}e^{-st}\,dt = g(t)e^{-st}\Big|_0^\infty - (-s)\int_0^\infty g(t)e^{-st}\,dt$$

$$= -g(0+) + sG(s).$$

By successive applications of this result, the general formula is obtained. For (10.171) we write $G(s) = \int_0^\infty g(t) e^{-st} dt$ and then differentiate with respect to s, giving

$$\frac{dG}{ds} = \int_0^\infty g(t) \frac{d}{ds} (e^{-st}) dt = \int_0^\infty (-t) g(t) e^{-st} dt,$$

and hence $-t g(t) \leftrightarrow dG(s)/ds$. Successive differentiations give (10.171).

To prove (10.172), we write $\int_0^t g(t) \, dt \leftrightarrow G_i(s)$, then differentiate, obtaining $g(t) \leftrightarrow sG_i(s) = G(s)$, hence $G_i(s) = (1/s)G(s)$. The converse is proved by writing

$$\int_s^\infty G(s) \, ds = \int_s^\infty \left\{ \int_0^\infty g(t) e^{-st} dt \right\} ds$$

$$= \int_0^\infty g(t) \left\{ \int_s^\infty e^{-st} ds \right\} dt$$

on changing the order of integration. Then

$$\int_s^\infty G(s) \, ds = \int_0^\infty g(t) \frac{e^{-st}}{-t} \bigg|_s^\infty dt = \int_0^\infty (1/t) g(t) e^{-st} dt;$$

hence $(1/t)g(t) \leftrightarrow \int_s^\infty G(s) \, ds$.

When Laplace transforms are being used, the convolution of two functions, $g_1(t)$ and $g_2(t)$, is usually defined by the expression

$$g_1(t) * g_2(t) = \int_0^t g_1(\tau) g_2(t - \tau) \, d\tau. \tag{10.174}$$

This definition appears to differ from that given in (10.123) because of the different limits. However, the difference is only apparent and is due to the fact that $g_i(t) = 0$, $t < 0$. As shown in fig. 10.8a, $g_1(\tau) = 0$, $\tau < 0$, while $g_2(t - \tau) = 0$, $\tau > t$. Thus changing the limits to $\pm\infty$ would not change the value of the integral.

The convolution theorem, (10.174), can be proven as follows:

$$g_1(t) * g_2(t) \leftrightarrow \int_0^\infty \left\{ \int_0^t g_1(\tau) g_2(t - \tau) d\tau \right\} e^{-st} dt$$

$$\leftrightarrow \int_0^\infty \left\{ \int_0^\infty g_1(\tau) g_2(t - \tau) \text{step}(t - \tau) d\tau \right\} e^{-st} dt$$

where the change of the upper limit and the insertion of $\text{step}(t - \tau)$ have not changed the value of the integral since $\text{step}(t - \tau)$ is unity for $\tau < t$ and zero for $\tau > t$. Changing

the order of integration gives

$$g_1(t) * g_2(t) \leftrightarrow \int_0^\infty g_1(\tau) \left\{ \int_0^\infty g_2(t - \tau) \text{step}(t - \tau) e^{-st} dt \right\} d\tau$$

$$\leftrightarrow \int_0^\infty g_1(\tau) e^{-\tau s} G_2(s) \, d\tau, \text{ using (10.166)},$$

$$\leftrightarrow G_1(s) G_2(s).$$

10.5 Linear systems
10.5.1 *Introduction*

We use the term *system* to denote a group of objects so related that, when an input is applied at one point, an output is generated at another point. We may know little or nothing of the detailed workings of the system.

A *linear system* is one in which the output $h(t)$ is proportional to the input, $g(t)$, that is, if $g(t) \to h(t)$, then $kg(t) \to kh(t)$, where k is a constant and the arrow represents the effect of the system. (Cheng, 1959, has a good account of linear systems.) By considering $g(t)$ to be the sum of two signals, we see that the *principle of superposition* holds, namely,

$$k_1 g_1(t) + k_2 g_2(t) \to k_1 h_1(t) + k_2 h_2(t). \tag{10.175}$$

A system is *time-invariant* if the same input produces the same output regardless of the time when the input is applied, that is,

$$kg(t - t_0) \to kh(t - t_0) \tag{10.176}$$

for all values of t_0.

In principle at least, systems can be described by means of differential equations relating output to input. The corresponding differential equations for linear systems are linear, and the coefficients of the equation are constants when the system is time-invariant. We shall study the properties of differential equations describing linear, time-invariant systems. Our discussion will not depend on the order of the equation and since many systems are represented by second-order equations, we consider the equation

$$\frac{d^2 h(t)}{dt^2} + a_1 \frac{dh(t)}{dt} + a_2 h(t) = g(t). \tag{10.177}$$

Using (10.170), we get

$$\frac{d^2 h(t)}{dt^2} \leftrightarrow s^2 H(s) - sh(0+) - h'(0+),$$

$$\frac{dh(t)}{dt} \leftrightarrow H(s) - h(0+)$$

where $h(0+)$ and $h'(0+)$ are the values of $h(t)$ and $\mathrm{d}h(t)/\mathrm{d}t$ at $t = 0$.

Substituting in the equation gives

$$(s^2 + a_1 s + a_2) H(s) - \{(s + a_1) h(0+) + h'(0+)\}$$
$$= G(s).$$

Solving for $H(s)$, we have

$$H(s) = \frac{G(s) + (s + a_1) h(0+) + h'(0+)}{(s^2 + a_1 s + a_2)}. \qquad (10.178)$$

The numerator in (10.178) is called the *total excitation transform*; it depends in part upon the excitation (input), $g(t)$, and in part upon the initial conditions of the system. When the initial output and its first derivative are zero, that is, $h(0+) = 0 = h'(0+)$, the system is said to be *initially relaxed* and the total excitation transform reduces to the input transform, $G(s)$. For convenience, we shall assume henceforth that the system is initially relaxed.

The quantity, $1/(s^2 + a_1 s + a_2)$ is known as the *transfer function*, $F(s)$. It is a function only of the properties of the system. Thus, the transform of the output corresponding to any input, $g(t)$, is given by

$$H(s) = F(s) G(s). \qquad (10.179)$$

We apply the convolution theorem to this equation and obtain

$$h(t) = f(t) * g(t) = \int_0^t g(\tau) f(t - \tau) \, \mathrm{d}\tau \qquad (10.180)$$

where $F(s) \leftrightarrow f(t)$.

If we apply a unit impulse $\delta(t)$ to a linear system, then $\delta(t) \leftrightarrow G(s) = +1$ from (10.159), so that $H(s) = F(s)$ and $h(t) = f(t)$. The response $f(t)$ to a unit impulse is the *impulse response* (often called the unit impulse response); it is the inverse Laplace transform of the transfer function. In principle we can predict the behavior of a linear system for an arbitrary input by applying a unit impulse and measuring the impulse response, then applying (10.180) to get $h(t)$.

Let us apply a unit step function to the input of a linear system. Then from (10.158) and (10.179), we obtain

$$H(s) = F_u(s) = (1/s) F(s)$$

where $F_u(s) \leftrightarrow f_u(t) = $ *step response* (also called 'unit step response'). We can find the relation between $f_u(t)$ and $f(t)$ as follows. From the above

$$F(s) = s F_u(s). \qquad (10.181)$$

But

$$\mathrm{d}\{f_u(t)\}/\mathrm{d}t \leftrightarrow s F_u(s) - f_u(0+) = F(s) - f_u(0+)$$

from (10.170); hence

$$f(t) = \mathrm{d}\{f_u(t)\}/\mathrm{d}t + f_u(0+)\delta(t). \qquad (10.182)$$

Since $f_u(t) = 0$, $t < 0$, the step response has a discontinuity at the origin of magnitude $f_u(0+)$. Thus the impulse response, $f(t)$, is equal to the derivative of the step response $f_u(t)$ plus an impulse at the origin of magnitude $f_u(0+)$.

10.5.2 *Superposition integral*

In §8.1.2a we showed that the output for a digital input to a linear system is equal to the superposition of a series of weighted impulse responses. In the limit as the sampling interval approaches zero, the sum becomes the convolution integral. A similar result holds when we use $f_u(t)$ instead of $f(t)$. To show this we write

$$f_u(t) * g(t) \leftrightarrow F_u(s) G(s),$$

$$\frac{\mathrm{d}}{\mathrm{d}t}\{f_u(t) * g(t)\} \leftrightarrow s F_u(s) G(s) - f_u(t) * g(t)|_{t=0}$$

from (10.170). Since $f_u(t) * g(t)|_{t=0} = \int_0^0 f_u(\tau) g(0 - \tau) \, \mathrm{d}\tau = 0$, we have from (10.181)

$$\frac{\mathrm{d}}{\mathrm{d}t}\{f_u(t) * g(t)\} \leftrightarrow s F_u(s) G(s) = F(s) G(s) = H(s),$$

where $H(s)$ is the output transform corresponding to the input $g(t)$. Taking the inverse transform gives

$$h(t) = \frac{\mathrm{d}}{\mathrm{d}t}\{f_u(t) * g(t)\}$$

$$= \frac{\mathrm{d}}{\mathrm{d}t}\left\{\int_0^t f_u(\tau) g(t - \tau) \, \mathrm{d}\tau\right\}$$

$$= \frac{\mathrm{d}}{\mathrm{d}t}\left\{\int_0^t g(\tau) f_u(t - \tau) \, \mathrm{d}\tau\right\}.$$

Applying Leibnitz' rule (10.131) to differentiate the integrals gives

$$h(t) = \int_0^t f_u(\tau) g'(t - \tau) \, \mathrm{d}\tau + f_u(t) g(0+)$$

$$= f_u(t) * g'(t) + f_u(t) g(0+)$$

$$= \int_0^t g(\tau) f_u'(t - \tau) \, \mathrm{d}\tau + g(t) f_u(0+) \qquad (10.183)$$

$$= f_u'(t) * g(t) + g(t) f_u(0+)$$

where the prime denotes differentiation with respect to t. These integrals are known as *Duhamel's integrals*.

The first integral in (10.183) is often referred to as the *Superposition Integral*. The name is due to the following concept. Let us approximate $g(t)$ by a series of step functions as in fig. 10.10, that is,

$$g(t) = g(0+) \operatorname{step}(t) + g'(\tau_1) \Delta\tau \operatorname{step}(t - \tau_1)$$
$$+ g'(\tau_2) \Delta\tau \operatorname{step}(t - \tau_2) + \dots$$

Using the principle of superposition the output will be the sum of the outputs for each of the step functions, hence

$$h(t) = g(0+) f_u(t) \operatorname{step}(t)$$
$$+ g'(\tau_1) f_u(t - \tau_1) \Delta\tau \operatorname{step}(t - \tau_1)$$
$$+ g'(\tau_2) f_u(t - \tau_2) \Delta\tau \operatorname{step}(t - \tau_2) + \dots$$

In the limit as $\Delta\tau \to 0$, we get

$$h(t) = g(0+) f_u(t) + \int_0^t g'(\tau) f_u(t - \tau) \, d\tau,$$
$$= f_u(t) g(0+) + \int_0^t f_u(\tau) g'(t - \tau) \, d\tau$$

which is the first equation in (10.183).

10.5.3 *Linear systems in series and parallel*
Assume that two linear systems with transfer functions $F_1(s)$ and $F_2(s)$ are connected in series so that the output of the first system is the input of the second system; then

$$H_1(s) = F_1(s) G(s);$$
$$h_1(t) = f_1(t) * g(t);$$
$$H_2(s) = F_2(s) H_1(s) = F_2(s) F_1(s) G(s);$$
$$h_2(t) = f_2(t) * \{f_1(t) * g(t)\}$$
$$= \{f_2(t) * f_1(t)\} * g(t). \tag{10.184}$$

Fig.10.10. Illustrating the superposition integral.

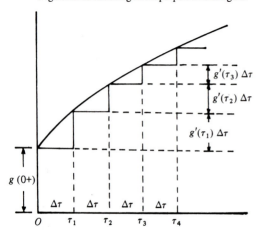

Thus two systems in series are equivalent to a single system whose transfer function is the product of the two individual transfer functions. Obviously this result holds for any number of systems in series.

When an input $g(t)$ is applied to two systems in parallel, the outputs will be superimposed so that

$$H(s) = H_1(s) + H_2(s) = G(s) F_1(s) + G(s) F_2(s)$$
$$= G(s) \{F_1(s) + F_2(s)\}. \tag{10.185}$$

Thus, the equivalent transfer function for systems in parallel is the sum of the individual transfer functions.

10.5.4 *Input–output relations for random functions*
Several important relations can be demonstrated between a random input to a linear system and the resulting output. Starting with (10.144), we write for the autocorrelation function of the output

$$\phi_{hh}(t) = \lim_{T \to \infty} (1/2T) \int_{-T}^{T} h(\tau) h(t + \tau) \, d\tau$$
$$= \lim_{T \to \infty} (1/2T) \int_{-T}^{T} \left\{ \int_{-\infty}^{\infty} f(x) g(\tau - x) \, dx \right\}$$
$$\times \left\{ \int_{-\infty}^{\infty} f(y) g(t + \tau - y) \, dy \right\} d\tau.$$

Interchanging the order of integration gives,

$$\phi_{hh}(t) = \int_{-\infty}^{\infty} f(x) \int_{-\infty}^{\infty} f(y) \left\{ \lim_{T \to \infty} (1/2T) \int_{-T}^{T} \right.$$
$$\left. \times g(\tau - x) g(t + x - y + \tau - x) \, d(\tau - x) \right\} dy \, dx$$

where we can change $d\tau$ to $d(\tau - x)$ in the last integral because x is constant in this integration. Thus,

$$\phi_{hh}(t) = \int_{-\infty}^{\infty} \int_{-\infty}^{\infty} f(x) f(y) \phi_{gg}(t + x - y) dx \, dy$$

using (10.144). This equation can be transformed to the frequency domain by using the Wiener autocorrelation theorem (10.145). Then

$$\Phi_{hh}(\omega) = \int_{-\infty}^{\infty} \phi_{hh}(t) e^{-j\omega t} \, dt$$
$$= \int_{-\infty}^{\infty} e^{-j\omega t} \, dt$$
$$\times \left\{ \int_{-\infty}^{\infty} \int_{-\infty}^{\infty} f(x) f(y) \phi_{gg}(t + x - y) \, dx \, dy \right\}.$$

Let $\mu = t + x - y$ so that

$$e^{-j\omega t} = e^{-j\omega(\mu - x + y)}, \quad d\mu = dt$$

since x and y are constant in the first integral. Then,

$$\Phi_{hh}(\omega) = \int_{-\infty}^{\infty} \phi_{gg}(\mu) e^{-j\omega\mu} d\mu$$

$$\times \int_{-\infty}^{\infty} f(x) e^{j\omega x} dx \int_{-\infty}^{\infty} f(y) e^{-j\omega y} dy$$

$$= \Phi_{gg}(\omega) \overline{F(\omega)} F(\omega) = |F(\omega)|^2 \Phi_{gg}(\omega). \quad (10.186)$$

Since $|F(\omega)|^2 = \Phi_{ff}(\omega) = $ energy spectrum of $f(t)$,

$$\Phi_{hh}(\omega) = \Phi_{ff}(\omega)\Phi_{gg}(\omega),$$

that is,

$$\phi_{hh}(t) = \phi_{ff}(t) * \phi_{gg}(t). \quad (10.187)$$

The cross-correlation of $g(t)$ and $h(t)$ becomes

$$\phi_{gh}(t) = \lim_{T \to \infty} (1/2T) \int_{-T}^{T} g(\tau) h(t + \tau) d\tau$$

$$= \lim_{T \to \infty} (1/2T) \int_{-T}^{T} g(\tau)$$

$$\times \left\{ \int_{-\infty}^{\infty} f(x)g(t + \tau - x) dx \right\} d\tau$$

$$= \int_{-\infty}^{\infty} f(x) \left\{ \lim_{T \to \infty} (1/2T) \right.$$

$$\left. \times \int_{-T}^{T} g(\tau)g(t + \tau - x) d\tau \right\} dx$$

$$= \int_{-\infty}^{\infty} f(x)\phi_{gg}(t - x) dx = f(t) * \phi_{gg}(t);$$

thus

$$\Phi_{gh}(\omega) = F(\omega)\Phi_{gg}(\omega). \quad (10.188)$$

10.6 Digital systems and z-transforms
10.6.1 *Sampling theorem*

When a continuous function, $g(t)$, is sampled at regular intervals, Δ, we obtain a series of values, $g(n\Delta)$. The sampled function, which we write g_t, can be regarded as the product of $g(t)$ and a comb (see (10.108)) with spacing Δ:

$$g_t = \sum_{n=-\infty}^{\infty} g(n\Delta)\delta(t - n\Delta). \quad (10.189)$$

(Although the comb extends from $-\infty$ to $+\infty$, we need to sum only over the range for which $g(t) \neq 0$.)

When we sample, we discard values of $g(t)$ in between the sampling points and hence apparently lose information. However, the Sampling Theorem states that $g(t)$ can be recovered exactly from the sampled values provided $g(t)$ has no frequencies above the Nyquist frequency, $\nu_N = \frac{1}{2}(\text{sampling frequency}) = 1/2\Delta$. The Fourier transform of $g(t)$, $G(\omega)$, thus must be zero for $|\omega| > \omega_N$, $\omega_N = 2\pi\nu_N = \pi/\Delta$, so that

$$g(t) = (1/2\pi) \int_{-\infty}^{\infty} G(\omega) e^{j\omega t} d\omega$$

$$= (1/2\pi) \int_{-\omega_N}^{\omega_N} G(\omega) e^{j\omega t} d\omega. \quad (10.190)$$

If we expand $G(\omega)$ in a Fourier series in the interval $-\omega_N$ to ω_N, the Fourier series will repeat $G(\omega)$ in each interval of width $2\omega_N$, although $G(\omega)$ is zero outside the interval $(-\omega_N, \omega_N)$. The fundamental period for the series is $1/2\omega_N$, hence $\omega_0 = 2\pi/T = \pi/\omega_N = \Delta$. Therefore, using (10.83) and (10.84) with ω replacing t, we get

$$G(\omega) = \sum_{r=-\infty}^{\infty} \alpha_r e^{jr\omega\Delta},$$

$$\alpha_r = (1/2\omega_N) \int_{-\omega_N}^{\omega_N} G(\omega) e^{-jr\omega\Delta} d\omega.$$

Comparison with (10.190) shows that if we take $t = r\Delta$, then

$$\alpha_{-r} = (\pi/\omega_N)g(r\Delta);$$

hence

$$G(\omega) = \sum_{r=-\infty}^{\infty} (\pi/\omega_N)g(r\Delta) \exp\{-jr\pi(\omega/\omega_N)\}.$$

This equation gives the correct values of $G(\omega)$ in the interval $-\omega_N$ to $+\omega_N$ but is not zero outside this interval as it should be to represent $g(t)$ exactly. To correct for this we can multiply by a boxcar, $\text{box}_a(\omega)$, extending from $-\omega_N$ to $+\omega_N$, that is, $G(\omega)$ is given exactly by

$$G(\omega) = \text{box}_a(\omega) \sum_{r=-\infty}^{\infty} (\pi/\omega_N)g(r\Delta)$$

$$\times \exp\{-jr\pi(\omega/\omega_N)\}. \quad (10.191)$$

Equation (10.130) gives

$$\omega_N \text{sinc} \omega_N t \leftrightarrow \pi \, \text{box}_{2\omega_N}(\omega).$$

Using (10.114), we find that

$$\omega_N \text{sinc}\{\omega_N(t - k)\} \leftrightarrow \pi \, \text{box}_{2\omega_N}(\omega) e^{-jk\omega}.$$

Using this result we can write the inverse transform of (10.191) in the form

$$g(t) = \sum_{r=-\infty}^{\infty} g(r\Delta) \text{sinc}(\omega_N t - r\pi). \quad (10.192)$$

This result shows that the function, $\mathrm{sinc}\,(\omega_N t - r\pi) = \sin \pi(t/\Delta - r)/\pi(t/\Delta - r)$, provides perfect interpolation to give $g(t)$ for all values of t, not merely for the sampling instants $r\Delta$.

When $G(\omega) \neq 0$ for values of $|\omega| > \omega_N$, the above proof breaks down and we are not able to recover $g(t)$ from the sampled values (see discussion in §8.1.2*b*).

10.6.2 *Convolution and correlation of sampled functions*

The integrals in (10.123), (10.125) and (10.139) involve continuous functions. When the functions are sampled functions, the integrals become summations. To show this, we start with continuous functions $f(t)$ and $g(t)$ so that $f(t) * g(t)$ is given by

$$f(t) * g(t) = \int_{-\infty}^{\infty} f(\tau) g(t - \tau)\,d\tau.$$

Replacing $f(\tau)$ by the sampled function

$$f_\tau = \sum_{k=-\infty}^{\infty} f(k\Delta)\,\delta(\tau - k\Delta)$$

(see (10.189)), we have

$$f_\tau * g(t) = \int_{-\infty}^{\infty} \left\{ \sum_k f(k\Delta)\,\delta(\tau - k\Delta) \right\} g(t - \tau)\,d\tau$$

$$= \sum_k \left\{ \int_{-\infty}^{\infty} \{f(k\Delta) g(t - \tau)\}\,\delta(\tau - k\Delta)\,d\tau \right\}.$$

Applying (10.102) to each integral, we get for each term in the sum the value $f(k\Delta)g(t - k\Delta)$. If we now sample $g(t)$, t becomes a multiple of Δ, hence $g(t - k\Delta) = g_{t-k}$, so that we have (note that $d\tau \to \Delta$)

$$f_t * g_t = \Delta \sum_k f_k g_{t-k}. \tag{10.193}$$

In the same way we find that

$$\phi_{fg}(\tau) = \Delta \sum_k f_k g_{k+\tau}. \tag{10.194}$$

Usually we set $\Delta = 1$ in these equations (e.g., (8.18) to (8.20), (8.35), (8.36), (8.38) and (8.40)) but care must be taken in some cases (as in problem 10.22*c*).

10.6.3 *Z-transforms*

Z-transforms are a special form of transform useful for calculations involving digital (sampled) functions. We take the Fourier transform of both sides of (10.189), using (10.132) and obtain

$$g_t \leftrightarrow G(\omega) = \sum_{n=-\infty}^{\infty} g(n\Delta)\,e^{-jn\omega\Delta}.$$

If we write $z = e^{-j\omega\Delta}$, we get

$$G(\omega) = \sum_{n=-\infty}^{\infty} g(n\Delta)z^n = G(z) \tag{10.195}$$

where $G(z)$ is the *z-transform* of g_t, that is, $g_t \leftrightarrow G(z)$. Thus, if $g_t = [1, 2, -5, 4, -6]$, $G(z) = 1 + 2z - 5z^2 + 4z^3 - 6z^4$. Negative powers of z denote values of time past; thus if $g_t = [2, 6, -\overset{\downarrow}{1}, 0, 5]$ (the superscribed arrow denotes $t = 0$), then $G(z) = 2z^{-2} + 6z^{-1} - 1 + 5z^2$. It is evident that multiplication by z is equivalent to delaying the time function by one sample interval and division by z to advancing it one sample interval.

[We could have taken the Laplace transform to get $G(z)$ in terms of $z = e^{-s\Delta}$. The Fourier form is more convenient for studying frequency characteristics, the Laplace form for examining stability, as when studying filters. A difference is that s ranges over that part of the complex plane to the right of the vertical line through σ, that is, $\mathrm{Re}[s] \geq \sigma(\S10.4.1)$, while the terminus of **z** lies on the unit circle with center at the origin.]

[In signal analysis z is often defined as $z = e^{j\omega\Delta}$, resulting in a polynomial in which the zs have negative exponents as compared with the above. The convention used here is more common in seismic data-processing.]

Clearly z, hence also $G(z)$, is a periodic function of ω with period $2\pi/\Delta$. As ω increases from c to $c + 2\pi/\Delta$, c being any real number, the terminus of z goes around the unit circle ($|z| = 1$) with center at the origin once in the clockwise direction; this follows from the relation

$$z = e^{-j\omega\Delta} = \cos \omega\Delta - j\sin \omega\Delta$$

(note that as ω increases from zero, $-\sin \omega\Delta$ increases in the negative direction). [If we had taken $z = e^{j\omega\Delta}$, z would rotate counterclockwise as ω increases.]

We shall calculate a few simple z-transforms. For example, let

$$g(t) = t\,\mathrm{step}\,(t) = \sum_{n=0}^{\infty} (n\Delta)\,\delta(t - n\Delta),$$

$$t\,\mathrm{step}\,(t) \leftrightarrow \sum_{n=0}^{\infty} (n\Delta)z^n = \Delta \sum_{n=1}^{\infty} nz^n$$

$$\leftrightarrow z\Delta \sum_{n=1}^{\infty} nz^{n-1} = z\Delta/(1 - z)^2. \tag{10.196}$$

Again, let

$$g(t) = e^{kt}\,\mathrm{step}\,(t) = \sum_{n=0}^{\infty} e^{kn\Delta}\,\delta(t - n\Delta);$$

$$e^{kt}\,\mathrm{step}\,(t) \leftrightarrow \sum_{n=0}^{\infty} e^{kn\Delta}z^n = \sum_{n=0}^{\infty} (e^{k\Delta}z)^n$$

$$\leftrightarrow 1 + e^{k\Delta}z + (e^{k\Delta}z)^2 + \cdots$$

$$= 1/(1 - e^{k\Delta}z). \tag{10.197}$$

Setting $k = 0$ gives $\mathrm{step}\,(t) \leftrightarrow 1/(1 - z)$. \hfill (10.198)

Setting $k = +j\theta$ gives

$$e^{j\theta t}\,\text{step}(t) \leftrightarrow \frac{1}{1 - z\,e^{j\theta\Delta}} \cdot = \frac{1}{(1 - z\cos\theta\Delta) - jz\sin\theta\Delta}$$

$$(10.199)$$

or

$$(\cos\theta t + j\sin\theta t)\,\text{step}(t) \leftrightarrow \frac{(1 - z\cos\theta\Delta) + j(z\sin\theta\Delta)}{(1 - z\cos\theta\Delta)^2 + (z\sin\theta\Delta)^2}.$$

Equating real and imaginary parts gives

$$\cos\theta t\,\text{step}(t) \leftrightarrow \frac{1 - z\cos\theta\Delta}{1 - 2z\cos\theta\Delta + z^2};$$

$$(10.200)$$

$$\sin\theta t\,\text{step}(t) \leftrightarrow \frac{z\sin\theta\Delta}{1 - 2z\cos\theta\Delta + z^2}.$$

As an example of the use of z-transforms in determining the frequency and phase characteristics of digital functions, we shall obtain the spectra of the digital function $g_t = [\ldots, 0, 0, -2, 0, \overset{\downarrow}{1}, 3, -2, 5, 0, 0, \ldots]$. The transform is $G(z) = -2z^{-2} + 1 + 3z - 2z^2 + 5z^3 = -2e^{2j\omega\Delta} + 1 + 3e^{-j\omega\Delta} - 2e^{-2j\omega\Delta} + 5e^{-3j\omega\Delta}$. Substituting values of ω gives values of $G(z)$ which are complex in general and from which we can get the amplitude and phase spectra as functions of ω. The following table gives typical values of $G(\omega)$ for a few values of ω.

$\omega\Delta$	$G(z)$	Amplitude	Phase
$0°$	5	5.00	$0.0°$
$45°$	$1 - \sqrt{2} - j4\sqrt{2}$	5.67	$85.8°$
$90°$	$5 + 2j$	5.39	$203.6°$
$135°$	$1 + \sqrt{2} - j4\sqrt{2}$	6.15	$293.0°$
$180°$	-11	11.00	$360.0°$
$225°$	$1 + \sqrt{2} + j4\sqrt{2}$	6.15	$67.0°$
$270°$	$5 - 2j$	5.39	$156.4°$
$315°$	$1 - \sqrt{2} + j4\sqrt{2}$	5.67	$274.2°$

10.6.4 *Calculation of z-transforms; Fast Fourier Transform*

Generally we must calculate z-transforms for many more values of the argument $\omega\Delta$ than we did in the last section to obtain the required precision. Consider the function

$$g_t = g_0, g_1, g_2, \ldots, g_{n-1};$$

$$G(z) = g_0 + g_1 z + g_2 z^2 + \ldots + g_{n-1}z^{n-1}.$$

It is convenient to take $n = 2^k$ where k is integral (this can always be achieved by adding zeros to g_t) and to calculate the transform for increments of $\omega\Delta$ equal to $2\pi/n$, that is, we take $\omega\Delta = r(2\pi/n)$, $r = 0, 1, 2, \ldots, n - 1$ (note that $G(z)$ repeats for $r \geq n$). If we let $q = e^{-j2\pi/n}$, then $z = e^{-jr(2\pi/n)} = q^r$. The various values of $G(z)$, G_r, can be written in matrix form

$$\begin{Vmatrix} G_0 \\ G_1 \\ G_2 \\ . \\ . \\ . \\ G_{n-1} \end{Vmatrix} = \begin{Vmatrix} 1 & 1 & 1 & \cdots & 1 \\ 1 & q & q^2 & \cdots & q^{(n-1)} \\ 1 & q^2 & q^4 & \cdots & q^{2(n-1)} \\ . & . & . & & . \\ . & . & . & & . \\ . & . & . & & . \\ 1 & q^{(n-1)} & q^{2(n-1)} & \cdots & q^{(n-1)^2} \end{Vmatrix} \begin{Vmatrix} g_0 \\ g_1 \\ g_2 \\ . \\ . \\ . \\ g_{n-1} \end{Vmatrix}$$

$$(10.201)$$

This method requires n^2 multiplications and n^2 additions. Since a seismic trace often has a few thousand values, millions of calculations are necessary. The Fast Fourier Transform (FFT) is an ingenious algorithm for calculating $G(z)$ with only $(n \log_2 n)$ calculations. For $n = 2^{10} = 1024$, the difference is between 10^4 and 2×10^6.

The Fast Fourier Transform (Cooley and Tukey, 1965) depends upon *doubling* processes by which a series is built up from (or decomposed into) shorter series. Let us take the time series

$$c_t = c_0, c_1, c_2, \ldots, c_{2n-1};$$

$$C(z) = c_0 + c_1 z + c_2 z^2 + \ldots + c_{2n-1}z^{2n-1},$$

and decompose it into two series:

$$c_t = x_0, y_0, x_1, y_1, \ldots, x_{n-1}, y_{n-1}.$$

We write

$$x_t = x_0, x_1, \ldots, x_{n-1};$$

$$X(z) = x_0 + x_1 z + \ldots + x_{n-1}z^{n-1};$$

$$y_t = y_0, y_1, \ldots, y_{n-1};$$

$$Y(z) = y_0 + y_1 z + \ldots + y_{n-1}z^{n-1},$$

where the values x_i, y_i occur at intervals of 2Δ, not Δ as in c_t.

We calculate $C(z)$ for $z = q^r$, $q = e^{-j2\pi/2n}$, $r = 0, 1, 2, \ldots (2n - 1)$, whereas $X(z)$, $Y(z)$ are calculated for the values $(q')^r$, $q' = e^{-j2\pi/n} = q^2$, $r = 0, 1, 2, \ldots, (n - 1)$. Writing x_r for the value of $X(z)$ for $z = q^{2r}$,

$$\left. \begin{aligned} X_r &= \sum_{i=0}^{n-1} x_i q^{2ri} \\ Y_r &= \sum_{i=0}^{n-1} y_i q^{2ri} \end{aligned} \right\} \quad r = 0, 1, \ldots, (n - 1). \quad (10.202)$$

But,

$$C_r = \sum_{i=0}^{2n-1} c_i q^{ri}, \quad r = 0, 1, \ldots, (2n-1).$$

For $r = 0, 1, \ldots, (n-1)$,

$$C_r = (x_0 + x_1 q^{2r} + x_2 q^{4r} + \ldots + x_{n-1} q^{2(n-1)})$$
$$+ q^r(y_0 + y_1 q^{2r} + y_2 q^{4r} + \ldots + y_{n-1} q^{2(n-1)})$$
$$= \sum_{i=0}^{n-1} x_i q^{2ri} + q^r \sum_{i=0}^{n-1} y_i q^{2ri}$$
$$= X_r + q^r Y_r. \qquad (10.203)$$

When $r = n, (n+1), \ldots, (2n-1)$, we must manipulate the exponents to express C_r in terms of X_r, Y_r. Thus,

$$C_r = \sum_{i=0}^{2n-1} c_i q^{ri}, \quad r = n, (n+1), \ldots, (2n-1).$$

We write $r = n + m$ so that $q^{ri} = q^{(n+m)i} = q^{ni} q^{mi} = (-1)^i q^{mi}, m = 0, 1, (n-1)$ since $q^{ni} = (e^{-j\pi})^i$. Hence,

$$C_r = \sum_{i=0}^{n-1} x_i q^{2mi} - q^m \sum_{i=0}^{n-1} y_i q^{2mi},$$

$$m = 0, 1, \ldots, (n-1),$$
$$= X_m - q^m Y_m$$
$$= X_{r-n} - q^{r-n} Y_{r-n},$$
$$r = n, (n+1), \ldots, (2n-1). \qquad (10.204)$$

To calculate C_r using (10.201) requires $2(2n)^2 = 8n^2$ arithmetical operations. To find X_r from (10.202) requires $2n^2$, hence to get C_r from (10.203) and (10.204) requires slightly more than $4n^2$ operations, a 50% saving. Since n is a multiple of 2, doubling can be continued until the sub-series consist of single elements of c_t with a tremendous saving in work when n is large.

10.6.5 *Application of z-transforms to digital systems*

By *digital systems* we refer to linear systems in which the input and output are sampled functions. If the system is analog, each element of the digital input g_t gives rise to a continuous output $h(t - n\Delta)$. In this case we sample the output in synchronism with the input to get h_t. Then,

$$F(\omega) = H(z)/G(z) = \text{polynomial in } z = F(z).$$

Thus,

$$H(z) = F(z)\, G(z) \leftrightarrow h_t = f_t * g_t. \qquad (10.205)$$

The utility of z-transforms for digital processing arises because they can be written by inspection and

manipulated as simple polynomials. For example, in §8.1.2a we convolved $f_t = (1, -1, \frac{1}{2})$ with $g_t = (1, \frac{1}{2}, -\frac{1}{2})$. The z-transforms are $F(z) = 1 - z + \frac{1}{2}z^2$ and $G(z) = 1 + \frac{1}{2}z - \frac{1}{2}z^2$; we have

$$f_t * g_t \leftrightarrow F(z)G(z) = (1 - z + \frac{1}{2}z^2)(1 + \frac{1}{2}z - \frac{1}{2}z^2)$$
$$= (1 - \frac{1}{2}z - \frac{1}{2}z^2 + \frac{3}{4}z^3 - \frac{1}{4}z^4);$$

thus, $f_t * g_t = (1, -\frac{1}{2}, -\frac{1}{2}, \frac{3}{4}, -\frac{1}{4})$.

As another example, in §8.1.2d, (8.32) gave the water reverberation filter for $n = 1$ as

$$f_t = (1, -2R, 3R^2, -4R^3, 5R^4, \ldots)$$

so that

$$f_t \leftrightarrow 1 - 2Rz + 3R^2z^2 - 4R^3z^3 + 5R^4z^4 + \ldots,$$

and the inverse filter, i_t, is such that $f_t * i_t = \delta_t \leftrightarrow 1$ (eq. (8.33)). We can solve for $I(z)$ by division since division by polynomials is a proper operation:

$$I(z) = 1/F(z) = 1 + 2Rz + R^2z^2.$$

Thus,

$$I(z) \leftrightarrow (1, 2R, R^2), \text{ which is (8.34).}$$

As a third example, in §8.1.3a we cross-correlated $x_t = (1, -1, \frac{1}{2})$ with $y_t = (1, \frac{1}{2}, -\frac{1}{2})$. From (8.38) we have $\phi_{xy}(\tau) = x_{-\tau} * y_\tau$. Hence,

$$x_{-\tau} \leftrightarrow \frac{1}{2}z^{-2} - z^{-1} + 1,$$
$$y_t \leftrightarrow 1 + \frac{1}{2}z - \frac{1}{2}z^2,$$
$$\phi_{xy}(\tau) \leftrightarrow \frac{1}{2}z^{-2} - \frac{3}{4}z^{-1} + \frac{1}{4} + z - \frac{1}{2}z^2$$

(see fig. 8.4f).

10.6.6 *Phase considerations*

(a) *Minimum-phase wavelets.* We define a causal function as a real function which is zero for negative time, i.e., $f(t) = 0, t < 0$. A physically-realizable function is a causal function which has finite energy, i.e. $\int_0^\infty |f(t)|^2 dt$ is finite. A minimum-delay function is a physically-realizable function whose transform has an inverse, that is, $f(t) \leftrightarrow F(z)$, and $1/F(z)$ is finite. $F(z)$ is said to be minimum-phase. The reasons for the terms 'minimum-delay' and 'minimum-phase' will be apparent later. (The literature is replete with different definitions of 'minimum-phase', the most common of which will be derived from the above definition.)

Consider a simple wavelet, $w_t = (a, -b)$ with transform $W(z) = (a - bz)$. Then,

$$\frac{1}{W(z)} = 1/(a - bz) = \frac{1}{a}\left(1 - \frac{b}{a}z\right)^{-1}.$$

If $|b/a| < 1$, we can expand (note that $|z| = 1$) and get

$$\frac{1}{W(z)} = \frac{1}{a}\left\{1 + \frac{b}{a}z + \left(\frac{b}{a}z\right)^2 + \ldots\right\}$$

$$= \text{convergent series.}$$

If $|b/a| > 1$, the series is divergent. Thus, for minimum-phase, $|a| > |b|$. The wavelet (b, a) is maximum-phase. When $a = \pm b$, the transform is $a(1 \pm z)$ and the inverse becomes infinite when $\omega\Delta$ equals $n\pi$ or $(2n+1)\pi$, hence the wavelet is not minimum-phase. When $(a - bz)$ is minimum-phase, so is $1/(a - bz)$ (see problem 10.31a).

Consider the wavelet $W(z) = (c - z) = c(1 - z/c)$, c being real. Then,

$$W(z) = c\{1 - (1/c)\cos\omega\Delta + j(1/c)\sin\omega\Delta\},$$

and the phase is $\gamma = \tan^{-1}\{\sin\omega\Delta/(c - \cos\omega\Delta)\}$. If $W(z)$ is minimum-phase, $|c| > 1$ and the denominator never becomes zero, hence $|\gamma| < \frac{1}{2}\pi$ and the curve of $\text{Im}\{W(z)\}$ versus $\text{Re}\{W(z)\}$ does not enclose the origin (see fig. 10.11a). Also, γ is periodic with frequency, values repeating each time $\omega\Delta$ increases by 2π (see fig. 10.11c).

Fig.10.11. Variation of phase γ for (a) minimum-phase wavelet and (b) maximum-phase wavelet as ω increases from 0 to 2π; (c) graphs of γ versus $\omega\Delta$ for (a) and (b).

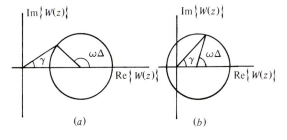

(a)   (b)

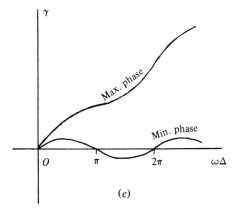

(c)

If $|c| < 1$, the denominator becomes zero twice as $\omega\Delta$ increases by 2π, hence γ assumes all values between 0 and 2π, that is, the curve of $\text{Im}\{W(z)\}$ versus $\text{Re}\{W(z)\}$ encloses the origin (fig. 10.11b). Each time $\omega\Delta$ increases by 2π, γ also increases by 2π (fig. 10.11c). The phase of the maximum-phase wavelet increases without limit while the phase of the minimum-phase wavelet is always between $-\frac{1}{2}\pi$ and $+\frac{1}{2}\pi$, hence is always less than that of the maximum-phase wavelet.

If $(a_i - b_i z)$, $i = 1, 2, \ldots, n$ are all minimum-phase, the sum $\sum_i(a_i - b_i z)$ may or may not be minimum-phase since $|\sum_i a_i|$ is not necessarily greater than $|\sum_i b_i|$. On the other hand the product $\prod_i(a_i - b_i z)$ is always minimum-phase since each term, hence the product also, is cyclic with frequency, hence the curve of $\text{Im}\{W(z)\}$ versus $\text{Re}\{W(z)\}$ cannot enclose the origin (if the curve encloses the origin, the phase increases without limit). If all of the terms $(a_i - b_i z)$ in the product are maximum-phase, the product is maximum-phase. If some of the factors are maximum-phase and some minimum-phase, the product is *mixed-phase*.

Consider the function $\prod_i(a_i - b_i z)/\prod_j(a_j - b_j z)$ where all the factors are minimum-phase. The phase of the function is the sum of the phases of the numerator minus the sum of the phases of the denominator; since all of the phases are cyclic, the phase of the function is also cyclic, hence the function is minimum-phase.

Since $(a_i - b_i z)$ is minimum-phase if $|a_i| > |b_i|$, the root of the equation $(a_i - b_i z) = 0$, namely, a_i/b_i, lies outside the unit circle. Similarly, the roots of a maximum-phase function lie inside the unit circle. When $\prod_i(a_i - b_i z)/\prod_j(a_j - b_j z)$ is minimum-phase, all of the roots of the numerator and denominator lie outside the unit circle; since roots of the denominator are called *poles*, we can say that all roots and poles of a minimum-phase function lie outside the unit circle. (When z is defined as $e^{+j\omega\Delta}$, these rules are reversed.)

The roots $z = 0, \pm 1$ follow the above rule. If $W(z)$ is multiplied by $z^{\pm m}$, m integral, the wavelet is shifted without change in shape; the factors and roots are those of $W(z)$ plus the factor $z^{\pm m}$ and root $z = 0$. However, the graph of $z^{\pm m}W(z)$ now encloses the origin as in fig. 10.11b, hence $z^{\pm m}W(z)$ is not minimum-phase. The convention when we are determining the phase of an isolated wavelet is to have $t = 0$ for the first non-zero element so that roots $z = 0$ do not occur.

Since the expansion of $1/(1 \pm z)$ is divergent, the expansion of $(1 \pm z)W(z)$ is also divergent, hence the product is not minimum-phase. When a root is only slightly larger than unity, problems are often encountered because the expansion converges slowly.

The terms 'minimum-phase' and 'maximum-phase' imply a comparison, and in fact they refer to the set of wavelets which have a given frequency spectrum. The four wavelets $(a - bz)$, $(b - az)$, $(\bar{a} - \bar{b}z)$, $(\bar{b} - \bar{a}z)$ all have the same frequency spectrum as is easily verified by multiplying by the conjugate complexes; two of the four are minimum-phase, two maximum. (Other wavelets with the same spectrum can be obtained by multiplying by any complex constant c where $|c| = 1$, so that there is an infinite number of wavelets with the same spectrum.) When a and b are real, as is most often the case, the four wavelets reduce to two.

(b) *Energy relations.* The intensity of a wave is proportional to the amplitude squared at any instant while the total energy is proportional to the sum of the amplitudes squared from $t = 0$ up to a given instant. For the wavelet (a, b) the total energy is a^2, $a^2 + b^2$ at $t = 0$ and $t = \Delta$, respectively. For the wavelet (b, a) the values are b^2, $b^2 + a^2$. If (a, b) is the minimum-phase wavelet, the energy builds up faster than for the maximum-phase wavelet (b, a). Since the two wavelets have the same frequency spectrum, the only difference is in the rate of build-up of energy. These results can be extended to more complex wavelets; for example, Robinson (1962) showed that the total energy at a time t for a minimum-phase wavelet is greater than or equal to that for any other wavelet with the same spectrum. Claerbout (1963) showed that the 'center of gravity' of the minimum-phase wavelet is closer to $t = 0$ than for any other wavelet with the same spectrum. These principles explain the origin of the term 'minimum-delay'.

(c) *Determining the minimum-phase wavelet for a given spectrum.* The spectrum of the wavelet $(a - bz)$, a and b being complex in general, is $(a - bz)(\bar{a} - \bar{b}z^{-1}) = -a\bar{b}z^{-1} + (a^2 + b^2) - \bar{a}bz$ (note that $\bar{z} = z^{-1}$). Thus the wavelet has the root a/b while the spectrum has this root plus the root $\bar{b}/\bar{a}$ (when a and b are real, the two roots are reciprocals). When a spectrum is of order $2n$, it must have n pairs of roots of the form $(z_i, 1/\bar{z}_i)$. Half the roots will be outside the unit circle, half inside. The minimum-phase wavelet is obtained by multiplying together n factors of the form $(z_i - z)$, for the z_i outside the unit circle.

The above method requires that we find the factors of the spectrum and this is time-consuming when n is large, hence other methods are used (see Claerbout, 1976, chapter 3). One method utilizes the special properties of the Toeplitz matrix to reduce the labor. Let $R(z)$ be a given spectrum and $W(z)$ the minimum-phase wavelet to be determined; $W(z)$ must have an inverse which we denote

$V(z)$. Then,

$$R(z) = W(z)W(\bar{z}), \quad \text{hence} \quad R(z)V(z) = W(\bar{z}).$$

Since $W(z)$ is minimum-phase, we can write $V(z) = b_0 + b_1 z + b_2 z^2 + \dots$.

Moreover, if

$$W(z) = a_0 + a_1 z + a_2 z^2 + \dots + a_n z^n,$$

then

$$W(\bar{z}) = \bar{a}_0 + \bar{a}_1 \bar{z} + \bar{a}_2 \bar{z}^2 + \dots + \bar{a}_n \bar{z}^n$$

and

$$R(z) = (r_{-n}z^{-n} + \dots + r_{-1}z^{-1} + r_0 + r_1 z + \dots + r_n z^n).$$

Then,

$$(r_{-n}z^{-n} + \dots + r_{-1}z^{-1} + r_0 + r_1 z + \dots + r_n z^n)$$
$$\times (b_0 + b_1 z + b_2 z^2 + \dots) = (\bar{a}_0 + \bar{a}_1 z^{-1}$$
$$+ \bar{a}_2 z^{-2} + \dots + \bar{a}_n z^{-n}).$$

Equating coefficients for positive powers of z gives:

$$z^0: \ r_0 b_0 + r_{-1}b_1 + \dots + r_{-n}b_n = \bar{a}_0,$$
$$z^1: \ r_1 b_0 + r_0 b_1 + \dots + r_{-n+1}b_n = 0,$$
$$z^2: \ r_2 b_0 + r_1 b_1 + \dots + r_{-n+2}b_n = 0,$$

$$\dots\dots\dots\dots\dots\dots\dots\dots\dots\dots\dots\dots$$

$$z^n: \ r_n b_0 + r_{n-1}b_1 + \dots + r_0 b_n = 0.$$

(We assume that $(n + 1)$ values of b_i are sufficient to give $W(z) = 1/V(z)$ with the required precision; if not, more terms can be obtained by equating coefficients of z^m to zero, $m > n$.) We note that $\bar{a}_0 = 1/\bar{b}_0$, hence we have $(n + 1)$ equations to solve for the $(n + 1)$ unknowns, b_i. The solution can be written in terms of a Toeplitz matrix $\mathcal{R}$ as $\mathcal{R}\mathcal{B} = \mathcal{W}$:

$$\begin{Vmatrix} r_0 & r_1 & \cdots & r_n \\ r_1 & r_0 & \cdots & \\ \cdot & \cdot & \cdots & \cdot \\ r_n & & \cdots & r_0 \end{Vmatrix} \begin{Vmatrix} 1 \\ b_1' \\ \cdot \\ b_n' \end{Vmatrix} = \begin{Vmatrix} w \\ 0 \\ \vdots \\ 0 \end{Vmatrix} \qquad (10.206)$$

where $r_n = r_{-n}$, $b_i' = b_i/b_0$, $w = \bar{a}_0/b_0 = 1/|b_0|^2$. Solving these equations for the $(n + 1)$ unknowns b_i in terms of the known r_i, we find $V(z)$, then $W(z)$ by inverting $V(z)$.

The straightforward solution of (10.206) requires computer time proportional to n^3 and memory proportional to n^2. The *Levinson recursion algorithm*, which we describe now, reduces these quantities by the factor n. The method is based on forming new equations by selecting a $(k \times k)$ matrix from the upper left-hand corner of $\mathscr{R}$ and the top k rows of $\mathscr{B}$ and $\mathscr{W}$, for example,

$$
\begin{Vmatrix} r_0 & r_1 & \cdots & r_{k-1} \\ r_1 & r_0 & \cdots & \cdot \\ \cdot & \cdot & \cdots & \cdot \\ r_{k-1} & \cdot & \cdots & r_0 \end{Vmatrix} \begin{Vmatrix} 1 \\ b_1^* \\ \vdots \\ b_{k-1}^* \end{Vmatrix} = \begin{Vmatrix} w^* \\ 0 \\ \vdots \\ 0 \end{Vmatrix}, \tag{10.207}
$$

where the asterisks signify that b_i^* and w^* are different from the corresponding quantities in (10.206) since they satisfy a different set of equations. The Levinson recursion algorithm shows how to obtain the solution for the $(k+1)$th case when we know the solution for the kth case. Thus, we can start with $k = 1$, that is $r_0 \times 1 = w$, then use the algorithm to get the solution for $k = 2$, that is for,

$$
\begin{Vmatrix} r_0 & r_1 \\ r_1 & r_0 \end{Vmatrix} \begin{Vmatrix} 1 \\ b_1^* \end{Vmatrix} = \begin{Vmatrix} w^* \\ 0 \end{Vmatrix},
$$

then continue the process until we get to $k = n + 1$; the final step gives the solution of (10.206) whereas intermediate solutions are discarded.

We start with (10.207) in which all of the b_i^*s and w^* are known since we have solved the equation, and we wish to solve the equation for $(k+1)$ in terms of these known quantities. We write the following equation which defines the quantity e:

$$
\begin{Vmatrix} r_0 & r_1 & \cdots & r_k \\ r_1 & r_0 & \cdots & \cdot \\ \cdot & & \cdots & \cdot \\ & & & b_{k-1}^* \\ r_k & \cdot & \cdots & r_0 \end{Vmatrix} \begin{Vmatrix} 1 \\ b_1^* \\ \vdots \\ b_{k-1}^* \\ 0 \end{Vmatrix} = \begin{Vmatrix} w^* \\ 0 \\ \vdots \\ 0 \\ e \end{Vmatrix} \tag{10.208}
$$

Obviously,

$$
e = r_k + r_{k-1} b_1^* + \ldots + r_1 b_{k-1}^* = \sum_{i=1}^{k} r_i b_{k-i}^*,
$$

hence is a known quantity. We now 'invert' $\mathscr{B}$ and $\mathscr{W}$ to get

$$
\begin{Vmatrix} r_0 & r_1 & \cdots & r_k \\ r_1 & r_0 & \cdots & \cdot \\ \cdot & & \cdots & \cdot \\ & & & b_1^* \\ r_k & \cdot & \cdots & r_0 \end{Vmatrix} \begin{Vmatrix} 0 \\ b_{k-1}^* \\ \vdots \\ b_1^* \\ 1 \end{Vmatrix} = \begin{Vmatrix} e \\ 0 \\ \vdots \\ 0 \\ w^* \end{Vmatrix} \tag{10.209}
$$

(one can verify that the last two equations are the same by direct expansion). We multiply (10.209) by a constant c_k

and subtract it from (10.206)

$$
\begin{Vmatrix} r_0 & r_1 & \cdots & r_k \\ r_1 & r_0 & \cdots & \cdot \\ \cdot & & \cdots & \cdot \\ r_k & \cdot & \cdots & r_0 \end{Vmatrix} \left\{ \begin{Vmatrix} 1 \\ b_1^* \\ \vdots \\ b_{k-1}^* \\ 0 \end{Vmatrix} - c_k \begin{Vmatrix} 0 \\ b_{k-1}^* \\ \vdots \\ b_1^* \\ 1 \end{Vmatrix} \right\}
$$

$$
= \begin{Vmatrix} w^* \\ 0 \\ \vdots \\ 0 \\ e \end{Vmatrix} - c_k \begin{Vmatrix} e \\ 0 \\ \vdots \\ 0 \\ w^* \end{Vmatrix} \tag{10.210}
$$

We wish this equation to reduce to the equation for the $(k+1)$th case, namely,

$$
\begin{Vmatrix} r_0 & r_1 & \cdots & r_k \\ r_1 & r_0 & \cdots & \cdot \\ \cdot & & \cdots & \cdot \\ r_k & \cdot & \cdots & r_0 \end{Vmatrix} \begin{Vmatrix} 1 \\ b_1^{**} \\ \vdots \\ b_k^{**} \end{Vmatrix} = \begin{Vmatrix} w^{**} \\ 0 \\ \vdots \\ 0 \end{Vmatrix} \tag{10.211}
$$

Comparing (10.210) and (10.211), we see that, to get 0 for the bottom element of $\mathscr{W}$, we must have $e - c_k w^* = 0$, that is, $c_k = e/w^*$. Also, w^{**} must equal $(w^* - c_k e) = w^*\{1 - (e/w^*)^2\}$. Finally, on the left, we have $b_i^{**} = b_i^* - c_k b_{k-i}^*$ (note that $b_k^{**} = -c_k$ since $b_k^* = 0$, $b_0^* = 1$). Finally, all of the quantities, b_i^{**} and w^{**}, in (10.211) can be found in terms of the known solution of (10.207).

(d) *Zero-phase and linear-phase wavelets.* We note that $(z^n + z^{-n}) = 2 \cos n\omega\Delta$, n integral; since the imaginary part is zero, the function has zero phase. Zero-phase wavelets can be obtained by multiplying pairs of elementary wavelets such as $(1 - az)(az^{-1} - 1) = az^{-1} - (1 + a^2) + az = W_Z(z)$. Since the phase is zero, $W_Z(z) = W_Z(\bar{z})$, hence the spectrum is

$$
W_Z^2(z) = \{az^{-1} - (1 + a^2) - az\}^2
$$
$$
= a^2 z^{-2} - 2a(1 + a^2)z^{-1} + (a^4 + 4a^2 + 1)
$$
$$
- 2a(1 + a^2)z + a^2 z^2.
$$

If $|a| < 1$, the minimum-phase wavelet with the same amplitude spectrum is $(1 - az)^2 = 1 - 2az + a^2 z^2$. Therefore if a minimum-phase wavelet can be written $\prod_i (1 - a_i z)^m$ where m is a multiple of 2, then an equivalent zero-phase wavelet can be found by replacing each pair of factors, $(1 - a_i z)^2$ by $(1 - a_i z)(a_i \bar{z} - 1)$.

A zero-phase wavelet is symmetrical about the origin, hence is neither causal nor physically realizable. The maximum amplitude is at $t = 0$. Roots of a zero-phase wavelet occur in pairs, $(a_i, 1/a_i)$, one of each pair being inside the unit circle, one outside.

If we multiply a zero-phase wavelet by z^n, that is, delay it by n sample intervals, we get

$$W_L(z) = z^n W_z(z).$$

Since $z^n = e^{-jc\omega}$ where c is constant, W_L has linear phase, that is $\gamma_L = c\omega$. Moreover, $W_L(z)$ is merely $W_z(z)$ displaced n time units and has the same roots (plus $z = 0$) and is symmetrical about $t = n\Delta$. Linear-phase wavelets are physically realizable if they start after $t = 0$. Zero-phase wavelets can be obtained from linear-phase wavelets by time shifting. Linear-phase wavelets are often called 'zero-phase.'

10.6.7 *Integral relations for inverse z-transforms*

Although z-transforms and inverse z-transforms can be written down by inspection when g_t is a time series, there are occasions when the equivalent of (10.87) is more convenient. As ω increases from 0 to $2\pi/\Delta$ (or from $-\pi/\Delta$ to $+\pi/\Delta$), z goes once around the unit circle in the clockwise direction (see §10.6.3). Therefore we have

$$g_n = \frac{1}{2\pi} \int_0^{2\pi/\Delta} G(\omega) e^{j\omega n\Delta} d\omega$$

$$= \frac{1}{2\pi} \oint G(z) z^{-n} \{e^{j\omega\Delta} dz/(-j\Delta)\},$$

$$= \frac{1}{2\pi j\Delta} \oint G(z) z^{-(n+1)} dz \quad (10.212)$$

where the integration is in the counterclockwise direction.

10.7 Cepstrum analysis

Transformation from the time domain into the frequency domain permits the equivalent of convolution to be carried out by the simpler operation of multiplication. Transformation from the frequency domain into the cepstrum domain permits such operations to be carried out by the even simpler process of addition. Moreover, in some cases frequencies which overlap in the frequency domain are separated sufficiently in the cepstrum domain that filtering can be carried out more efficiently (Ulrych, 1971.)

The *cepstrum*, $\hat{g}(\zeta)$, is given by an inverse transform of the log of the frequency spectrum:

$$\hat{g}(\zeta) = (1/2\pi) \int_{-\infty}^{\infty} \ln\{G(\omega)\} e^{j\omega\zeta} d\omega. \quad (10.213)$$

The transformation from the time-domain is usually carried out in three steps:

$$\left. \begin{aligned} g(t) &\leftrightarrow G(\omega) = |G(\omega)| e^{j\gamma(\omega)}, \\ \hat{G}(\omega) &= \ln\{G(\omega)\} = \ln|G(\omega)| + j\gamma(\omega), \\ \hat{G}(\omega) &\leftrightarrow \hat{g}(\zeta) \\ &= (1/2\pi) \int_{-\infty}^{\infty} \{\ln|G(\omega)| + j\gamma(\omega)\} e^{j\omega\zeta} d\omega. \end{aligned} \right\} \quad (10.214)$$

Thus the essential feature which characterizes cepstrum analysis is taking the logarithm before making the inverse transformation. To return to the time-domain, the above three steps are reversed:

$$\left. \begin{aligned} \hat{g}(\zeta) &\leftrightarrow \hat{G}(\omega) = \int_{-\infty}^{\infty} \hat{g}(\zeta) e^{-j\omega\zeta} d\zeta, \\ G(\omega) &= \exp \hat{G}(\omega), \\ G(\omega) &\leftrightarrow g(t) = (1/2\pi) \int_{-\infty}^{\infty} G(\omega) e^{j\omega t} d\omega. \end{aligned} \right\} \quad (10.215)$$

(Other definitions of cepstrum are also used (e.g., Ulrich, 1971; Båth, 1974) but the definition above is the most common in seismic data analysis.)

For a discrete function g_t, we use the z-transform and the above steps become

$$\left. \begin{aligned} G(z) &= \sum_i g_i z^i, \\ \hat{G}(z) &= \ln\left(\sum g_i z^i\right) \\ \hat{g}_\zeta &= (1/2\pi j\Delta) \oint \hat{G}(z) z^{-(\zeta+1)} dz, \end{aligned} \right\} \quad (10.216)$$

using (10.212). To return to the time-domain, we have

$$\left. \begin{aligned} \hat{G}(z) &= \sum \hat{g}_\zeta z^\zeta = \ln\{G(z)\}, \\ G(z) &= \exp\{\hat{G}(z)\} = \exp\left(\sum \hat{g}_\zeta z^\zeta\right), \\ g_t &\leftrightarrow G(z). \end{aligned} \right\} \quad (10.217)$$

The variable ζ is called the *quefrency*, a permutation of the letters in frequency, just as 'cepstrum' is a permutation of the letters in 'spectrum.' The cepstrum can be expressed in terms of its *lampitude* $\hat{a}(\zeta)$ and its *saphe* $\hat{\gamma}(\zeta)$,

$$\hat{g}(\zeta) = \hat{a}(\zeta) e^{j\hat{\gamma}(\zeta)}. \quad (10.218)$$

The equivalent of filtering in the time or frequency domains is called *liftering* when performed in the cepstrum domain.

An essential step in going to the cepstrum domain is finding the phase $\gamma(\omega)$, usually by means of (10.91). Since $\tan(\theta + \pi n) = \tan\theta$, each value calculated for $\gamma(\omega)$ is ambiguous by πn. This ambiguity must be removed before transforming to the cepstrum domain, an operation called 'uncracking' the phase ambiguity. One method is to utilize the fact that $\gamma(\omega)$ is continuous and add π to $\gamma(\omega)$ in a trial-and-error approach, the values adopted being those which make the slope of $\gamma(\omega)$ as smooth as possible. An alternative is to take the derivative of the expression for $\gamma(\omega)$ in (10.91), a procedure which does not introduce the ambiguities:

$$\frac{d\gamma(\omega)}{d\omega} = \frac{d}{d\omega}\left[\tan^{-1}\{X(\omega)/R(\omega)\}\right]$$

$$= \frac{R(\omega)dX(\omega)/d\omega - X(\omega)dR(\omega)/d\omega}{\{R(\omega)\}^2 + \{X(\omega)\}^2}. \qquad (10.219)$$

Stoffa *et al.* (1974) discuss the deconvolution of marine reverberatory noise, starting by weighting an observed time series g_i to give another series g_t, whose z-transform is

$$G(z) = \sum_i g_i a^i z^i,$$

where a is a constant slightly smaller than one which makes g_t minimum-phase (see also §8.1.4). We may then associate the slowly-varying components with the source and the reverberation, and the rapidly-varying components with the reflector series. The non-linear operation of taking the logarithm results in undersampling of $G(z)$ but the weighting lessens alias effects. (Stoffa suggests starting with $a = 0.94$, presumably for $\Delta = 0.004\,\text{s}$, and then increasing a until aliasing begins to create problems, to ascertain the largest value of a which can be used.)

10.8 Filtering

10.8.1 *Introduction*

Filters are devices which pass or fail to pass information based on some measurable discriminant. Usually the discriminant is frequency and the filter alters the amplitude and/or phase spectra of signals which pass through it. Analog filtering was discussed briefly in §5.4.5 and digital filtering in §8.2.1. Although some of the following discussion is applicable to analog filtering, emphasis will be on digital filtering. The literature dealing with filtering is vast; many references are given in Kulhánek (1976), Båth (1974), Blackman and Tukey (1958), Finetti *et al.* (1971), Lee (1960). The following discussion is based to a considerable extent on Kulhánek.

Most of the filters with which we deal are assumed to be linear to facilitate calculation, and so we here assume linearity. Digital filters are more versatile than analog filters, partly because we are not restricted to physically-realizable components such as capacitors, inductances and resistances, partly because with digital filters we know future values of the signal as well as the present and past values upon which analog filters must act.

The output of a physical system cannot precede the input; thus, when $g(t)$ and $h(t)$ are respectively the input and output signals, if $g(t) = 0$ for $t < 0$, then $h(t) = 0$, $t < 0$ for analog filters. This is not necessarily the case for digital filters.

A filter is *stable* if the output is finite for any finite input. The output in the time-domain for a linear system is given by (10.180). A filter $f(t)$ will be stable provided that

$$\int_{-\infty}^{\infty} |f(t)|\,dt < +\infty. \qquad (10.220)$$

This requires that $f(t)$ be finite everywhere and approach zero as t approaches $\pm\infty$ (Treitel and Robinson, 1964).

We may write (10.179) in the form

$$F(s) = H(s)/G(s).$$

The right-hand side is almost invariably expressible as the ratio of two polynomials in s with the numerator of lower order than the denominator (if this is not the case, long division leads to terms in s^n where n is positive, and these usually give rise to instability). Applying the method of partial fractions we can write

$$F(s) = \sum_{i=1}^{n} A_i/(s - s_i),$$

where s_i is one of the n roots of the equation $G(s) = 0$. Taking the inverse transform gives

$$f(t) = \sum_{i=1}^{n} A_i \exp\{s_i t\} = \sum_{i=1}^{n} A_i \exp\{(a_i + jb_i)t\}$$

since in general s_i is complex. For $f(t)$ to remain finite as t approaches infinity, all of the a_i must be negative; thus the roots s_i, usually called the *poles* of $H(s)$, must lie on the left-hand side of the complex plane.

The above discussion applies equally well to digital filters, in which case (10.220) becomes

$$\sum_{-\infty}^{\infty} |f_t| < \infty. \qquad (10.221)$$

10.8.2 *Filter synthesis and analysis*

Filters can be designed either by requiring that a given input produce a desired output (*filter synthesis*) or by investigating the effects of a given filter on various input signals (*filter analysis*). As an example of filter synthesis, we design a filter to transform a sampled input g_t into a desired output h_t where

$$g_t = \sum_{k=0}^{m-1} g_k \delta(t - k\Delta), \quad h_t = \sum_{k=0}^{n-1} h_k \delta(t - k\Delta).$$

For simplicity we assume $m = n$, zeros being added to g_t or h_t to achieve this. From (10.179) we obtain

$$F(z) = \mathcal{H}(z)/G(z). \tag{10.222}$$

Since $\mathcal{H}(z)$ and $G(z)$ are polynomials in z, long division gives a polynomial which may be of infinite order. In practice, infinite polynomials must be truncated to a reasonable number of terms. Once we have $F(z)$, we can get f_t; then the output y_t for any input x_t can be found by convolution:

$$y_t = f_t * x_t.$$

We could also get y_t by using (10.222) to write

$$Y(z) = F(z)X(z) = \{\mathcal{H}(z)/G(z)\} X(z),$$

or

$$Y(z)G(z) = X(z)\mathcal{H}(z)$$

hence

$$y_t * g_t = x_t * h_t,$$

or by (10.193),

$$\sum_{k=0}^{r} y_{r-k} g_k = \sum_{k=0}^{r} x_{r-k} h_k, \quad r = 0, 1 \ldots n - 1,$$

which is a set of n equations in the n unknowns, y_k. Setting $r = 0, 1, \ldots$, we obtain the solutions

$$y_0 = x_0 h_0/g_0,$$

$$y_1 = (x_1 h_0 + x_0 h_1)/g_0 - y_0 g_1/g_0,$$

hence

$$y_r = \left(\sum_{k=0}^{r} x_{r-k} h_k/g_0 \right) - \left(\sum_{k=1}^{r} y_{r-k} g_k/g_0 \right). \tag{10.223}$$

Since the system is linear, there is no loss of generality by setting $g_0 = +1$. We also note that the initial value of k in the second summation means that g_t has been delayed one time unit (see §10.6.3, also problem 10.37). Therefore we can write

$$y_t = x_t * h_t - (y_t * g_t)', \tag{10.224}$$

where the prime means that g_t is delayed one unit.

Equations (10.223) and (10.224) give a solution in terms of present (x_r) and past inputs (x_0 to x_{r-1}) and past

outputs (y_0 to y_{r-1}). Filters of this type are *predictive*, *recursive* or *feedback*. Such equations can be solved iteratively, y_0 being found first, then y_1, y_2, etc., a type of calculation convenient with digital computers.

As an example of filter analysis, we take a specific case of (10.223),

$$y_r = ax_r - by_{r-1}, \quad r = 0, 1, 2, \ldots, n - 1 \tag{10.225}$$

(corresponding to $g_t = [1, b, 0, 0, \ldots]$, $h_t = [a, 0, 0, \ldots]$), a and b being real, and we determine the filter properties. Taking z-transforms of the sequences obtained by giving r the values $0, 1, 2, \ldots, n - 1$ in (10.225), we get

$$Y(z) = aX(z) - bzY(z),$$

where the factor z takes into account the delay of g_t by one unit. Solving for $Y(z)$ gives

$$Y(z) = aX(z)/(1 + bz),$$

$$F(z) = Y(z)/X(z) = a/(1 + bz). \tag{10.226}$$

Provided $|bz| < 1$, we can expand the right-hand side and obtain

$$F(z) = a(1 + bz)^{-1} = a \sum_{r=0}^{\infty} (-bz)^r, \tag{10.227}$$

hence

$$f_t = (a, -ab, ab^2, -ab^3, \ldots).$$

Equation (10.221) shows that $|b| < 1$ for the filter to be stable. The series in (10.227) may converge very slowly and a recursive solution may be better than finding f_t as above, then calculating $f_t * x_t$.

A second-order recursive filter can be defined by

$$y_r = ax_r - by_{r-1} - cy_{r-2}. \tag{10.228}$$

Then,

$$Y(z) = aX(z) - bzY(z) - cz^2Y(z),$$

$$F(z) = \frac{Y(z)}{X(z)} = \frac{a}{1 + bz + cz^2} = \frac{a}{c(z - z_1)(z - z_2)}$$

$$= \frac{(a/cz_1z_2)}{(1 - z/z_1)(1 - z/z_2)} \tag{10.229}$$

where z_1, z_2 are roots of the denominator.

Reference to (10.227) shows that the second-order recursive filter is equivalent to two first-order filters in series, the transfer functions being

$$\left(\frac{1}{z_i} \frac{(a)^{\frac{1}{2}}}{c} \right) \Big/ \left(1 - \frac{z}{z_i} \right), \quad i = 1, 2.$$

Comparison with (10.221), shows that the filter will be stable provided that $|z/z_i| < 1$, that is, $|b \pm (b^2 - 4c)^{\frac{1}{2}}| > 2|c|$. In general recursive filters of any order can be replaced by first-order filters in series.

Equations (10.228) and (10.229) can be generalized as follows:

$$y_r = x_r - f_1 y_{r-1} - \ldots - f_r y_0,$$
$$r = 0, 1, 2, \ldots, n-1$$

where we have set $a = 1 = f_0$ (this can always be done by introducing a scale factor). Then,

$$x_r = y_r + f_1 y_{r-1} + \ldots + f_r y_0,$$

and on taking z-transforms, we have

$$X(z) = Y(z)(1 + f_1 z + \ldots + f_{n-1} z^{n-1})$$

or

$$Y(z) = X(z)/(1 + f_1 z + \ldots + f_{n-1} z^{n-1}). \quad (10.230)$$

Thus the general form of the feedback filter of order n is

$$F(z) = 1 \bigg/ \bigg(\sum_{r=0}^{n} f_r z^r \bigg). \quad (10.231)$$

10.8.3 *Frequency filtering*

Frequency filters are classified as *low-pass, high-pass* or *bandpass* according as they discriminate against frequencies above or below a certain limiting frequency or outside of a given band of frequencies. 'Ideal' filters of these types are the following:

Low-pass $\quad F_L(\omega) = +1, |\omega| < |\omega_0|,$
$$\left.\begin{array}{l} \\ = 0, |\omega| > |\omega_0|; \end{array}\right\} \quad (10.232)$$

High-pass $\quad F_H(\omega) = 0, |\omega| < |\omega_0|,$
$$\left.\begin{array}{l} \\ = +1, |\omega| > |\omega_0|; \end{array}\right\} \quad (10.233)$$

Bandpass $\quad F_B(\omega) = +1, |\omega_1| < |\omega| < |\omega_2|,$
$$\left.\begin{array}{l} \\ = 0, |\omega_1| > |\omega| \text{ or } |\omega| > |\omega_2|. \end{array}\right\} \quad (10.234)$$

These filters are discontinuous at $\omega_0, \omega_1, \omega_2$. Obviously a bandpass filter is equivalent to a low-pass filter with $|\omega_0| = |\omega_2|$ in series with a high-pass filter with $|\omega_0| = |\omega_1|$.

The low-pass filter can be obtained from (10.130):

$$F_L(\omega) = \text{box}_{2\omega_0}(\omega) \leftrightarrow (\omega_0/\pi) \, \text{sinc}(\omega_0 t) = f_L(t). \quad (10.235)$$

For digital functions, provided $|\omega_0| < \omega_N = \pi/\Delta$, (10.235) becomes

$$f_t^L = (1/\pi) \sum_{n=-\infty}^{\infty} \omega_0 \, \text{sinc}(n\omega_0 \Delta). \quad (10.236)$$

Since $F_L(\omega)$ does not change the signal amplitude or phase within the passband, the filter is distortionless.

For continuous functions a high-pass filter with cut-off frequency ω_0 is given by

$$f_H(t) = (1/2\pi) \int_{-\infty}^{-\omega_0} e^{j\omega t} \, d\omega + (1/2\pi) \int_{+\omega_0}^{\infty} e^{j\omega t} \, d\omega$$

$$= (1/\pi) \int_{\omega_0}^{\infty} \cos \omega t \, d\omega \quad (10.237)$$

since $e^{j\omega t} = \cos \omega t + j \sin \omega t$ and $\sin \omega t$ is odd. Then

$$f_H(t) = (1/\pi t) \sin \omega t \big|_{+\omega_0}^{\infty} = -(\omega_0/\pi) \, \text{sinc}(\omega_0 t) \quad (10.238)$$

since $\lim_{\omega \to \infty} \sin \omega t = 0$ (see Papoulis, 1962, p. 278). Thus $f_H(t) = -f_L(t)$ provided both filters have the same cutoff frequency ω_0.

For a digital filter, the response should be zero above the Nyquist frequency to avoid aliasing. Changing the limit in (10.238) from $\pm\infty$ to $\pm\omega_N$ gives

$$f_H(t) = (1/\pi t)(\sin \omega_N t - \sin \omega_0 t). \quad (10.239)$$

Changing to digital functions, since $\omega_N n \Delta = n\pi$,

$$f_t^H = (1/\pi) \sum_{n=-\infty}^{\infty} \{\omega_N \, \text{sinc}(n\omega_N \Delta) - \omega_0 \, \text{sinc}(n\omega_0 \Delta)\}$$

$$\left.\begin{array}{l} = -(\omega_0/\pi) \sum \text{sinc}(\omega_0 n \Delta), \quad n \neq 0 \\ = (1/\pi)(\omega_N - \omega_0), \quad n = 0. \end{array}\right\} \quad (10.240)$$

As for $f_H(t)$, $f_t^H = -f_t^L$ (except at $t = 0$) when both filters have the same cutoff frequency, ω_0. Thus, the design of high-pass and bandpass filters is essentially the same as that of low-pass filters.

To achieve an ideal low-pass filter requires an infinite series for f_t^L, which is impossible. The result of using a finite series for f_t^L is to introduce ripples, both within and without the passband, the effect being especially noticeable near the cutoff frequency (Gibbs' phenomenon – see §10.3.5). These effects result from the discontinuity and can be partially overcome by multiplying f_t^L by a smoothing window function (see §10.8.5).

A noisy signal may be cross-correlated with a Vibroseis-type signal,

$$g_v(t) = \sin \{\omega_0 + (\omega_1 - \omega_0)t/L\}t, \quad 0 < t < L,$$
$$= 0, \quad t < 0 \text{ and } t > L,$$

where ω_0, ω_1 are positive constants (compare (5.3));

$$G_v(\omega) \approx \text{constant}, \quad \omega_0 < \omega < \omega_1,$$
$$\approx 0, \, \omega < \omega_0, \quad \omega > \omega_1.$$

Such an operation, called *chirp filtering*, is roughly equivalent to bandpass filtering, as shown in fig. 10.12.

If $g(t)$ is the input to a filter whose impulse response, $f_M(t)$, is given by

$$f_M(t) = g(-t),$$

then clearly $F_M(\omega) = \overline{G(\omega)}$ and output of the filter is given by

$$H(\omega) = F_M(\omega)\,G(\omega) = |G(\omega)|^2,$$

hence $h(t)$ = autocorrelation of $g(t)$. Filters of this type are called *matched* or *conjugate filters*. When the input consists of $g(t)$ plus random noise, the output is mainly the autocorrelation of $g(t)$ since the cross-correlation of $g(t)$ and the noise will be approximately zero for all shifts.

10.8.4 *Butterworth filters*

The *Butterworth filter* is a common form of low-pass filter; it can be defined by

$$|F(\omega)|^2 = 1/\{1 + (\omega/\omega_0)^{2n}\} \tag{10.241}$$

where ω_0 is the 'cutoff' frequency and n determines the sharpness of the cutoff. Curves of $|F(\omega)|$ for various values of n are shown in fig. 10.13.

To investigate the stability of the filter we use a Laplace-transform definition:

$$|F(s)|^2 = 1/\{1 + (-1)^n s^{2n}\} \tag{10.242}$$

where $s = \sigma + j(\omega/\omega_0)$. This function has no zeros but has $2n$ poles given by the roots (see fig. 10.3) of

$$s^{2n} = -1,\ n\ \text{even},$$
$$= +1,\ n\ \text{odd}.$$

These roots are of the form $(\pm a \pm jb)$, a and b being real and positive. Thus, the roots are symmetrical about the real and imaginary axes. When n is odd, two roots reduce to ± 1. Since $|F(s)|^2 = F(s)\overline{F(s)}$, if the roots $-a \pm jb$ (and -1 when n is odd) are assigned to $F(s)$, and $+a \pm jb$

Fig. 10.12. Chirp filter. (After Kulhánek, 1976.)
(a) Impulse response and (b) amplitude spectrum.
The frequency increases linearly from 10 to 30 Hz in 2 s.

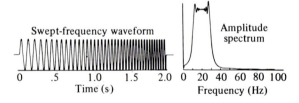

Swept-frequency waveform

Amplitude spectrum

| 0 | .5 | 1.0 | 1.5 | 2.0 | 0 | 20 40 60 80 100 |

Time (s)

Frequency (Hz)

(and $+1$) to $\overline{F}(s)$, $F(s)$ is stable. For the nth order digital filter

$$F_n(s) = 1/\{(s+1)(s + a_1 + jb_1)(s + a_1 - jb_1)$$
$$\times (s + a_2 + jb_2)(s + a_2 - jb_2)\ldots\}$$

(the factor $(s+1)$ is omitted when n is even).

The expression for $F_n(s)$ can be expressed as partial fractions (§10.1.7), transformed to the time-domain and then to the z-domain to get the digital filter. Thus, for $n = 2$, the roots of $s^4 = -1$ are $(\pm 1 \pm j)/(2)^{\frac{1}{2}}$; hence

$$F_2(s) = \cfrac{1}{\{s + (1 + j)/\sqrt{2}\}\{s + (1 - j)/\sqrt{2}\}}$$

$$= \frac{j}{\sqrt{2}}\left[\cfrac{1}{\{s + (1 + j)/\sqrt{2}\}} - \cfrac{1}{\{s + (1 - j)/\sqrt{2}\}}\right].$$

Using (10.161) we get

$$f(t) = (j/\sqrt{2})\{e^{-(1+j)t/\sqrt{2}} - e^{-(1-j)t/\sqrt{2}}\}\,\text{step}\,(t)$$

$$= \{\sqrt{2}\,e^{-t/\sqrt{2}}\sin(t/\sqrt{2})\}\,\text{step}\,(t).$$

Applying (10.200) and the results of problem 10.26, we obtain

$$F(z) = \sqrt{2}\left\{\cfrac{ze^{-\Delta/\sqrt{2}}\sin(\Delta/\sqrt{2})}{1 - 2ze^{-\Delta/\sqrt{2}}\cos(\Delta/\sqrt{2}) + z^2 e^{-\Delta\sqrt{2}}}\right\}.$$

Equation (10.241) gives no information about the phase characteristics of the filter. We can construct

Fig. 10.13. Amplitude response of a Butterworth filter.

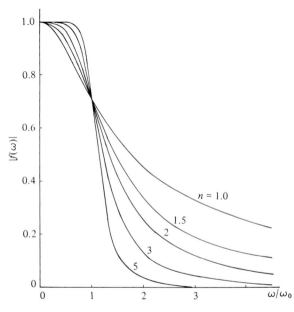

$|F(\omega)|$

$n = 1.0$

1.5

2

3

5

ω/ω_0

Butterworth filters with different phase characteristics, for example,

$$F_n(\omega) = \frac{e^{-j\omega t_0}}{\{1 + (\omega/\omega_0)^{2n}\}^{\frac{1}{2}}}:$$

for $n = \frac{1}{2}$,

$$F_{\frac{1}{2}}(\omega) = \frac{1}{1 + j(\omega/\omega_0)} = \frac{1 - j(\omega/\omega_0)}{1 + (\omega/\omega_0)^2}.$$

The first filter has a linear phase while the second has the phase

$$\gamma(\omega) = -\tan^{-1}(\omega/\omega_0).$$

10.8.5 *Windows*

We often wish to select a portion of a signal for study, or we may wish to 'smooth' a function such as a transform. We can achieve these objectives by multiplying the signal or transform by a *window* or *gate*, a function which varies in a more-or-less convenient manner within an interval and is zero outside this interval. We represent a window by $w(t) \leftrightarrow W(\omega)$, the symbolism emphasizing that we can think of the window as being in either the time or frequency domain ('window' is also used in a third sense to denote merely an interval of time, especially an interval in which data can be recorded free from interference by noise such as ground roll). The result of applying a window to a signal, $g(t)$, in the time domain is to give

$$h_w(t) = g(t)w(t) \leftrightarrow (1/2\pi)G(\omega) * W(\omega).$$

Using a time-domain window is a form of frequency filtering so that the transform of the part of the signal selected by the window is distorted in both amplitude and phase in comparison with $G(\omega)$.

The following list includes the more commonly-used time-domain windows (see Båth, 1974, pp. 157–64). The formulas give $w(t)$ in the interval $0 < |t| < T$, $w(t)$ being zero for $|t| > T$ (except for the Gaussian window, (h), where the range is $\pm\infty$). Obviously similar windows can be applied in the frequency domain, the inverse transforms being obtained using (10.118).

(a)   Boxcar window:

$$w(t) = \text{box}_{2T}(t) \leftrightarrow W(\omega) = 2T \, \text{sinc} \, \omega T; \qquad (10.243)$$

(b)   Sinc window:

$$w(t) = \text{sinc}(\pi t/T) \leftrightarrow W(\omega)$$

$$= \frac{T}{\pi} \sum_{n=0}^{\infty} \frac{(-1)^n}{(2n+1)(2n+1)!}$$

$$\{(\pi - \omega T)^{2n+1} + (\pi + \omega T)^{2n+1}\} \qquad (10.244)$$

(c)   Fejér kernel window:

$$w(t) = \text{sinc}^2(\pi t/T) \leftrightarrow W(\omega)$$

$$\left.\begin{aligned}
&= (T/2\pi^2) \sum_{n=0}^{\infty} \frac{(-1)^n}{(2n+1)(2n+1)!} \\
&\quad \times \{(2\pi + \omega T)^{2n+2} \\
&\quad - 2(\omega T)^{2n+2} + (2\pi - \omega T)^{2n+2}\}.
\end{aligned}\right\} \qquad (10.245)$$

(d)   Cosine window:

$$\left.\begin{aligned}
w(t) &= +1, \quad 0 < |t| < 4T/5, \\
&= \frac{1}{2} + \frac{1}{2}\cos(5\pi t/T), \quad 4T/5 < |t| < T, \\
w(t) &\leftrightarrow W(\omega) = \frac{(\sin\omega T + \sin 4\omega T/5)}{\omega\{1 - (\omega T/5\pi)^2\}}.
\end{aligned}\right\} \qquad (10.246)$$

(e)   Hanning window:

$$\left.\begin{aligned}
w(t) &= \frac{1}{2} + \frac{1}{2}\cos(\pi t/T), \\
w(t) &\leftrightarrow W(\omega) = T[\text{sinc}\,\omega T \\
&\quad + (2\omega T\sin\omega T)/\{\pi^2 - (\omega T)^2\}].
\end{aligned}\right\} \qquad (10.247)$$

(f)   Hamming window:

$$\left.\begin{aligned}
w(t) &= 0.54 + 0.46\cos(\pi t/T), \\
w(t) &\leftrightarrow W(\omega) = T[1.08\,\text{sinc}\,\omega T \\
&\quad + (0.92\omega T\sin\omega T)/\{\pi^2 - (\omega T)^2\}].
\end{aligned}\right\} \qquad (10.248)$$

(g)   Triangular window:

$$w(t) = (1 - |t|/T) \leftrightarrow W(\omega) = T\,\text{sinc}^2(\omega T/2). \qquad (10.249)$$

(h)   Gaussian window:

$$\left.\begin{aligned}
w(t) &= \exp(-at^2), \quad a > 0, \\
w(t) &\leftrightarrow W(\omega) = \sqrt{\pi/a}\,\exp(-\omega^2/4a).
\end{aligned}\right\} \qquad (10.250)$$

Combinations of windows are also used, for example the discontinuous sides of a boxcar can be modified with a cosine taper. A common technique is to apply a window, for example, a boxcar, in the time-domain, then smooth the resulting transform by applying a second window in the frequency domain.

The effect on the spectrum of $g(t)$ of applying a window in the time-domain is determined by $W(\omega)$. Thus, curves of $W(\omega)$ give some idea of the effect of using the window. In general the wider the window and the gentler the fall-off, the less effect on the spectrum. Båth (1974, pp. 164–71), Blackman and Tukey (1958) and Kurita (1969) discuss practical aspects of 'window carpentry'.

10.8.6 *Optimum filters*

(a) *Introduction.* The filters discussed up to this point have been based on periodic or aperiodic inputs. If the input noise is a stationary random function, we may design a filter that will give an 'optimum' output according to some criterion. Among the most important and most widely used of such criteria is the *Wiener* or *least-squares* criterion. To apply this criterion, we compare the output of the filter with some 'desired' output, the difference being the 'error' in the output; we then design the filter to minimize the power (or energy) of the error by applying the principle of least squares. The Wiener criterion is also called the ℓ_2 *norm* (see Claerbout, 1976, p. 121, 123); it provides the maximum likelihood estimate if the errors have a *Gaussian* or *normal distribution* where $P(e_j)$, the probability of an error e_j, equals

$$P(e_j) = \{1/\sigma(2\pi)^{\frac{1}{2}}\} \exp\left[-\tfrac{1}{2}(e_j/\sigma)^2\right] \qquad (10.251)$$

where σ is the standard deviation (that is, the square root of the variance).

Another criterion sometimes used to find f_t is the ℓ_1 *norm* which minimizes $\sum|e_j|$ (Claerbout, 1976, p. 123; Claerbout and Muir, 1973; Taylor, 1981). It provides the maximum likelihood estimate if the errors have a *Laplacian* or *one-sided exponential distribution* where probability of an error e_j is

$$\begin{aligned} P(e_j) &= (1/\mu)\exp\left[-e_j/\mu\right], & e_j \geqslant 0 \\ &= 0 & e_j < 0, \end{aligned} \qquad (10.252)$$

where μ is the mean absolute error and the variance is μ^2. Use of the ℓ_1 norm is less sensitive to errors than the Wiener or ℓ_2 norm.

The ℓ_4 *norm* which minimizes $\sum e_j^4$ is used in minimum-entropy filtering (§10.8.6e). *Parsimonious deconvolution* (Postic *et al.*, 1980) minimizes $(\sum e_j^p)^{1/p}/(\sum e_j^q)^{1/q}$ where p is very slightly larger than q. Occasionally the minimax, Chebyshev or ℓ_∞ norm is used, for example in array design (Rietsch, 1979).

(b) *Wiener (least-squares) filtering.* Let us write for the input

$$g(t) = g_s(t) + g_n(t) \qquad (10.253)$$

where the subscripts refer to 'signal' and 'noise'. If $h(t)$ and $\hbar(t)$ denote the 'actual' and 'desired' outputs, their difference, the error, is $e(t)$. Then, the energy of the error, E, is

the sum of the squares of $e(t)$. For continuous functions, this gives

$$\begin{aligned} E &= \lim_{T\to\infty} (1/2T) \int_{-T}^{T} \{h(t) - \hbar(t)\}^2 dt \\ &= \lim_{T\to\infty} (1/2T) \int_{-T}^{T} \left[\left\{\int_{-\infty}^{\infty} f(\tau)g(t-\tau)d\tau\right\} - \hbar(t)\right]^2 dt \\ &= \lim_{T\to\infty} (1/2T) \left[\int_{-T}^{T} \left\{\int_{-\infty}^{\infty} f(\tau)g(t-\tau)d\tau \right.\right. \\ &\quad \left.\left. \times \int_{-\infty}^{\infty} f(\sigma)g(t-\sigma)d\sigma\right\} dt \right. \\ &\quad \left. - 2\int_{-T}^{T}\int_{-\infty}^{\infty} \{f(\tau)g(t-\tau)d\tau\}\hbar(t)dt \right. \\ &\quad \left. + \int_{-T}^{T} \hbar^2(t)\,dt \right] \end{aligned}$$

where the square of the integral giving $h(t)$ has been written as the product of two integrals which are identical except for the dummy variables of integration. Interchanging the order of integration gives

$$\begin{aligned} E &= \int_{-\infty}^{\infty} f(\tau)d\tau \int_{-\infty}^{\infty} f(\sigma)d\sigma \left\{\lim_{T\to\infty} (1/2T)\right. \\ &\quad \left. \times \int_{-T}^{T} g(t-\tau)g(t-\sigma)dt\right\} \\ &\quad - 2\int_{-\infty}^{\infty} f(\tau)d\tau \left\{\lim_{T\to\infty}(1/2T)\int_{-T}^{T}\hbar(t)g(t-\tau)dt\right\} \\ &\quad + \lim_{T\to\infty}(1/2T)\int_{-T}^{T}\hbar(t)^2\,dt \\ &= \int_{-\infty}^{\infty} f(\tau)d\tau \int_{-\infty}^{\infty} f(\sigma)\phi_{gg}(\tau-\sigma)d\sigma \\ &\quad - 2\int_{-\infty}^{\infty} f(\tau)\phi_{g\hbar}(\tau)d\tau + \phi_{\hbar\hbar}(0), \end{aligned}$$

using (10.144) and (10.147).

Since we have specified $g(t)$ and $\hbar(t)$, E is a function of $f(t)$ only and so the problem reduces to that of finding the function $f(t)$ which will minimize E. Determination of the form of $f(t)$ in general involves the calculus of variations and we only state the result:

$$\int_{-\infty}^{\infty} f(\sigma)\phi_{gg}(\tau-\sigma)d\sigma = \phi_{g\hbar}(\tau), \qquad \tau \geqslant 0. \qquad (10.254)$$

This integral equation, known as the *Wiener–Hopf equation*, holds for a causal linear system, the function $f(t)$ satisfying this equation being the impulse response of the desired optimum filter. The solution of (10.254) is lengthy and complicated, mainly because of the requirement that $\tau \geqslant 0$. For details the reader is referred to Lee (1960, pp. 360–7, 389–92).

(c) *Prediction-error filtering.* Consider a filter $f_t = f_1$, $f_2,\ldots,f_n$ which is designed to predict the causal time series $g_t = g_0, g_1,\ldots,g_m$, $m > n$, one time unit ahead based on current and past values of g_t. For example, at $t = 3\Delta$, the filter predicts g_4 based on the current value g_3 and past values g_0, g_1, g_2. In general the predicted value of g_j is

$$g_j = f_1 g_{j-1} + f_2 g_{j-2} + \ldots + f_j g_0$$
$$= \sum_{k=1}^{j} f_k g_{j-k} = \sum_{k=1}^{n} f_k g_{j-k},$$
$$j = 1, 2,\ldots, m \qquad (10.255)$$

where we can use the upper limit n since $g_{j-k} = 0, k > j$. Note that the above implies that the predicted value of g_0 is zero. The error in the prediction of g_j is e_j where

$$e_j = \sum_{k=1}^{n} f_k g_{j-k} - g_j = \sum_{k=0}^{n} f_k g_{j-k} \qquad (10.256)$$

where $f_0 = -1$. We call the filter $-1, f_1, f_2,\ldots, f_n$ the *prediction-error filter of length* $(n + 1)$ *for unit prediction distance.*

In the foregoing section the prediction distance was one unit but it is possible to design filters for prediction distance (*span*) p, p integral (Robinson and Treitel, 1980). Moreover, the above discussion dealt with prediction based on current and past values, that is, *forward prediction*. Whenever the entire set of values of g_t has been recorded, *backward prediction* based on future values is also possible, as is prediction based on combinations of the two. The data for adjacent traces may also be available and these can be used in multichannel prediction methods (Claerbout, 1976, pp. 139–40).

The prediction concept implies that all of the data involved are parts of the same ensemble, hence the statistics of the ensemble are central to the prediction concept. Whether or not the ensemble is ergodic and stationary (§10.3.10) and the nature of the distribution are clearly relevant to the prediction method.

If we use the Wiener (least-squares) criterion to determine the filter f_t, we obtain the following normal equations (see problem 10.42a, also compare with (8.58),

(10.27) to (10.29)):

$$\sum_{k=1}^{n} f_k \phi_{gg}(r - k) = \phi_{gg}(r), \quad r = 1, 2,\ldots, n. \quad (10.257)$$

A simple expression can be obtained for the error power E by using (10.256) and (10.257). Noting that $e_0 = -g_0$, we have

$$E = \sum_{j=0}^{m} e_j^2 = \sum_{j=0}^{m} \left(\sum_{k=1}^{n} f_k g_{j-k} - g_j \right)^2$$
$$= \sum_{j=0}^{m} \left\{ \left(\sum_{k=1}^{n} f_k g_{j-k} \sum_{\ell=1}^{n} f_\ell g_{j-\ell} \right) \right.$$
$$\left. - 2 g_j \left(\sum_{k=1}^{n} f_k g_{j-k} \right) + g_j^2 \right\}$$

where we use different summation indices k, ℓ to get the square of the first term. Interchanging the order of summation gives

$$E = \sum_k f_k \left\{ \sum_\ell f_\ell \left(\sum_j g_{j-k} g_{j-\ell} \right) \right\}$$
$$- 2 \sum_k f_k \left(\sum_j g_j g_{j-k} \right) + \phi_{gg}(0),$$
$$= \sum_k f_k \left\{ \sum_\ell f_\ell \phi_{gg}(k - l) \right\} - 2 \sum_k f_k \phi_{gg}(-k) + \phi_{gg}(0)$$
$$= \sum_k f_k \phi_{gg}(k) - 2 \sum_k f_k \phi_{gg}(-k) + \phi_{gg}(0)$$

on using the normal equations. Recalling that $f_0 = -1$, we have

$$E = - \sum_{k=0}^{n} f_k \phi_{gg}(k). \qquad (10.258)$$

Equations (10.257) and (10.258) can be combined to give the following matrix equation

$$\begin{Vmatrix} \phi_{gg}(0) & \phi_{gg}(-1) & \cdots & \phi_{gg}(-n) \\ \phi_{gg}(1) & \phi_{gg}(0) & \cdots & \phi_{gg}(1-n) \\ \cdots & & & \\ \phi_{gg}(n) & \phi_{gg}(n-1) & \cdots & \phi_{gg}(0) \end{Vmatrix} \begin{Vmatrix} f_0 \\ f_1 \\ \vdots \\ f_n \end{Vmatrix} = \begin{Vmatrix} -E \\ 0 \\ \vdots \\ 0 \end{Vmatrix}$$

$$(10.259)$$

(d) *Maximum-entropy filtering.* In thermodynamics, entropy is a measure of the disorder (unpredictability) of molecular motion. In information theory, Shannon and Weaver (1949) regard *entropy* as a measure of the unpredictability of a time series. The amount of information which can be extracted increases with the entropy. At one extreme a perfectly predictable series, such as a sine wave, bears no information and at the other extreme white noise is completely unpredictable and hence potentially carries maximum information. *Maximum-entropy filtering* attempts to produce a filtered output which is as unpredictable as possible while still having the same autocorrelation function as the original time series, that is, of all the time series which have a given autocorrelation function, maximum-entropy filtering selects the one which has the maximum unpredictability.

We could regard the number of digits required to encode information as a measure of entropy. For example, if we have four equally probable events, the various possibilities can be encoded as 00, 01, 10, 11, which requires two digits where $2 = -\log_2(\frac{1}{4})$; for eight equally probable events we would need three digits $(3 = -\log_2(\frac{1}{8}))$. In general, for equally probable events the entropy is measured by $-\log_2(1/P)$ where P is the probability of each event. When the events are not equally probable, the entropy S_n is the average:

$$S_n = \sum_j \log(1/P_j)/(1/P_j) = -\sum_{j=1}^{n} P_j \log P_j. \tag{10.260}$$

(The base of the logarithms is arbitrary except when comparing entropies with different bases.) For a signal of infinite length, we define the entropy density S as

$$S = \lim_{n \to \infty} \{S_n/(n+1)\} \tag{10.261}$$

The matrix ϕ_{gg} in (10.259) is of Toeplitz form and Smylie *et al.* (1973) deduce the relation

$$E_\infty = (\omega_N/\pi) \exp\left[(1/2\omega_N) \int_{-\omega_N}^{+\omega_N} \ln\{\Phi_\infty(\omega)\} d\omega \right] \tag{10.262}$$

where E_∞ = prediction-error power for $n = \infty$, ω_N = Nyquist frequency and $\Phi_\infty(\omega)$ is the spectral density of g_t, that is

$$\Phi_\infty(\omega) = \Phi_\infty(z) = \sum_{k=-\infty}^{+\infty} \phi_{gg}(k)z^k. \tag{10.263}$$

Smylie *et al.* show that for n infinite

$$S = \frac{1}{2} \ln E_\infty$$

$$= \frac{1}{2} \ln(\omega_N/\pi) + (1/4\omega_N) \int_{-\omega_N}^{+\omega_N} \ln\{\Phi_\infty(\omega)\} d\omega. \tag{10.264}$$

In practical problems we work with a finite signal g_t which we assume to be a sample of one member of an ensemble. Although we have only a finite number of values of $\phi_{gg}(j)$, theoretically an infinity of values of $\phi_{gg}(j)$ exists outside the range of our measurements. Usually operations on data assume that $\phi_{gg}(j)$ is zero outside the

range of measurements. However, Burg (1972, 1975) suggested that a more reasonable choice of the unknown values of $\phi_{gg}(j)$ is one which adds no information, hence adds no entropy so that S is stationary with respect to $\phi_{gg}(k)$, $|k| > n$. Thus, using (10.263) and (10.264), we get the result

$$\partial S/\partial \phi_{gg}(k) = 0 = \int_{-\omega_N}^{\omega_N} \{z^k/\Phi_\infty(z)\} d\omega, \quad |k| > n. \tag{10.265}$$

Although $\Phi_\infty(z)$ is an infinite series, (10.265) implies that $1/\Phi_\infty(z)$ is a finite series of the form

$$1/\Phi_\infty(z) = \sum_{r=-n}^{n} c_r z^r. \tag{10.266}$$

Since $\Phi_\infty(z)$ is a real function of z, $1/\Phi_\infty(z)$ must also be real, hence $c_k = \bar{c}_{-k}$ and therefore

$$1/\Phi_\infty(z) = \sum_{r=-n}^{n} c_r z^r = \mathscr{G}(z)\mathscr{G}(z^{-1}) \tag{10.267}$$

where we can take $\mathscr{G}(z)$ as minimum-phase, $\mathscr{G}(z^{-1})$ as maximum-phase.

In addition to satisfying (10.267), $\Phi_\infty(z)$ must be consistent with the known autocorrelation values, $\phi_{gg}(j)$, $|j| \leqslant n$. Let

$$\Phi_n(z) = \phi_{gg}(-n)z^{-n} + \dots + \phi_{gg}(-1)z^{-1}$$
$$+ \phi_{gg}(0) + \phi_{gg}(1)z + \dots + \phi_{gg}(n)z^n;$$

then the terms in $\Phi_\infty(z)$ between z^{-n} and z^n must be the same as those of $\Phi_n(z)$.

Equation (10.259) can be written

$$\sum_{s=0}^{n} f_s \phi_{gg}(j-s) = -E\delta_j^0, \quad j = 0, 1, \dots, n, \tag{10.268}$$

where δ_j^0 is the Kronecker delta ($\delta_i^j = 1$, $i = j$; $\delta_i^j = 0$, $i \neq j$). This can be expressed in terms of $f * \phi_{gg}$ as follows:

$$(f * \phi_{gg})_j = -E\delta_j^0 + p_j, \quad |j| \leqslant n \tag{10.269}$$

p_j being zero for $j \geqslant 0$; thus the convolution is zero for j positive, equals E for $j = 0$, and has unspecified values p_j for j negative.

Taking z-transforms, we get

$$F(z)\Phi_n(z) = -E + P(z) \tag{10.270}$$

where

$$P(z) = p_{-1}z^{-1} + \dots + p_{-n}z^{-n}.$$

Factorization of $\Phi_n(z)$ gives

$$\Phi_n(z) = G(z)G(z^{-1})$$

$$= (g_0 + g_1 z + \dots + g_n z^n)(g_n z^{-n} + \dots + g_0).$$

Substituting in (10.270) gives

$$F(z)G(z)G(z^{-1}) = -E + P(z),$$

or

$$F(z)G(z) = \{-E + P(z)\}/G(z^{-1})$$
$$= (1/g_0)\{-E + (p_{-1}z^{-1} + \ldots + p_{-n}z^{-n})\}$$
$$\times \{1 + (g_1/g_0)z^{-1} + \ldots + (g_n/g_0)z^{-n}\}^{-1}.$$

Therefore,

$$(-1 + f_1 z + \ldots + f_n z^n)(g_0 + g_1 z + \ldots + g_n z^n)$$
$$= (-E/g_0 + \text{negative powers of } z).$$

Since the left-hand side has no negative powers of z, it follows that

$$(-1 + f_1 z + \ldots + f_n z^n)(g_0 + g_1 z + \ldots + g_n z^n)$$
$$= -E/g_0,$$

or

$$F(z)G(z) = -E/g_0 = -g_0, E = g_0^2,$$

on equating powers of z (including z^0). Thus,

$$G(z) = -g_0/F(z) \qquad (10.271)$$

and

$$1/\Phi_n(z) = \{G(z)G(z^{-1})\}^{-1} = F(z)F(z^{-1})/g_0^2$$
$$= F(z)F(z^{-1})/E. \qquad (10.272)$$

Taking $\mathscr{G}$ as the value of G as $n \to \infty$, comparison of (10.272) and (10.267) shows that

$$\mathscr{G}(z) = F(z)/E^{\frac{1}{2}}. \qquad (10.273)$$

Note that if g_t is minimum-delay, both $\mathscr{G}(z)$ and $F(z)$ can be taken as minimum-phase, $\mathscr{G}(z^{-1})$ and $F(z^{-1})$ being maximum-phase. Finally, we have from (10.267) and (10.273)

$$\Phi_\infty(z) = \{\mathscr{G}(z)\mathscr{G}(z^{-1})\}^{-1} = E\{F(z)F(z^{-1})\}^{-1}.$$
$$(10.274)$$

Thus, to determine the maximum-entropy spectral density, we first find the prediction-error filter f_t and the associated error power E by solving (10.259), then use (10.274) to find the maximum-entropy spectral density, $\Phi_\infty(z)$; the methods of §10.6.6c then enable us to find the desired maximum-entropy signal. Equation (10.259) can be solved by a recursive method similar to the Levinson algorithm; details are given in Smylie *et al.* (1973), Robinson and Treitel (1980) and Andersen (1974).

(e) *Minimum-entropy filtering* (deconvolution). Minimum-entropy filtering (Wiggins, 1977, 1978) attempts to find a linear filter which maximizes the 'spiky' characteristics of a signal, thereby reducing the disorder of the signal, hence minimizing the entropy. Maximizing the spikyness of a signal is equivalent to finding the smallest number of large spikes consistent with the observed signal. One way of increasing the spikyness of a signal g_t is to raise the values to some positive power, for example, the fourth power, since this makes the difference between large and small values much greater. Since this criterion is especially sensitive to high amplitudes, it tends to focus attention on the strongest events, which we assume are reflections standing out against the background noise.

We assume that we have N traces, each covering the same time interval from $t = 0$ to $t = n\Delta$. We write g_{ij} for the value of the ith trace at $t = j\Delta$. The filter coefficients are f_k, $k = 1, 2, \ldots, N_f$, and the filter output is h_{ij}. Then,

$$h_{ij} = \sum_k f_k g_{i,j-k}. \qquad (10.275)$$

As the number of spikes in the outputs h_{ij} decreases, the results become simpler. Wiggins (1978) takes as a measure of 'simplicity' the quantity Γ defined by the equation

$$\Gamma = \sum_i \Gamma_i, \quad \Gamma_i = \sum_j h_{ij}^4 \Big/ \Big(\sum_j h_{ij}^2 \Big)^2, \qquad (10.276)$$

then seeks a maximum of Γ by varying the filter coefficients, f_k. This leads to N_f equations obtained in the usual way:

$$\frac{\partial \Gamma}{\partial f_k} = 0 = \sum_i \frac{\partial \Gamma_i}{\partial f_k}$$

$$= \sum_i \left\{ 4 \sum_j h_{ij}^3 \Big/ \Big(\sum_j h_{ij}^2 \Big)^2 \right.$$

$$\left. - 4 \Big(\sum_j h_{ij}^4 \Big) \Big(\sum_j h_{ij} \Big) \Big/ \Big(\sum_j h_{ij}^2 \Big)^3 \right\} \frac{\partial h_{ij}}{\partial f_k}.$$

Writing $u_i = \sum_j h_{ij}^2 = n \times$ (variance of the ith output) (since $(h_{ij})_{\text{av}} \approx 0$), we have

$$\sum_i \left(u_i^{-2} \sum_j h_{ij}^3 - u_i^{-1} \Gamma_i \sum_j h_{ij} \right) g_{i,j-k} = 0,$$

$$k = 1, 2, \ldots, N_f.$$

Thus,

$$\sum_i \left(u_i^{-1} \Gamma_i \sum_j h_{ij} g_{i,j-k} \right) = \sum_i \left(u_i^{-2} \sum_j h_{ij}^3 g_{i,j-k} \right);$$

using (10.275), we find

$$\sum_i \left\{ u_i^{-1} \Gamma_i \sum_j \left(\sum_l f_l g_{i,j-l} g_{i,j-k} \right) \right\}$$

$$= \sum_i \left(u_i^{-2} \sum_j h_{ij}^3 g_{i,j-k} \right).$$

Interchanging the order of summation on the left-hand side gives

$$\sum_l f_l \left\{ \sum_i u_i^{-1} \Gamma_i \left(\sum_j g_{i,j-l} g_{i,j-k} \right) \right\}$$

$$= \sum_i \left(u_i^{-2} \sum_j h_{ij}^3 g_{i,j-k} \right), \quad k = 1, 2, \dots, N_f. \quad (10.277)$$

The summation over j on the left-hand side of the equation is the autocorrelation of the ith trace, so that the expression in curly brackets is a weighted sum of the autocorrelations of the observed signals. The summation over j on the right-hand side of the equation is a cross-correlation of the inputs and the outputs cubed, the effect of the cubing being to give great weight to the spiky components.

Since the filter coefficients enter into the calculation of u_i, Γ_i and h_{ij} in (10.277), the equations cannot be solved directly. However, we can assume values for f_k initially, use these to obtain the quantities u_i, Γ_i, h_{ij}, then solve for f_k. These values can then be used to determine u_i, Γ_i and h_{ij} again and so the equations can be solved for a second filter. Wiggins (1978) states that about 4–6 iterations are usually enough to determine the filter with sufficient precision.

Problems

10.1. In mechanics, moment or torque $\mathbf{M}$ of a force $\mathbf{F}$ about a point O is equal to the product of the magnitude of the vector $\mathbf{r}$ from O to the point of application of $\mathbf{F}$ and the component of $\mathbf{F}$ perpendicular to $\mathbf{r}$. Show that $\mathbf{M} = \mathbf{r} \times \mathbf{F}$.

10.2. Verify (10.10). (Hint: Write the vectors in terms of components and use the relations between vector products of the unit vectors.)

10.3. Three vectors $\mathbf{A}$, $\mathbf{B}$ and $\mathbf{C}$ can be multiplied together in three ways: $(\mathbf{A} \cdot \mathbf{B})\mathbf{C}$, $\mathbf{A} \cdot (\mathbf{B} \times \mathbf{C})$, and $\mathbf{A} \times (\mathbf{B} \times \mathbf{C})$. (a) Which of these three are scalars and which are vectors, and what are the directions of the vectors? (b) Show that

$$\mathbf{A} \cdot (\mathbf{B} \times \mathbf{C}) = \mathbf{B} \cdot (\mathbf{C} \times \mathbf{A}) = \mathbf{C} \cdot (\mathbf{A} \times \mathbf{B}) = \begin{vmatrix} a_x & a_y & a_z \\ b_x & b_y & b_z \\ c_x & c_y & c_z \end{vmatrix},$$

that this gives the volume of the parallelopiped defined by $\mathbf{A}$, $\mathbf{B}$ and $\mathbf{C}$, and that changing the cyclic order changes the sign:

$$\mathbf{A} \cdot (\mathbf{B} \times \mathbf{C}) = -\mathbf{A} \cdot (\mathbf{C} \times \mathbf{B}) = -\mathbf{B} \cdot (\mathbf{A} \times \mathbf{C})$$

$$= -\mathbf{C} \cdot (\mathbf{B} \times \mathbf{A}).$$

(c) Show that $\mathbf{A} \times (\mathbf{B} \times \mathbf{C}) = (\mathbf{A} \cdot \mathbf{C})\mathbf{B} - (\mathbf{A} \cdot \mathbf{B})\mathbf{C}$. [Hint: use (10.8) and (10.10) to expand both sides.] (d) Why are parentheses necessary in writing $\mathbf{A} \times (\mathbf{B} \times \mathbf{C})$ but not for $\mathbf{A} \cdot \mathbf{B} \times \mathbf{C}$?

10.4. Show that (a) the vector $\nabla \psi$ is perpendicular to the contours $\psi = $ constant; (b) $\nabla \psi$ is in the direction of, and equal in magnitude to, the maximum rate of increase of ψ; (c) the rate of increase of ψ in any direction is equal to the projection of $\nabla \psi$ in that direction.

10.5. By direct expansion using (10.8), (10.10) and the definition of ∇, verify the following identities:

$$\nabla \times \nabla \psi = 0 = \nabla \cdot \nabla \times \mathbf{A},$$

$$\nabla \times (\nabla \times \mathbf{A}) = \nabla(\nabla \cdot \mathbf{A}) - \nabla^2 \mathbf{A}$$

(the latter being valid only in rectangular coordinates).

10.6. Use (10.15) to (10.17) to verify the following expressions for $\nabla \psi$, $\nabla \cdot \mathbf{A}$, $\nabla^2 \psi$ for (a) cylindrical coordinates and for (b) spherical coordinates (see fig. 2.31). In cylindrical coordinates, $x = r \cos \theta$, $y = r \sin \theta$, $z = z$; and

$$\nabla \psi = \frac{\partial \psi}{\partial r} \mathbf{i}_1 + \frac{1}{r} \frac{\partial \psi}{\partial \theta} \mathbf{i}_2 + \frac{\partial \psi}{\partial z} \mathbf{i}_3,$$

$$\nabla \cdot \mathbf{A} = \frac{1}{r} \frac{\partial}{\partial r}(r A_r) + \frac{1}{r} \frac{\partial A_\theta}{\partial \theta} + \frac{\partial A_z}{\partial z},$$

$$\nabla^2 \psi = \frac{1}{r} \frac{\partial}{\partial r}\left(r \frac{\partial \psi}{\partial r} \right) + \frac{1}{r^2} \frac{\partial^2 \psi}{\partial \theta^2} + \frac{\partial^2 \psi}{\partial z^2},$$

where $\mathbf{i}_1$, $\mathbf{i}_2$, $\mathbf{i}_3$ are unit vectors in the direction of increasing r, θ, z, and A_r, A_θ, A_z are components of $\mathbf{A}$ in the r-, θ-, z-directions. In spherical coordinates, $x = r \sin \theta \cos \phi$, $y = r \sin \theta \sin \phi$, $z = r \cos \theta$; and

$$\nabla \psi = \frac{\partial \psi}{\partial r} \mathbf{i}_1 + \frac{1}{r} \frac{\partial \psi}{\partial \theta} \mathbf{i}_2 + \frac{1}{r \sin \theta} \frac{\partial \psi}{\partial \phi} \mathbf{i}_3,$$

$$\nabla \cdot \mathbf{A} = \frac{1}{r^2} \frac{\partial}{\partial r}(r^2 A_r) + \frac{1}{r \sin \theta} \frac{\partial}{\partial \theta}(A_\theta \sin \theta)$$

$$+ \frac{1}{r \sin \theta} \frac{\partial A_\phi}{\partial \phi},$$

$$\nabla^2\psi = \frac{1}{r^2}\frac{\partial}{\partial r}\left(r^2\frac{\partial\psi}{\partial r}\right) + \frac{1}{r^2\sin\theta}\frac{\partial}{\partial\theta}\left(\sin\theta\frac{\partial\psi}{\partial\theta}\right)$$

$$+ \frac{1}{r^2\sin^2\theta}\frac{\partial^2\psi}{\partial\phi^2},$$

where $\mathbf{i}_1$, $\mathbf{i}_2$, $\mathbf{i}_3$ are unit vectors in the directions of increasing r, θ, ϕ, and A_r, A_θ, A_ϕ are components of $\mathbf{A}$ in the r-, θ- and ϕ-directions.

10.7. (a) Prove that

$$l^2 + m^2 + n^2 = 1$$

where (l, m, n) are the direction cosines of a vector. [Hint: Start with a vector $\mathbf{A}$ with direction cosines (l, m, n) and find $\mathbf{A}^2$.] (b) The perpendicular from the origin to a plane has length h and direction cosines (l, m, n). Show that the equation of the plane is

$$lx + my + nz = h.$$

10.8. (a) Prove the following corollary of (10.2):

$$\sum_j(-1)^{k+j}a_{ji}M_{jk} = 0 = \sum_j(-1)^{j+k}a_{ij}M_{kj}, \quad i \ne k.$$

(b) verify (10.20) using (10.2) and part (a).

10.9. (a) When more than two matrices are multiplied together, show that the order of multiplying adjacent pairs is arbitrary; thus

$$\mathscr{ABC} = (\mathscr{AB})\mathscr{C} = \mathscr{A}(\mathscr{BC}).$$

(b) Prove that $(\mathscr{ABC})^{\mathrm{T}} = \mathscr{C}^{\mathrm{T}}\mathscr{B}^{\mathrm{T}}\mathscr{A}^{\mathrm{T}}$ by applying the basic law of matrix multiplication. (c) Show that the multiplication of partitioned matrices gives the same result as the basic law of multiplication by setting up matrices $\mathscr{A}$ and $\mathscr{B}$ of sizes (3×4) and (4×5) and carrying out the multiplication for the unpartitioned matrices and for $\mathscr{A}$ partitioned with a (2×3) matrix in the upper left corner and $\mathscr{B}$ with a (3×3) matrix in the upper left corner.

10.10. (a) Referring to §8.2.1d, show that the error $e_t = h_t - g_t * f_t$ can be written in matrix form as $\mathscr{E} = \mathscr{H} - \mathscr{GF}$ and that the normal equations become

$$\mathscr{F} = (\mathscr{G}^{\mathrm{T}}\mathscr{G})^{-1}\mathscr{G}^{\mathrm{T}}\mathscr{H} = \phi_{gg}^{-1}\phi_{gh}$$

where $\mathscr{G}$ has the same form as $\mathscr{A}$ in (10.23) and $\mathscr{F}$, $\mathscr{E}$, $\mathscr{H}$ are column matrices of orders $(n + 1) \times 1$, $(2n + 1) \times 1$, $(2n + 1) \times 1$ respectively. (b) Show that v in (10.42) is the minimum value of E. (c) Show that v^* in (10.47) is the same as v.

10.11. (a) The following series expressions can be found in any calculus text:

$$e^x = 1 + x + \frac{x^2}{2!} + \frac{x^3}{3!} + \cdots,$$

$$\sin x = x - \frac{x^3}{3!} + \frac{x^5}{5!} - \frac{x^7}{7!} + \cdots,$$

$$\cos x = 1 - \frac{x^2}{2!} + \frac{x^4}{4!} - \frac{x^6}{6!} + \cdots.$$

Show that

$$\left.\begin{array}{l}\cos x = \frac{1}{2}(e^{jx} + e^{-jx}), \\[2mm] \sin x = \frac{1}{2j}(e^{jx} - e^{-jx}), \\[2mm] (\cos x \pm j\sin x) = e^{\pm jx}.\end{array}\right\} \text{ Euler's formulas}$$

$$(\cos x \pm j\sin x)^n = e^{\pm jnx}$$
$$= (\cos nx \pm j\sin nx) \quad \text{de Moivre's theorem}$$

(b) Evaluate z_1^2, $1/z_1$, $z_1 z_2$ for $z_1 = 2 - 3j$, $z_2 = 4 + 9j$; express z_1 and z_2 in polar form and repeat the above, verifying that the results are the same in both cases. (c) Verify the following formulas. [Hint: Start with the third Euler formula, form the sum, then equate real and imaginary parts.]

$$\sum_{r=0}^{n-1}\cos(x + r\gamma) = \frac{\sin\frac{1}{2}n\gamma}{\sin\frac{1}{2}\gamma}\cos\{x + \frac{1}{2}(n - 1)\gamma\},$$

$$\sum_{r=0}^{n-1}\sin(x + r\gamma) = \frac{\sin\frac{1}{2}n\gamma}{\sin\frac{1}{2}\gamma}\sin\{x + \frac{1}{2}(n - 1)\gamma\}.$$

10.12. (a) Use the method of least squares to fit the line $V = V_0 + az$ to the following data; (b) fit the curve $V = V_0 + az + bz^2$ to the same data.

z	V	z	V
0.50 km	2.02 km/s	2.00 km	2.50 km/s
1.00	2.16	2.50	2.58
1.50	2.32	3.00	2.60

10.13. Show that (10.37) has the matrix solution

$$\mathscr{A} = (\mathscr{D}\mathscr{D}^{\mathrm{T}})^{-1}\mathscr{D}\mathscr{Y}$$

where

$$\mathscr{D} = \begin{Vmatrix} 1 & 1 & 1 & & 1 \\ x_1 & x_2 & x_3 & & x_n \\ x_1^2 & x_2^2 & x_3^2 & & x_n^2 \\ \cdots & & & & \\ x_1^m & x_2^m & x_3^m & \cdots & x_n^m \end{Vmatrix},$$

Fig.10.14. Boxcar filter.

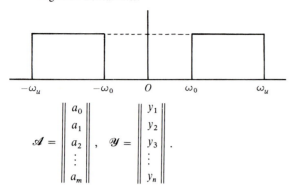

$$\mathscr{A} = \begin{Vmatrix} a_0 \\ a_1 \\ a_2 \\ \vdots \\ a_m \end{Vmatrix}, \quad \mathscr{Y} = \begin{Vmatrix} y_1 \\ y_2 \\ y_3 \\ \vdots \\ y_n \end{Vmatrix}.$$

10.14. A periodic function $g(t)$ can be represented by a finite series of the form

$$g(t) \approx S_n(t)$$

$$= \tfrac{1}{2}p_0 + \sum_{r=1}^{n} (p_r \cos r\omega_0 t + q_r \sin r\omega_0 t).$$

(a) Show that, if $S_n(t)$ gives a least-squares 'best fit' to $g(t)$, p_r and q_r must be equal to a_r and b_r in (10.77) and (10.78). (b) If we calculate a_r, b_r up to $r = 5$, then decide that we need a better approximation by extending up to $r = 8$, must we recalculate a_r, b_r for $r = 1, \dots 5$?

10.15. Verify (10.82).

10.16 (a) Writing $g(t) = r(t) + jx(t)$, $G(\omega) = R(\omega) + jX(\omega)$, where $g(t) \leftrightarrow G(\omega)$, and $r(t)$, $x(t)$, $R(\omega)$ and $X(\omega)$ are real, derive the following:

$$R(\omega) = \int_{-\infty}^{\infty} \{r(t) \cos \omega t + x(t) \sin \omega t\} \, dt;$$

$$X(\omega) = \int_{-\infty}^{\infty} \{r(t) \sin \omega t - x(t) \cos \omega t\} \, dt;$$

$$r(t) = (1/2\pi) \int_{-\infty}^{\infty} \{R(\omega) \cos \omega t - X(\omega) \sin \omega t\} \, d\omega;$$

$$x(t) = (1/2\pi) \int_{-\infty}^{\infty} \{R(\omega) \sin \omega t + X(\omega) \cos \omega t\} \, d\omega.$$

(b) When $g(t)$ is a real function,

$$R(\omega) = R(-\omega), X(\omega) = -X(-\omega), G(-\omega) = \overline{G(\omega)};$$

$$g(t) = (1/2\pi) \operatorname{Re} \left\{ \int_{-\infty}^{\infty} G(\omega) e^{j\omega t} \, d\omega \right\}$$

$$= (1/\pi) \int_{0}^{\infty} \{R(\omega) \cos \omega t - X(\omega) \sin \omega t\} \, d\omega.$$

(c) When $g(t)$ is real and even, $G(\omega)$ is real and even.
(d) When $g(t)$ is real and odd, $G(\omega)$ is imaginary and odd.
(e) When $g(t)$ is real and causal,

$$R(\omega) = \int_{0}^{\infty} g(t) \cos \omega t \, dt,$$

$$X(\omega) = -\int_{0}^{\infty} g(t) \sin \omega t \, dt.$$

10.17. (a) Verify (10.101) by subtracting the transforms of two step functions. (b) Show that the waveshape in the time-domain corresponding to a boxcar bandpass filter (fig. 10.14) is: (i) the modulated sinusoid

$$f(t) = (2/\pi t) \sin \{\tfrac{1}{2}(\omega_u - \omega_0)t\} \cos \{\tfrac{1}{2}(\omega_u + \omega_0)t\};$$

(ii) the difference between two sinc functions.

10.18. Prove (10.124) by finding the inverse transform of $G_1(\omega) * G_2(\omega)$.

10.19. (a) Verify the four relationships in (10.136) by drawing curves to represent $g_1(t)$ and $g_2(t)$, carrying out the required reflections and translations and comparing the results. (b) Verify the last three relationships of (10.136) by substitution in the integral expressions.

10.20. Prove the second relationship in (10.137).

10.21. (a) Show that the autocorrelation of an aperiodic function has its greatest value for zero shift. [Hint: Start with the identity

$$\int_{-\infty}^{\infty} \{g_1(\tau) - g_1(t + \tau)\}^2 \, d\tau > 0,$$

expand and identify the various integrals as $\phi_{11}(0)$ or $\phi_{11}(t)$.] (b) Same as (a) except for a random function.

10.22. Two functions, $X(\omega)$ and $R(\omega)$, related by equations (10.151) and (10.152), are known as a *Hilbert transform pair*. (a) Verify the following Hilbert transform pairs:

$$\delta(\omega) \leftrightarrow -1/\pi\omega,$$
$$\cos \omega \leftrightarrow -\sin \omega,$$
$$\sin \omega \leftrightarrow \cos \omega$$
$$\operatorname{sinc} \omega \leftrightarrow (\cos \omega - 1)/\omega.$$

(b) The treatment of quadrature filters in §10.3.11 deals with continuous functions; discuss the case of digital functions. (c) Show that (10.154) becomes for digital functions

$$y_t = (1/\pi) \sum_{n=-\infty}^{+\infty} x_{t-n}(e^{jn\pi} - 1)/n.$$

10.23. (a) Verify the following Laplace transform pairs:

$$t^n \leftrightarrow n!/s^{n+1}, \quad n = \text{positive integer};$$

$$\cos^2 \omega t \leftrightarrow \tfrac{1}{2}\left(\frac{1}{s} + \frac{s}{s^2 + 4\omega^2}\right);$$

$$(t-2)^5 \text{ step}\,(t-2) \leftrightarrow e^{-2s}(5!/s^6);$$

$$(t-2)^2 \text{ step}\,(t-3) \leftrightarrow e^{-3s}(s^2 + 2s + 2)/s^3.$$

(b) Find the inverse transforms of:

$$1/(s^2 + 9),\ e^{-4s}/(s^2 + 9),\ \int_s^{+\infty} ds/(s^2 + 9),\ 1/s(s^2 + 9).$$

10.24. Solve the following differential equations using Laplace transforms (see §10.1.7 for part (b)):

(a) $\dfrac{dy}{dx} + 3y = 5e^{-2x}, \quad y = 4 \text{ at } x = 0;$

(b) $\dfrac{d^2y}{dt^2} + 5\dfrac{dy}{dt} + 4y = \sinh 2t,$

$\dfrac{dy}{dt} = 0 \text{ and } y = \tfrac{1}{36} \text{ at } t = 0.$

10.25. Show that

$$\sinh kt \leftrightarrow (z \sinh k\Delta)/(z^2 - 2z \cosh k\Delta + 1),$$

$$\cosh kt \leftrightarrow (1 - z \cosh k\Delta)/(z^2 - 2z \cosh k\Delta + 1).$$

10.26. Derive the z-transform equivalents of (10.114) and (10.115); also (10.166) and (10.167).

10.27. (a) Show that the wavelet $z(2 - z)$ is not minimum-phase by considering the variation of the phase γ as in §10.6.6a. (b) Generalize this result for the wavelet $z^n(c - z)$, then to the wavelet $z^n W(z)$ where $n =$ integer and $W(z)$ is any minimum-phase wavelet.

10.28. Using the digital functions $a_t = [a_0, a_1, a_2, a_3, a_4]$ and $b_t = [b_0, b_1, b_2]$, work out the equivalents of (10.123) and (10.125) in terms of z-transforms.

10.29. The z-transform of a wavelet is

$$[\{1 + (0.5 + 0.5j)z\}\{1 + (0.5 - 0.5j)z\}(1 - 0.5z)]^2.$$

(a) Plot the waveshape. (b) Plot the roots with respect to the unit circle; what can you say about the phase of the wavelet? (c) What is the zero-phase wavelet with the same spectrum (see §10.6.6d)? Plot it. (d) Plot the roots of the zero-phase wavelet.

10.30. Show that a zero-phase wavelet has its maximum amplitude at the origin. [Hint: Consider the energy spectrum.]

10.31. (a) If $(a + bz)$ is minimum-phase, show that $1/(a + bz)$ is also minimum-phase. (b) Show that the convolution of two minimum-delay wavelets is also minimum-delay. (c) Given that $A(z)$ is minimum-phase while $B(z)$ is causal but not minimum-phase, under what conditions will the sum, $A(z) + B(z)$, be minimum-phase? (d) Writing $\mathscr{B}^*$, $\mathscr{B}^\dagger$ and $\mathscr{B}^{**}$ for the column matrices on the left sides of (10.210) and (10.211), we have

$$\mathscr{B}^* - c_k \mathscr{B}^\dagger = \mathscr{B}^{**}.$$

Show that $\mathscr{B}^{**}$ is minimum-phase if $\mathscr{B}^*$ is. [Hint: Note that $\mathscr{B}^\dagger = z^k \mathscr{B}^*$.]

10.32. Show that for a minimum-phase wavelet of the form

$$W(z) = \frac{\displaystyle\prod_{i=1}^{m}(a_i - z)}{\displaystyle\prod_{k=1}^{n}(b_k - z)}, \quad a_i \neq 0, \quad b_k \neq 0:$$

(a) the initial amplitude cannot be zero; (b) the maximum amplitude is not necessarily the first element.

10.33. Show that when we use the definition $z = e^{+j\omega\Delta}$, (10.212) becomes

$$g_n = (1/2\pi j\Delta)\oint G(z)z^{n-1}\,dz$$

where the integration is along the unit circle in the counterclockwise direction.

10.34. If a spectrum $R(z)$ does not have roots $z = 0, \pm 1$, show that there is one and only one minimum-phase wavelet corresponding to $R(z)$ (ignore multiplicative constants c where $|c| = 1$).

10.35. (a) Given the spectrum of a wavelet (assumed to be minimum-phase) $-6z^{-2} - 5z^{-1} + 38 - 5z - 6z^2$, use (10.206) to find the wavelet; (b) find the wavelet using the Levinson algorithm (see (10.207) to (10.211)).

10.36 Given the wavelet $w_t = [1\overset{\downarrow}{4}4, -96, -56, 48, 1, -6, 1, 0]$, show that: (a) the z-transform of the autocorrelation (the spectrum) is

$$R(z) = 144z^{-6} - 960z^{-5} + 664z^{-4} + 7200z^{-3}$$
$$- 13015z^{-2} - 11100z^{-1} + 35430$$
$$- 11100z - 13015z^2 + 7200z^3$$
$$+ 664z^4 - 960z^5 + 144z^6;$$

(b) w_t is minimum-delay (find the factors of $W(z)$); (c) the corresponding zero-phase wavelet is $[12, -40, -39, \overset{\downarrow}{170}, -39, -40, 12]$; (d) what is the earliest causal linear-phase wavelet corresponding to (c)?

10.37. Show that the right-hand expression in (10.223) corresponds to g_t delayed by one time unit, hence verify (10.224). [Hint: Investigate the right-hand term in (10.223) graphically, then study the summation in relation to the graph.]

10.38. (a) Verify the relation $f_H(t) = -f_L(t)$, $t \neq 0$ when both filters have the same cutoff frequency, by starting from the relation $f_L(t) + f_H(t) \leftrightarrow F_L(\omega) + F_H(\omega)$. (b) Compare the values of $f_L(t), f_H(t), f_t^L, f_t^H$.

10.39. Show that the filter

$$F(\omega) = e^{-jk\omega}, \quad |\omega| < \omega_0,$$
$$= 0, \qquad |\omega| > \omega_0,$$

is identical with that of (10.232) except that $f_L(t)$ is shifted by k time units.

10.40. Obtain the digital filter corresponding to the Butterworth filter for $n = 3$.

10.41. Verify the transforms in (10.243) to (10.250). [Hint: In (10.244) multiply sinc $(\pi t/T)$ by box$_{2T}(t)$ and use (10.124); for (10.245) note the result in (10.249); for (10.250), note problem 4.11a.]

10.42. (a) Verify (10.257). [Hint: Follow the same procedure as in the derivation of (8.58) except that there is no equation corresponding to f_0 since it is constant.] (b) Verify (10.259).

Appendices

A List of abbreviations used in volume 2

AAPG American Association of Petroleum Geologists
AGC automatic gain control
AGIP AGIP Petroleum Company
CDP common-depth-point
CGG Compagnie Générale de Géophysique
GSI Geophysical Service Inc.
GTS Geoscience Technology Services Corp.
NMO normal moveout
OTC Offshore Technology Conference
rms root mean square
SEG Society of Exploration Geophysicists
3-D three-dimensional

B Trade names and proper names used

Name   *Whose tradename*
Seiscrop   Geophysical Service Inc.
Sosie   Société Nationale Elf-Aquitaine
Vibroseis   Conoco Inc.

C A seismic report

(a) *Title page*

List: for whom work was done, name of project or area, dates of project, name of contractor making report, individuals responsible for report.

(b) *Enclosures*

List: attachments, documents which go with report. Figures, maps and sections should be used where they are able to convey information more clearly than textual description. Enclose only relevant data. Consider photo-reducing maps or sections for inclusion. Label enclosures so they will be identifiable if separated from report.

(c) *Abstract*

Briefly state why work was done, what was done and how results are to be used. No longer than a half-page.

(*d*) **Introduction**

(1) Briefly state objectives. If report covers only processing, review relevant information about field operations or previous processing.

(2) Describe location of work, usually with a map. Distinguish data being discussed from other data shown on map.

(3) Describe data quality in general terms and nature of problems encountered (multiples, static problems, line misties, structure problems, etc.)

(*e*) **Processing procedures and analysis**

(1) Standard processing sequence used (often shown by flow chart). Parameter values, method of datum correction. Discuss processes by trade name and describe objective and methods which unusual programs employ.

(2) Describe testing done to determine processing sequence and parameters. Location of test points. Describe (often include example of) displays used to determine parameters for muting, filtering, determining stack response, velocity, static corrections, residual statics, etc.

(3) Experimentation done, where, and conclusions.

(4) Discuss velocities; data from previous work, wells, other sources. How often were velocity analyses run, datum used. How much velocity variation?

(*f*) **Results**

(1) Include copies of sections and list of data processed.

(2) List problems encountered, including where unable to read tapes, poor documentation, survey, elevation, uphole problems, etc.

(3) Special problems observed.

(*g*) **Conclusions**

Did processing meet objectives? How could objectives have been better met?

(*h*) **Recommendations**

Reprocessing, further testing, for next work in this area.

(*i*) **Appendix**

(1) Copy of lines processed.

(2) Statistics.

(3) Special studies not relevant to main objectives.

(4) Personnel list.

(5) References.

D   Symbols used in mapping

(*a*)   **Structure symbols**

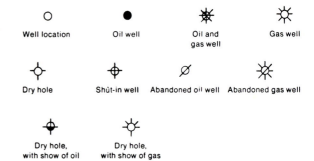

Apparent dip

Anticlinal axis, reversal of dip direction

Anticline plunging to the left

Synclinal axis plunging to the left

Normal fault with upthrown side to the north or hachures on downthrown side

Thrust or reverse fault; barbs are on side of upper block; contours in lower may be dashed when underneath the fault

Strike-slip fault showing sense of movement

Dashed or dotted contours indicate inferred or doubtful structure or sometimes an alternate interpretation or other kind of data (perhaps outline of gravity anomaly, inferred subcrops, etc.)

Strike and dip of bedding, the number indicating the amount of dip (usually in degrees)

(*b*)   **Well symbols**

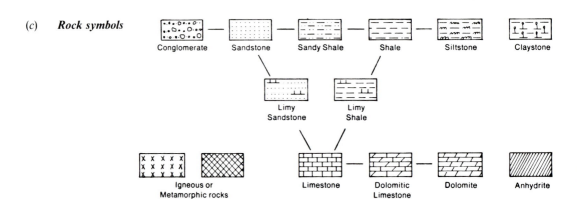

Note: Intermediate rock compositions are represented by combining symbols..

References

The numbers in square brackets are the section numbers in which references are cited.

Andersen, N. O. (1974) On the calculation of filter coefficients for maximum entropy spectral analysis: *Geophysics*, **39**, 69–72.   [10.8.6*d*]

Anstey, N. A. (1964) Correlation techniques – a review: *Geophy. Prosp.*, **12**, 355–82.   [8.1.3*a*]

Anstey, N. A. (1970) Signal characteristics and instrument specifications; Vol. 1 of *Seismic Prospecting Instruments*: Berlin, Gebrüder Borntraeger.   [8.0]

Anstey, N. A. (1973) How do we know we are right? *Geophys. Prosp.*, **21**, 407–11.   [9.0]

Anstey, N. A. (1977), *Seismic Interpretation – the Physical Aspects*: Boston, International Human Resources Development Corp.   [9.0]

Aud, B. W. (1976) History of abnormal pressure determinations from seismic data: Offshore Technology Conference Preprints, paper 2611; Dallas, OTC.   [7.2.4]

Backus, M. M. (1959) Water reverberations – their nature and elimination: *Geophysics*, **24**, 233–61.   [8.1.2*d*]

Balch, A. H. (1971) Color sonagrams: A new dimension in seismic data interpretation: *Geophysics*, **36**, 1074–98.   [9.2.7]

Bates, R. L., and Jackson, J. A. (1980) *Glossary of Geology*, 2nd ed.: Falls Church, Va., AGI.   [9.0]

Båth, M. (1974) *Spectral analysis in geophysics*: Amsterdam, Elsevier.   [10.0, 10.7, 10.8.1, 10.8.5]

Bendat, J. S. and Piersol, A. G. (1966) *Measurement and analysis of random data*: New York, Wiley.   [10.3.10]

Blackman, R. B., and Tukey, J. W. (1958) *The Measurement of Power Spectra*: New York, Dover.   [10.8.1, 10.8.5]

Bois, P and la Porte, M. (1970) Pointe automatique: *Geophys. Prosp.*, **18**, 489–504.   [8.4.1]

Bracewell, R. (1965) *The Fourier Transform and its Applications*: New York, McGraw-Hill.   [8.4.2]

Brown, L. F. and Fisher, W. L. (1980) *Seismic Stratigraphic Interpretation and Petroleum Exploration*: Tulsa, AAPG Continuing Education Course Note Series # 16.
                                                                                                       [9.3.5, 9.7.1]

Brown, A. R., Dahm, C. G. and Graebner, R. J. (1981) Stratigraphic case history using three-dimensional seismic data in the Gulf of Thailand: *Geophys. Prosp.* **29**, 327–49.   [9.3.5]

Bubb, J. N. and Hatlelid, W. G. (1977) Seismic recognition of carbonate buildups; in *Seismic Stratigraphy – Applications to Hydrocarbon Exploration*, pp. 185–204 (ed. C. E. Payton): Tulsa, AAPG Memoir 26.   [9.3.4]

Burg, J. P. (1972) The relationship between maximum entropy spectra and maximum likelihood spectra: *Geophysics*, **37**, 375–76.   [10.8.6d]

Burg, J. P. (1975) *Maximum entropy spectral analysis*: Ph.D. thesis, Dept. of Geophysics, Stanford University, Palo Alto, Calif.   [10.8.6d]

Buxtorf, A. (1916) Prognosen und Befunden beim Hauensteinbasis und Grenchenberg Tunnel und die Bedeutung der letzern für die Geologie des Juragebirges: *Verh. Naturf. Gesell. Basel*, **27**, 185–254.   [9.3.1b]

Campbell, F. F. (1965) Fault criteria: *Geophysics*, **30**, 976–97.   [9.3.2a]

Cassand, J., Damotte, B, Fontanel, A., Grau, G., Hemon, C. and Lavergne, M. (1971) *Seismic filtering*: Tulsa, SEG. (Translated by N. Rothenburg from *Le Filtrage en Sismique*, 1966: Paris, Editions Technip.)   [10.0]

Cheng, D. K. (1959), *Analysis of Linear Systems*: Reading, Mass., Addison-Wesley.   [10.5.1]

Chun, J. H. and Jacewitz, C. A. (1981) Fundamentals of frequency-domain migration: *Geophysics*, **46**, 717–33.
                                                                                                        [8.3.3]

Churchill, R. V. (1963) *Fourier Series and Boundary Value Problems*, 2nd. ed.: New York, McGraw-Hill.   [10.2.2]

Claerbout, J. F. (1963) *Digital Filtering and Applications to Seismic Detection and Discrimination*: Cambridge, Mass., M.I.T. M.Sc. thesis.   [10.6.6b]

Claerbout, J. F. (1976) *Fundamentals of Geophysical Data Processing*: New York, McGraw-Hill.
                                              [8.0, 8.2.2b, 8.3.4, 10.0, 10.1.5, 10.6.6c, 10.8a, c]

Claerbout, J. F. and Muir, F. (1973) Robust modeling with erratic data: *Geophysics*, **38**, 826–44.   [10.8.6a]

Clarke, G. K. C. (1968) Time-varying deconvolution filters: *Geophysics*, **33**, 936–44.   [8.2.1]

Clement, W. A. (1977) Case history of geoseismic modeling of basal Morrow–Springer sandstones, Watonga–Chickasha trend, Geary, Oklahoma; in *Seismic Stratigraphy – Applications to Hydrocarbon Exploration*, pp. 451–76 (ed. C. E. Payton): Tulsa, AAPG Memoir 26.   [9.7.4]

Cook, E. E. and Taner, M. T. (1969) Velocity spectra and their use in stratigraphic and lithologic differentiation: *Geophys. Prosp.*, **17**, 433–48.   [8.2.3a]

Cook, F. A., Albaugh, D. S., Brown, L. D., Kaufman, S., Oliver, J. E., and Hatcher, R. D. (1979) Thin-skinned tectonics in the crystalline Southern Appalachians: *Geology*, **7**, 563–7.   [9.9]

Cook, F. A., Brown, L. D., and Oliver, J. E. (1980) The Southern Appalachians and the growth of continents: *Scientific American*, **243**, No. 4. 156–68.   [9.9]

Cooley, J. W. and Tukey, J. W. (1965) Algorithm for the machine calculation of complex Fourier series: *Math. Computation*, **19**, 297–301.   [10.6.4]

Crump, M. D. (1974) A Kalman filter approach to the deconvolution of seismic signals: *Geophysics*, **39**, 1–13.   [8.2.1f]

Dahlstrom, C. D. A. (1970) Structural geology in the eastern margin of the Canadian Rocky Mountains: *Canadian Petrol. Geol. Bull.*, **18**, 332–406.   [9.3.1b]

Dahm, C. G. and Graebner, R. J. (1982) Field development with three-dimensional seismic methods in the Gulf of Thailand – a case history: *Geophysics*, **47**, 149–176.   [9.6]

Daly, R. A., Manger, G. E. and Clark, S. P. (1966) Density of rocks; in *Handbook of Physical Constants*, pp. 19–26 (ed. S. P. Clark): Geological Society of America Memoir 97, New York.   [7.1.3]

Davis, T. L. (1972) Velocity variations around Leduc reefs, Alberta: *Geophysics*, **37**, 584–604.   [9.3.4]

Dix, C. H. (1955) Seismic velocities from surface measurements: *Geophysics*, **20**, 68–86.   [7.3.3a]

Dobrin, M. B. (1977) Seismic exploration for stratigraphic traps; in *Seismic stratigraphy – Applications to Hydrocarbon Exploration*, pp. 329–52 (ed. C. E. Payton): Tulsa, AAPG Memoir 26.   [9.3.6]

Domenico, S. N. (1974) Effect of water saturation on seismic reflectivity of sand reservoirs encased in shale: *Geophysics*, **39**, 759–69.   [7.1.7]

Domenico, S. N. (1976) Effect of brine-gas mixture on velocity in an unconsolidated sand reservoir: *Geophysics*, **41**, 882–94.   [7.1.7]

Domenico, S. N. (1977) Elastic properties of unconsolidated porous sand reservoirs: *Geophysics*, **42**, 1339–68.   [7.1.7]

Duska, L. (1963) A rapid curved-path method for weathering and drift corrections: *Geophysics*, **28**, 925–47.   [7.2.2]

Fail, J. P. and Grau, G. (1963) Les filtres en eventail: *Geophys. Prosp.*, **11**, 131–63.   [8.2.7, 10.3.2]

Faust, L. Y. (1951) Seismic velocity as a function of depth and geologic time: *Geophysics*, **16**, 192–206.   [7.1.6]

Faust, L. Y. (1953) A velocity function including lithologic variation: *Geophysics*, **18**, 271–88.   [7.1.5]

Feagin, F. J. (1981) Seismic data display and reflection perceptability: *Geophysics*, **46**, 106–20.   [9.2.7]

Finetti, I., Nicolich, R. and Sancin, S. (1971) Review on the basic theoretical assumptions in seismic digital filtering: *Geophys. Prosp.*, **19**, 292–320.   [8.0, 10.8.1]

Flinn, E. A., ed., Robinson, E. A. and Treitel, S. (1967) Special issue on the MIT Geophysical Analysis Group reports: *Geophysics*, **32**, 411–525. [8.0]

Fitch, A. A. (1976) *Seismic Reflection Interpretation*: Berlin, Gebrüder Borntraeger. [9.0]

Galbraith, R. M. and Brown, A. R. (1982) Field appraisal with three-dimensional seismic surveys offshore Trinidad: *Geophysics*, **47**, 177–95. [9.6]

Gallup, W. B. (1951) Geology of Turner Valley oil and gas field, Alberta: *AAPG Bull.*, **35**, 797–821. [9.3.1*b*]

Gardner, G. H. F., Gardner, L. W. and Gregory, A. R. (1974) Formation velocity and density – the diagnostic basics for stratigraphic traps: *Geophysics*, **39**, 770–80. [7.1.2, 7.1.3, 7.1.5]

Garotta, R. (1971) Selection of seismic picking based upon the dip, move-out and amplitude of each event: *Geophys. Prosp.*, **19**, 357–70. [8.4.1]

Garotta R. and Michon, D. (1967) Continuous analysis of the velocity function and of the moveout corrections: *Geophys. Prosp.*, **15**, 584–97. [8.2.3*a*]

Gassmann, F. (1951) Elastic waves through a packing of spheres: *Geophysics*, **16**, 673–85. [7.1.5]

Gazdag, J. (1981) Modeling of the acoustic wave equation with transform methods: *Geophysics*, **46**, 854–9. [9.4.3]

Geertsma, J. (1961) Velocity log interpretation: the effect of rock bulk compressibility: *Soc. Petroleum Engineers AIMME Trans.* **222**, 235–53. [7.1.7]

Goguel, J. (1962) *Tectonics*: San Francisco, W. H. Freeman. [9.3.1*b*]

Gregory, A. R. (1976) Fluid saturation effects on dynamic elastic properties of sedimentary rocks: *Geophysics*, **41**, 895–921. [7.1.7]

Gregory, A. R. (1977) Aspects of rock physics from laboratory and log data that are important to seismic interpretation; in *Seismic Stratigraphy – Applications to Hydrocarbon Exploration*, pp. 15–46 (ed. C. E. Payton): Tulsa, AAPG Memoir 26. [7.1.1, 7.1.5]

Gretener, P.E. (1979) *Pore Pressure: fundamentals, general ramifications and implications for structural geology*: Education course note series #4, Tulsa, AAPG. (Revised). [7.2.4, 9.3.1*b*]

Hagedoorn, J. G. (1954) A process of seismic reflection interpretation: *Geophys. Prosp.*, **2**, 85–127. [8.3.2]

Handbook of Chemistry and Physics; (1975): Cleveland, CRC Press. [10.3.3]

Harding, T. P. and Lowell, J. D. (1979) Structural styles, their plate-tectonic habitats, and hydrocarbon traps in petroleum provinces: *AAPG Bull.*, **63**, 1016–58. [9.3.1*b*]

Harms, J. C. and Tackenberg, P. (1972) Seismic signatures of sedimentation models: *Geophysics*, **37**, 45–58. [9.7.4]

Harris, L. D. and Milici, R. C. (1977) *Characteristics of thin-skinned styles of deformation in the Southern Appalachians and potential hydrocarbon traps*: US Geological Survey Prof. Paper 1018. [9.3.1*b*]

Hatton, L., Larner, K. and Gibson, B. S. (1981) Migration of seismic data from inhomogeneous media: *Geophysics*, **46**, 751–67. [8.3.5*c*]

Hilterman, F. J. (1970) Three-dimensional seismic modeling: *Geophysics*, **35**, 1020–37. [9.4.3]

Hobbs, B. E., Weams, W. D. and Williams, P. F. (1976) *Outline of Structural Geology*: New York, Wiley. [9.3.1*b*]

Hofer, H. and Varga, W. (1972) Seismogeologic experience in the Beaufort Sea: *Geophysics*, **37**, 605–19. [7.4]

Hubbert, M. K. (1937) Scale models and geologic structures: *Geol. Soc. Am. Bull.*, **48**, 1459–1520. [9.4.2]

Hun, F. (1978) Correlation between seismic reflection amplitude and well productivity – a case study: *Geophys. Prosp.*, **26**, 157–62. [9.7.4]

Isaacs, B., Oliver, J. and Sykes, L. R. (1968) Seismology and the new global tectonics: *J. Geophysical Research*, **73**, 5855–99. [9.3.1*a*]

Jains, S. and Wren, A. E. (1980) Migration before stack: procedure and significance: *Geophysics*, **45**, 204–12. [8.3.5*d*]

Jankowsky, W. (1970) Empirical investigation of some factors affecting elastic wave velocities in carbonate rocks: *Geophys. Prosp.*, **18**, 103–18. [7.2.1]

Judson, D. R., Lin, J., Schultz, P. S. and Sherwood, J. W. C. (1980) Depth migration after stack: *Geophysics*, **45**, 361–75. [8.3.5*c*]

Kanasewich, E. R. (1973) *Time Sequence Analysis in Geophysics*: Edmonton, Univ. of Alberta Press. [8.0, 10.0]

Kaplan, W. (1952) *Advanced calculus*: Reading, Mass., Addison-Wesley. [10.3.4]

Kleyn, A. H. (1977) On the migration of reflection-time contour maps; *Geophys. Prosp.*, **25**, 125–40. [8.3.6]

Kokesh, F. P. and Blizard, R. B. (1959) Geometrical factors in sonic logging: *Geophysics*, **24**, 64–76. [7.3.2]

Kulhánek, O. (1976) *Introduction to Digital Filtering in Geophysics*: Amsterdam, Elsevier. [10.0, 10.8.1, 10.8.3]

Kunetz, G. and Fourmann, J. M. (1968) Efficient deconvolution of marine seismic records: *Geophysics*, **33**, 412–23. [8.2.1*e*]

Kuo, S. S. (1965) *Numerical Methods and Computers*: Reading, Mass., Addison-Wesley. [10.1.6]

Kurita, T. (1969) Spectral analysis of seismic waves, Part I, Data windows for the analysis of transient waves: *Spec. Contrib. Geophys. Inst., Kyoto Univ.*, **9**, 97–122. [10.8.5]

Larner, K. L., Hatton, L., Gibson, B. S. and Hsu, I. C. (1981) Depth migration of imaged time sections: *Geophysics*, **46**, 734–50. [8.3.5*c*, 9.5.1]

Lee, Y. W. (1960) *Statistical Theory of Communication*: New York, Wiley. [8.0, 10.3.10, 10.8.1, 10.8.6*b*]

Levin, F. K. and Shah, P. M. (1977) Peg-leg multiples and dipping reflectors: *Geophysics*, **42**, 957–81. [8.2.5*a*]

Lindseth, R. O. (1979) Synthetic sonic logs – a process for stratigraphic interpretation: *Geophysics*, **44**, 3–26. [7.1.2, 9.4.5, 9.7]

Lowell, J. D. (1972) Spitzbergen Tertiary orogenic belt and the Spitzbergen fracture zone: *Geol. Soc. Am. Bull.*, **83**, 3091–102. [9.3.1*b*]

Lyons, P. L. and Dobrin, M. B. (1972) Seismic exploration for stratigraphic traps; in *Stratigraphic Oil and Gas Fields – Classification, Exploration Methods and Case Histories*, pp. 225–43 (ed. R. E. King): Tulsa, AAPG Memoir 16. [9.3.6]

Marr, J. D. (1971) Seismic stratigraphic exploration – part I: *Geophysics*, **36**, 311–29; part II: *Geophysics*, **36**, 533–53; part III; *Geophysics*, **36**, 676–89. [9.3.6]

Matsuzawa, A., Tamano, T., Aoki, Y. and Ikawa, T. (1979) Structure of the Japan Trench subduction zone from multi-channel seismic reflection records: pp. 171–82 in *Marine Geology*: Amsterdam, Elsevier. [9.9]

Maureau, G. T. and van Wijhe, D. H. (1979) Prediction of porosity in the Permian carbonate of eastern Netherlands using seismic data: *Geophysics*, **44**, 1502–17. [9.7.4]

May, B. T. and Covey, J. D. (1981) An inverse ray method for computing geologic structures from seismic reflections: zero-offset case: *Geophysics*, **46**, 268–87. [9.4.6]

McQuillin, R., Bacon, M. and Barclay, W. (1979) *An Introduction to Seismic Interpretation*: Houston, Gulf Publishing Co. [9.0]

Meckel, L. D., Jr and Nath, A. K. (1977) Geologic considerations for stratigraphic modeling and interpretation; in *Seismic Stratigraphy – Applications to Hydrocarbon Exploration*, pp. 417–38 (ed. C. E. Payton): Tulsa, AAPG Memoir 26. [7.1.3]

Middleton, D. and Wittlesey, J. R. B. (1968) Seismic models and deterministic operators for marine reverberation: *Geophysics*, **33**, 557–83. [8.1.2*d*]

Millahn, K. O. (1980) In-seam seismics: position and development: *Prakla–Seismos Report*, **80**, no. 2 + 3, 19–30. [8.2.8]

Mitchum, R. M., Vail, P. R. and Thompson, S. (1977) The depositional sequence as a basic unit for stratigraphic analysis: in *Seismic Stratigraphy – Applications to Hydrocarbon Analysis*, pp. 53–62 (ed. C. E. Payton): Tulsa, AAPG Memoir 26. [9.7.2, 9.7.3]

Musgrave, A. W. and Bratton, R. H. (1967) Practical application of Blondeau weathering solution; in *Seismic Refraction Prospecting*, pp. 231–46 (ed. A. W. Musgrave): Tulsa, SEG. [7.2.2]

Neidell, N. S. and Poggiagliolmi, E. (1977) Stratigraphic modeling and interpretation; in *Seismic Stratigraphy – Applications to Hydrocarbon Exploration*, pp. 389–416 (ed. C. E. Payton) Tulsa, AAPG Memoir 26. [9.7.4]

Neidell, N. S. and Taner, M. T. (1971) Semblance and other coherency measures for multichannel data: *Geophysics*, **36**, 482–97. [8.1.3*f*]

Otis, R. M. and Smith, R. B. (1977) Homomorphic deconvolution by log spectral averaging: *Geophysics*, **42**, 1146–57. [8.2.1*f*]

Papoulis, A. (1962) *The Fourier Integral and its Applications*: New York, McGraw-Hill. [10.3.1, 10.3.3, 10.3.6, 10.3.11, 10.8.3]

Paturet, D. (1971) Different methods of time–depth conversion with and without migration: *Geophys. Prosp.*, **19**, 27–41. [8.4.1]

Paulson, K. V. and Merdler, S. C. (1968) Automatic seismic reflection picking: *Geophysics*, **33**, 431–40. [8.4.1]

Payton, C. E., ed. (1977) *Seismic Stratigraphy – Applications to Hydrocarbon Exploration*: Tulsa, AAPG Memoir 26. [9.7.1]

Peacock, K. L. and Treitel, S. (1969) Predictive deconvolution: theory and practice: *Geophysics*, **34**, 155–69. [8.2.1*e*]

Pickett, G. R. (1963) Acoustic character logs and their applications in formation evaluation: *Jour. Pet. Tech.*, 659–67. [7.1.2]

Pipes, L. A. and Harvill, L. R. (1970) *Applied Mathematics for Engineers and Physicists*, 3rd ed.: New York, McGraw-Hill. [10.0, 10.1.1, 10.1.2*d*, 10.1.3*b*, 10.1.5]

Plumley, W. J. (1980) Abnormally high fluid pressure: survey of some basic principles: *AAPG Bull.*, **64**, 414–22. [7.2.4]

Postic, A., Fourmann, J., and Claerbout, J. (1980) Parsimonious deconvolution: preprint of paper at SEG 50th Annual Meeting in Houston. [10.8.6*a*]

Press, F. (1966) Seismic velocities: in *Handbook of Physical Constants*, pp. 195–218 (ed. S. P. Clark, Jr): Geological Society of America, Memoir 97. [7.1.1, 7.1.2, 7.1.5]

Rackets, H. M. (1971) A low-noise seismic method for use in permafrost regions: *Geophysics*, **36**, 1150–61. [7.2.3]

Ramsayer, G. R. (1979) Seismic stratigraphy, a fundamental exploration tool: OTC Paper 3568. [9.7.2]

Reynolds, E. B. (1970) Predicting over-pressured zones with seismic data: *World Oil* **171**, # 5. [7.2.4]

Rietsch, E. (1979) Geophone sensitivities for Chebyshev optimized arrays: *Geophysics*, **44**, 1142–3. [10.8.6*a*]

Rittenhouse, G. (1972) Stratigraphic trap classification; in *Stratigraphic Oil and Gas Fields – Classification, Exploration Methods and Case Histories*, pp. 14–28 (ed. R. E. King): Tulsa, AAPG Memoir 16. [9.3.6]

Robie, R. A., Bethke, P. M., Toulmin, M. S. and Edwards, J. L. (1966) X-ray crystallographic data, densities and molar volumes of minerals: in *Handbook of physical constants*, pp. 27–73 (ed. S. P. Clark, Jr): Geological Society of America, Memoir 97. [7.1.3]

Robinson, E. A. (1962) *Random wavelets and cybernetic systems*: London, Griffin. [10.6.6*b*]

Robinson, E. A. (1967*a*) *Multichannel Time Series Analysis with Digital Computer Programs*: San Francisco, Holden-Day. [10.0]

Robinson, E. A. (1967*b*) Predictive decomposition of time series with application to seismic exploration: *Geophysics*, **32**, 418–84. [10.0]

Robinson, E. A. (1967*c*) *Statistical Communication and Detection*: London, Griffin. [10.0]

Robinson, E. A. and Treitel, S. (1964) Principles of digital filtering: *Geophysics*, **29**, 395–404. [8.0]

Robinson, E. A. and Treitel, S. (1967) Principles of digital Wiener filtering: *Geophys. Prosp.*, **15**, 311–33. [8.2.1*d*]

Robinson, E. A. and Treitel, S. (1973) *The Robinson–Treitel reader*: Tulsa, Seismograph Service.　　　[10.0]

Robinson, E. A., and Treitel, S. (1980) *Geophysical Signal Analysis*: Englewood Cliffs, N. J., Prentice-Hall.　　　[8.0, 10.0, 10.8.6*c*, *d*]

Roksandic, M. M. (1978) Seismic facies analysis concepts: *Geophys. Prosp.*, **26**, 383–98.　　　[9.7.3]

Sangree, J. B. and Widmier, J. M. (1979) Interpretation of depositional facies from seismic data: *Geophysics*, **44**, 131–60.　　　[9.7.3]

Sattlegger, J. W. and Stiller, P. K. (1974) Section migration before stack, after stack or in-between: *Geophys. Prosp.*, **22**, 297–314.　　　[8.3.5*d*]

Sattlegger, J. W., Stiller, P. K., Echterhoff, J. A. and Hentschke, M. K. (1980) Common-offset-plane migration: *Geophys. Prosp.*, **28**, 859–71.　　　[8.3.5*d*]

Schneider, W. A. (1978) Integral formulation for migration in two dimensions and three dimensions: *Geophysics*, **43**, 49–76.　　　[8.3.2]

Schneider, W. A. and Backus, M. M. (1968) Dynamic correlation analysis: *Geophysics*, **33**, 105–26.　　　[8.2.3*a*]

Schramm, M. W., Dedman, E. V. and Lindsey, J. P. (1977) Practical stratigraphic modeling and interpretation; in *Seismic Stratigraphy – Applications to Hydrocarbon Exploration*, pp. 477–502 (ed. C. E. Payton): Tulsa, AAPG Memoir 26.　　　[9.8]

Schultz, P. S. and Sherwood, J. W. C. (1980) Depth migration before stack: *Geophysics*, **45**, 376–93.　　　[8.3.5*c*, *d*]

Shannon, C. E. and Weaver, W. (1949) *The Mathematical Theory of Communications*: Urbana, Ill., Univ. of Illinois Press.　　　[10.8.6*d*]

Sheriff, R. E. (1973) *Encyclopedic Dictionary of Exploration Geophysics*: Tulsa, SEG.　　　[preface]

Sheriff, R. E. (1975) Factors affecting seismic amplitudes: *Geophys. Prosp.*, **23**, 125–38.　　　[8.2.4]

Sheriff, R. E. (1977*a*) Using seismic data to deduce rock properties; in *Developments in Petroleum Geology – I*, pp. 243–74 (ed. G. D. Hobson): London, Applied Science Publishers.　　　[7.1.4]

Sheriff, R. E. (1977*b*) Limitations on resolution of seismic reflections and geologic detail derivable from them; in *Seismic Stratigraphy – Applications to Hydrocarbon Exploration*, pp. 3–14 (ed. C. E. Payton): Tulsa, AAPG Memoir 26.　　　[9.7.1]

Sheriff, R. E. (1978) *A first Course in Geophysical Exploration and Interpretation*: Boston, International Human Resources Development Corp.
　　　[7.1.2, 7.1.5, 8.3.2, 9.0, 9.5.1, 9.6]

Sheriff, R. E. (1980) *Seismic Stratigraphy*: Boston, International Human Resources Development Corp.
　　　[9.3.6, 9.7.1, 9.7.3, 9.8]

Sheriff, R. E. and Farrell, J. (1976) Display parameters of marine geophysical data: OTC paper 2567.　　　[9.2.7]

Sherwood, J. W. C. and Trorey, A. W. (1965) Minimum-phase and related properties of the response of a horizontally stratified absorptive earth to plane acoustic waves:

Geophysics, **30**, 191–7.　　　[8.1.4]

Shipley, T. H., Houston, M. K., Buffler, R. T., Shaub, F. J., McMillen, K. J., Ladd, J. W. and Worzel, J. L. (1979) Seismic evidences for widespread possible gas hydrate horizons on continental slopes and rises: *AAPG Bull.*, **63**, 2204–13.　　　[7.2.5]

Silverman, D. (1967) The digital processing of seismic data: *Geophysics*, **32**, 988–1002.　　　[8.0]

Silvia, M. T. and Robinson, E. A. (1979) *Deconvolution of Geophysical Time Series in the Exploration for Oil and Natural Gas*: Amsterdam, Elsevier.　　　[10.0]

Smylie, D. E., Clarke, C. K. G. and Ulrych, T. J. (1973) Analysis of irregularities in the earth's rotation; in *Methods in Computational Physics, vol. 13, Geophysics*, pp. 391–430 (ed. B. A. Bolt): New York, Academic Press.　　　[10.8.6*d*]

Stoffa, P. L., Buhl, P. and Bryan, G. M. (1974) The application of homomorphic deconvolution to shallow-water marine seismology: *Geophysics*, **39**, 401–26.
　　　[8.2.1*f*, 10.7]

Stommel, H. E. and Graul, M. (1978) Current trends in geophysics: *Indonesian Petroleum Association Proceedings*, Jakarta, Indonesia.　　　[9.4.4*a*]

Swan, B. G. and Becker, A. (1952) Comparison of velocities obtained by delta-time analysis and well velocity surveys: *Geophysics*, **17**, 575–85.　　　[7.3.3*b*]

Taner, M. T. (1976) Simplan: simulated plane-wave exploration: SEG 46th Meeting (abstract in *Geophysics*, **42**, 186–7)　　　[8.2.6]

Taner, M. T. (1980) Long-period sea-floor multiples and their suppression: *Geophys. Prosp.*, **28**, 30–48.　　　[8.2.1*c*]

Taner, M. T., Cook, E. E. and Neidell, N. S. (1970) Limitations of the reflection seismic method, lessons from computer simulations: *Geophysics*, **35**, 551–73.
　　　[7.4, 8.2.5*a*, 9.4.6]

Taner, M. T. and Koehler, F. (1969) Velocity spectra; digital computer derivation and applications of velocity functions: *Geophysics*, **34**, 859–81.　　　[8.2.3*a*]

Taner, M. T. and Koehler, F. (1981) Surface consistent corrections: *Geophysics*, **46**, 17–22.　　　[8.2.4]

Taner, M. T. and Sheriff, R. E. (1977) Application of amplitude, frequency and other attributes to stratigraphic and hydrocarbon determination; in *Seismic Stratigraphy – Applications to Hydrocarbon Exploration*, pp. 301–28 (ed. C. E. Payton): Tulsa, AAPG Memoir 26.　　　[8.4.2, 9.2.7, 9.7.4]

Taner, M. T. Koehler, F. and Alhilali, K. A. (1974) Estimation and correction of near-surface time anomalies: *Geophysics*, **39**, 441–63.　　　[8.2.2*b*]

Taner, M. T., Koehler, F. and Sheriff, R. E. (1979) Complex seismic trace analysis: *Geophysics*, **44**, 1041–63.
　　　[8.4.2, 9.2.7, 9.7.4, 9.8]

Taylor, H. (1981) The l_1 norm in seismic data distribution: in *Developments in Geophysical Exploration Methods – 2*, pp. 53–76 (ed. A. A. Fitch): London, Applied Science Publishers.　　　[10.8.6*a*]

Telford, W. M., Geldart, L. P., Sheriff, R. E. and Keys, D. A. (1976) *Applied Geophysics*: Cambridge, England, Cambridge Univ. Press. [9.2.5]

Timoshenko, S. and Goodier, J. N. (1951) *Theory of Elasticity*, 2nd ed.: New York, McGraw-Hill. [7.1.5]

Timur, A. (1968) Velocity of compressional waves in porous media at permafrost temperatures: *Geophysics*, 33, 584–95. [7.2.3]

Timur, A. (1977) Temperature dependence of compressional and shear wave velocities in rocks: *Geophysics*, 42, 950–6. [7.1.6]

Treitel, S. and Robinson, E. A. (1964) The stability of digital filters: *IEEE Transactions on Geoscience Electronics* vol. GE–2, pp. 6–18. [10.8.1]

Treitel, S. and Robinson, E. A. (1981) Maximum entropy spectral decomposition of a seismogram into its minimum entropy component plus noise: *Geophysics*, 46, 1108–15. [10.8.6d]

Treitel, S., Shanks, J. L. and Frasier, C. W. (1967) Some aspects of fan filtering: *Geophysics*, 32, 789–800. [8.2.7, 10.3.2]

Trorey, A. W. (1970) A simple theory for seismic diffractions: *Geophysics*, 35, 762–84. [9.4.4b]

Trorey, A. W. (1977) Diffractions for arbitrary source-receiver locations: *Geophysics*, 42, 1177–82. [9.4.4b]

Trusheim, F. (1960) Mechanism of salt migration in Northern Germany: *AAPG Bull.*, 44, 1519–41. [9.3.1b]

Tucholke, B. E., Bryan, G. M. and Ewing, J. I. (1977) Gas-hydrate horizons in seismic-profiler data from the Western North Atlantic: *AAPG Bull.*, 61, 698–707. [7.2.5]

Tucker, P. M. and Yorston, H. J. (1973) *Pitfalls in Seismic Interpretation*: Tulsa, SEG. [9.0]

Ulrych, T. J. (1971) Application of homomorphic deconvolution to seismology: *Geophysics*, 36, 650–60. [10.7]

Vail, P. R., Mitchum, R. M. and Thompson, S. (1977), Relative changes of sea level from coastal onlap; in *Seismic Stratigraphy – Applications to Hydrocarbon*

Exploration, pp. 63–81 (ed. C. E. Payton): Tulsa, AAPG Memoir 26. [9.7.2]

Vail, P. R., Todd, R. G. and Sangree, J. B. (1977), Chronostratigraphic significance of seismic reflections: pp. 99–116 in Payton, C. E. (ed.), *Seismic Stratigraphy – Applications to Hydrocarbon Exploration*: AAPG Memoir 26. [9.7.2]

Vetter, W. J. (1981) Forward-generated synthetic seismogram for equal-delay layered media models: *Geophys. Prosp.*, 29, 363–73. [9.4.4a]

Waters, K. H. (1978) *Reflection Seismology*: New York, Wiley. [7.3.3a]

Watkins, J. S., Walters, L. A. and Godson, R. H. (1972) Dependence of *in-situ* compressional-wave velocity on porosity in unsaturated rocks: *Geophysics*, 37, 29–35. [7.2.2]

Webster, G. M., ed. (1978) *Deconvolution*: Tulsa, SEG, *Geophysical Reprint Series*, 1, vol. 1 and 2. [8.2.1a]

White, J. E. (1965) *Seismic Waves: Radiation, Transmission and Attenuation*: New York, McGraw-Hill. [7.1.5]

Wiggins, R. A. (1977) Minimum entropy deconvolution; in *Proc. Int. Symp. Computer-aided Seismic Analysis and Discrimination*, pp. 7–14: IEEE Computer Society. [10.8.6e]

Wiggins, R. A. (1978) Minimum entropy deconvolution: *Geoexploration*, 16, 21–35. [10.8.6]

Wylie, Jr, C. R. (1966) *Advanced engineering mathematics*, 3rd ed: New York, McGraw-Hill. [10.0, 10.1.1, 10.1.6, 10.2.1]

Wyllie, M. R. J., Gregory, A. R. and Gardner, L. W. (1956) Elastic wave velocities in heterogeneous and porous media: *Geophysics*, 21, 41–70. [7.1.4]

Wyllie, M. R. J., Gregory, A. R. and Gardner, G. H. F. (1958) An experimental investigation of factors affecting elastic wave velocities in porous media: *Geophysics*, 23, 459–93. [7.1.4]

Zieglar, D. L. and Spotts, J. H. (1978) Reservoir and source-bed history of Great Valley of California: *AAPG Bull.*, 62, 813–26. [7.1.4]

Index

In this index, I refers to volume 1
and II refers to volume 2